W9-AWT-505

RATIONAL MECHANICS
The Classic Notre Dame Course

R. Catesby Taliaferro

Edited by
A. Hahn, T. Banchoff,
and F. Crosson

DOVER PUBLICATIONS, INC.
Mineola, New York

Bibliographical Note

Rational Mechanics: The Classic Notre Dame Course is a new work,
first published by Dover Publications, Inc., in 2014.

International Standard Book Number

ISBN-13: 978-0-486-49982-6
ISBN-10: 0-486-49982-0

Manufactured in the United States by Courier Corporation
49982001   2014
www.doverpublications.com

Contents

R. Catesby Taliaferro

Autobiographical Remarks

I taught at Portsmouth Priory from September 1944 to July 1952, except for the year 1947-1948 when I was on leave of absence at St. Johns College, Annapolis. So I arrived as the war was drawing to a close. As might be expected at such a time, while the heads of the principal departments, English, Classics, Modern Languages, History, and Mathematics (Kelly, Ansgar, Flower, Lally, and Brady) were competent and seriously committed instructors, the younger masters, with the exception of certain young monks, were often not suited to their jobs. That was this not just a result of the war, I gathered from a bitter remark later made to Schehl and myself by Dom Richard Flower: I have never been given any but amateurs for my department. This problem of masters who have graduated with a nondescript undergraduate degree and little intellectual interest and elect to teach for several years before moving on to something else is common to many (perhaps most) private schools. I should better say that to many it is a regular practice rather than a problem, which vitiates their actions from the very beginning.

At my arrival, Dom Gregory Borgstedt was Prior and Headmaster, offices which had been held by Father Hugh Diman, the founder of the School. Since Father Hugh was quite old at that time, I am unable to report on his ideas of a school, but I would guess he wished a Catholic adaptation of the venerable New England institutions like Groton, with its emphasis on character building as well as a classical curriculum, if not like Exeter, with its more purely intellectual purpose (which, of course, forms character too, but indirectly). In any case, the ideal of Dom Gregory was fairly clear. He was interested in a Benedictine liturgical climate with which the young boys would be imbued for life. But circumstances were to push the school in another direction, as we shall see. He certainly never had intended that mathematics should be very important, but, after all, Dr. Brady, a physicist, was his assistant headmaster and, even more decisive, the details of the running of the school were in the hands of a young mathematician, Dom Andrew Jenks, certainly the most brilliant member of the Community and Faculty, learned in many fields, who would have been remarkable in any group. He had been a graduate student of the older Birkhoff at Harvard. Unfortunately, he was later put in charge of a house which was not at all in his line of work.

The regime of discipline at Portsmouth was the most casual I could have imagined. This, I was told, was in reaction to Jesuit practice. The classroom took care of itself, and two of the houses, the New under Kelley and the Red for the two years that Dom Ansgar kept it, were managed in an orderly way, the New rather conventionally, the Red most unconventionally, which resulted in bad feeling between the two. I lived for the first three years in the Red before moving off campus. During the first year, Father Ansgar allowed two ranking seniors, Tom Cuddihy and d'Uzès to drink a glass of beer with me before retiring. But the School dining room was an impossible mess. Monks were never seen there except on very few occasions. This had led at some time or other (or at least everyone supposed so) to a double standard of menu for the Monastery and School. I finally suggested that I might ask to be served in the Faculty Room. But happily Dom Aelred Wall (as I remember) and Bede Gorman were moved into the dining room, with the former at the high table, and civil society was restored.

The academic standards were low, as I judged. This was certainly true in mathematics. Grades were too high and failures too few in all subjects. And this was the judgment of at least one prestigious university which advised prospective applicants for admission not to prepare at Portsmouth, advice they reversed several years later. There was a seeming exception in the Classics Department, where Dr. Schehl, a former professor of Greek and Latin in Austria, had for a short time startled the School by demanding a very high performance. He was temporarily in Washington, but returned my second year in time to join the Mathematics Department in raising the academic standards to such a pitch that a few years later a dean of a New England college was to say that a 60 at Portsmouth was worth an 80 anywhere else.

Schehl was a most difficult man to work with (only Father Ansgar could manage him); but, although I could never fully share his Weltanschauung, we could agree on what we considered the ideal school, the nineteenth-century classical Lycée of France and classical Gymnasium of Germany, with their emphasis on languages and mathematics and their high demands on performance, which had produced such a wealth of great scientists, scholars, and professional men of all kinds. Dom Andrew Jenks was of this tradition. No one was fully conscious at the time that, in the forties and fifties, the problem of the prep school or high school would pass from the purely educational plane to that of world politics. The New England prep school was beginning to soften, even introducing sociology in some cases into the curriculum, while the American public high school was impotent from a heavy dose of Dewey-ism given it in the nineteen-twenties and the doctrine of non-transferability of disciplines added on by a Georgia cracker become professor of education at Columbia. The falsity of this doctrine as well as that of the fixed IQ was apparent to anyone who inspected with intelligence the records of students, but you were treated as a simpleton if you stated it publicly: a low IQ could be raised by faithful work at a demanding course in a fundamental subject. The subsequent politicalizing of these schools was to destroy them after a brief revival in mathematics and languages brought on by Sputnik, when, in the words of a Harvard dean of education, Khrushchev became our Minister of Education. For, barely noticed, Kapitza, a Russian mathematical physicist, a decade or so before, had returned to Russia from Cambridge to rescue the Russian schools from the progressivism introduced by the Revolution and return them to that ideal of the nineteenth century which permitted the Soviet Union to become a superpower.

In effecting the change of Portsmouth's standards, the Classics Department and the Mathematics Department pursued two different paths. The Classics Department simply upped its demands for mastery of the usual classical texts, Caesar, Cicero, etc. But the Mathematics Department innovated by directing the whole four years to the calculus and the foundations of analysis. Exceptionally, in the last year of my tenure at Portsmouth, after Schehl's final departure, Dom David Hurst read Plato's Meno in Greek with James Murphy. It should also be mentioned that Dom Alban Baer introduced the sonnets of John Donne into his Junior English.

It was, I believe, in my second year that Dom Gregory reacted and, in his address to the Faculty, insisted that Portsmouth was a school for the average boy and remarked that some of the Faculty were speaking over the heads of these students, which prompted Schehl to chant in the corridors of the Red: "Domine, super capita eorum loquamur." But Dom Gregory soon shifted an, in one of the rare conversations I had with him, he asked me if I thought a school as difficult as Portsmouth was getting to be could recruit enough students. I replied that, since in the matter of high academic standards, there would be no Catholic competitor and few, if any, non-Catholic, there was no reason to have any fear. Indeed, this advice very soon proved to be correct. For, at least during my first two years, undesirable students were brought in late to fill up the School's quota; this practice was soon obsolete, since the School had more qualified applicants than it could accept. Later parents of some of my students told me that they had removed their sons from other schools to Portsmouth because of its high standards and because it did not restrict its graduates to Catholic colleges.

There are two stories connected with this shift which are very likely true. The first is that Dom Gregory asked an influential father of several Portsmouth students who had been helpful to the Priory and School what were the necessary qualities of a good prep school. He replied, only two: high academic standards and good food. And somewhat later, when the superior of a New England congregation wrote to ask how he could raise the standards of his school, the Prior replied, with disarming candor, that he should employ plenty of lay masters who do not have to obey him. I myself observed, during the Christmas vacation of 1947, the Prior crossing the Manor House Hall, where a little boy was sitting waiting for an interview in the office. Just barely stopping, the Prior said: Be sure and make very good grades this spring so we can consider your admission. The tide had turned.

But, several years later, Dom Gregory and several other monks felt that the School was pre-empting too much of the Community's time and energy and applied to Ampleforth for permission to make the Priory more purely contemplative. It was assumed this meant suppression of the School. The story goes that Ampleforth consulted some university or other authority as to the rank of Portsmouth among American Catholic schools and received the reply that it was first by a large margin. This prompted Ampleforth to send Dom Aelred Graham to replace Dom Gregory as Prior when the latter retired to Elmira. Don Aelred Wall was already headmaster.

I would never have succeeded in teaching the ambitious program of the Mathematics Department without the cooperation of the majority of the students and the exceptional achievement of a few. Of course, one never asks for cooperation; one either gets it or doesn't. Above all, the student should not be too self-conscious about what he is doing. In my first year, Pruszanowski was outstanding in the third form (although there were others like Lummis and McCormick who did very well), Price and Palffy in the fourth form. In succeeding years, there were Denney, Richard Madden, Belt, Fred Fisher, Putnam, Stern, James Murphy, Vicini, and Daniel Childs. There were also those like William Ruckelshaus and Spaeth whose achievement was not so remarkable but who took the advanced courses beyond requirements. I was also lucky in arriving with a new football coach, Ralph Hewitt, who understood very well the priorities and was a help rather than a hindrance that many coaches would have been.

During 1948-1949, I was allowed by Dom Gregory, at my request, to teach informally to the novices of the Community a course in philosophy, which consisted of some Greek mathematics and some of the later dialogues of Plato. Most of the students were graduates of secular colleges, among them Arthur Gardiner, who, attacked by polio, was soon to finish much of his life in an artificial respirator. I was told later that they did exceptionally well on an examination set and corrected by Ampleforth. The examiner wrote that they had obviously not learned all this from a Thomist textbook. But their official instructor, a zealous Dominican Thomist from the Philippines via Providence College, was not pleased. Of him, Dom Hilary Martin said: he is like a Viollet-le-Duc gothic church. At that time Portsmouth was under attack from Father Feeney, and I was not allowed to teach philosophy again. Two years later, however, Dom Aelred Graham came for tea and before leaving quizzed me on my metaphysics and seemed quite satisfied I was no dangerous heretic. The American Church, not to speak of Laval in Canada, was so bewitched by Thomism that they forgot another powerful tradition in the Church, that of the fourteenth-century Scotists and especially that of the fifteenth-century cardinal Nicolas Cusanus, which underlies the most spiritual side of modern physics and mathematics. They were, therefore, totally unprepared for Vatican II and became the easy prey to all kinds of irrational existentialisms. There had been warning signals, for, even under Pius XII, a monsignor, dispatched from Rome to Laval, had suggested in a public lecture that the Thomists were going too far. I met the extreme Thomism again when I left Portsmouth for Notre Dame and watched with astonishment its rapid disintegration in the nineteen sixties.

At the time I left Portsmouth there were in my mind two subjects under consideration. The first, the problem of the second form, which seemed to me neglected. And, second, the introduction of a rigorous course in rational mechanics (in vectors, naturally) for the final year in physics, using all the mathematics the student now possessed.

To the suggestion often agitated of introducing two standards into the School, I was and am unalterably opposed. In one sense Father Gregory was right, the School is for the average boy, granted one knows what that means. But the average boy must never set the pace. The same deep and rigorous studies should be given to all, some more slowly than others. In mathematics this was usually achieved by having the student repeat courses. The average man must never be allowed to dominate; there is nothing more destructive to society than the consciously powerful average man.

R. Catesby Taliaferro

Catesby Taliaferro: A Memoir

It was in the fall of 1936 at the University of Chicago that I first met Catesby Taliaferro. I had gone there after taking an M.A. in mediaeval studies at the University of Michigan to take part in the revival of Aristotelianism that was occurring there under the aegis of Mortimer Adler and Richard McKeon. Catesby was there along with his friend and mentor the philosopher Scott Buchanan, as a member of the Committee on the Liberal Arts, then newly established by the president of the university, Robert Hutchins.

This committee consisted of Adler, Buchanan, and McKeon as its leading members and two or more younger assistants for each of them. Its task was to draw up the ideal curriculum of a liberal arts college based on reading and discussing the great books. However, the committee hardly got off the ground. Strong disagreements among its members over the selection of the books to be read and the best method of reading them led after a few meetings to its break-up, and it was reduced to a rump session including Buchanan and his associates. Adler continued his efforts through teaching his Trivium, an honors course for pre-law students based on reading the Great Books, of which Catesby was an assistant and which Buchanan also frequently attended. It was there that I became acquainted with Catesby.

These various efforts produced a plan of studies which quickly and entirely unexpectedly became one that could be put into execution. For during 1936-37 it turned out that the centuries old St. John's College at Annapolis, Md. was on the verge of bankruptcy and in need of strenuous help if it were to survive. Fortunately Buchanan with the help of friends was able to obtain sponsors so that he could take over the college. The fall of 1937 saw the beginning of the New Program at St. John's College based on the Great Books with Buchanan as its dean and Taliaferro one of its tutors.

R. Catesby Taliaferro was born in New York, N.Y. April 3, 1907. The R. stood for Robert a name he never used. Yet Robert Catesby was an old family name of the Taliaferros in colonial Virginia, and it would be interesting to know if there was any relation to Robert Catesby (1573-1605), the main leader of the Gun Powder Plot, the Roman Catholic conspiracy to blow up King James I and the Parliament on Nov. 5, 1605.

Catesby did his undergraduate studies at the University of Virginia, receiving the A.B. degree in 1928. He then went to France, where he first attended the University of Lyons from 1929-31 and also taught English at the nearby Lycée of Rochefort, and then in 1931-32 studied philosophy at the Sorbonne in Paris. The following year he was back at the University of Virginia to pursue the doctorate and in 1936 received the Ph.D. with special competence in Plato and ancient Greek science and mathematics. In the summer and fall of 1942 he studied at the Advanced School of Mechanics, Brown University.

Taliaferro's publications fully demonstrate his special competence. There is first the translation of The Mathematics-Composition of Claudius Ptolemy, known also by the title given it by its Arabic translators as Almagest (The Greatest). Catesby's translation of this massive and comprehensive treatise of ancient Greek astronomy consists of thirteen books along with an introduction and an appendix (The Passage from the Ptolemaic to the Copernican System and thence to that of Kepler) amounts to 478 pages. Hardly less daunting is his translation of The Conics by Apollonius of Perga, a treatise in three books of 307 pages. These two works were published together by Encyclopaedia Britannica in its set Great Books of the Western World in 1952. His translation of St. Augustine's De Musica appeared in 1947; it is a mathematical work on the theory of proportions that underlies the acoustical expression of them in what is more commonly called music. Besides these major works, Catesby also wrote a number of essays, including and introduction to Plato's Timaeus, another on Plato and the liberal arts, and one on the concept of matter in Descartes and Leibniz.

Taliaferro was known and appreciated as an excellent and devoted teacher, also a very demanding one. He entered the academic profession, when as already mentioned, he became a tutor in Buchanan's New Program at St. John's College, where he taught from 1937-42. Then after a year at Hamilton College, teaching meteorology, he became Master in Mathematics at Portsmouth Priory School, a Catholic Benedictine boarding school in Rhode Island, where he taught 1944-47. Then after a year back at St. John's, he returned to the Priory and remained there teaching until 1952: In that year he came to the University of Notre Dame, first as a professor in the Great Books program of Liberal Studies, and soon thereafter in the Mathematics Department, until his retirement in 1972. He then moved to Rocky Hill, N.J. near Princeton. There he died July 12, 1987, and was buried in Hollywood Cemetery, Richmond, Va.

Otto Bird

Further Biographical Remarks

I came to know Catesby Taliaferro (pronounced Tolliver) in the early sixties when we each taught a section of the first-year Honors Calculus course for Arts and Letters students, a course which he had developed and whose topics, pace and level were set by him. The course started out with elementary number theory, including such topics as the Euclidean algorithm and linear congruences, and proof by induction was studied and used. A student was later quoted in the Notre Dame yearbook as saying that although the material was called number theory, it was really numberless theory. Catesby liked to begin the course this way because there was usually a large enrollment, and by the end of the first week or two those who did not have a serious interest were gone. The number theory was followed by a rigorous treatment of limits, continuity, derivatives and integrals. I greatly enjoyed teaching the course, and did so several times after that.

Catesby did his undergraduate work at the University of Virginia, and then spent time working with Mortimer Adler on the development of the Great Books program at the Universiy of Chicago. After that he was instrumental in setting up the Great Books curriculum at St. John's in Annapolis. He taught English and studied philosophy in France, and finished a Ph.D. in philosophy at the University of Virginia. He was fluent in French and did several major translations of classical works from Greek and Latin into English. He found time one summer to study rational mechanics at Brown. The teaching he talked about most to me took place at Portsmouth Priory prep school in Rhode Island, from whence he came to Notre Dame's Great Books Program (now called the Program of Liberal Studies) in 1952. In the mid-fifties he moved to the Mathematics Department in order to be able to teach courses at a level of rigor and abstraction that were deemed unsuitable for the Great Books Program.

In the Mathematics Department he developed a junior-level course in rational mechanics and became the academic advisor to all the Arts and Letters mathematics majors. His students loved this course because, although it was basically a physics course, the mathematical and historical niceties were emphasized. In those days undergraduate academic advisors had full power over course choices for their advisees, and Catesby used this power relentlessly to be sure that his advisees achieved a broad level of liberal education, and in particular that they were enrolled in the best and most demanding philosophy courses. His advisees revered him as an intellectual leader, and often went to his home for sherry and conversation.

Catesby was a very conservative educator and a very conservative Catholic. Rumor had it that he had been disowned by his family when he converted to Roman Catholicism. He did not like the results of Vatican II, and he felt that Notre Dame had relaxed both its religious and educational standards in the sixties. He frequently referred to football coaches as "those thugs." He also dressed conservatively. I recall only once having seen him without a buttoned-up three-button dark tweedish sport coat over a flannel shirt and tie. He told me once that he wore flannel shirts because they didn't have to be laundered as often.

Around 1965 Catesby began to suffer from Parkinson's disease, which soon progressed to the point that he could not control the motions of his arms and other parts of his body. He said that if he took his medicine he could not think straight, and if he did not take it, his body was in such motion that he could not teach; he was embarrassed to be seen in public, or sometimes even in private. He felt forced to retire, and he went to live with a sister in Princeton.

John Derwent

On Being a Student of Dr. R. Catesby Taliaferro

Dr. R. Catesby Taliaferro (pronounced "Tolliver") was a remarkable teacher at the University of Notre Dame in the 1950s and '60s. He had a great influence on a number of students who went on to become professional mathematicians, many of whom recall with appreciation his unique teaching style and his demanding reputation. His colleagues and students in the General Program on Liberal Education also appreciated his attention to the philosophy and history that he included in his courses in Number Theory and Rational Mechanics. I'd like to recount some of the highlights from my time with him, first as an undergraduate in both of his courses, then as a novice teacher in two summer programs at Notre Dame, and finally as a friend during his retirement in New Jersey, a span of more than twenty years.

In the 1950s, Notre Dame attracted many top students, mainly from Catholic high schools throughout the country. Based on their standing in secondary school and their scores on the newly established SAT exams, a hundred or so of the incoming students in the Arts and Letters College were placed in honors courses in Mathematics and in Rhetoric and Composition. In those days, there were no elective choices made by students, except for the foreign language they would pursue, so most of the students in the two honors courses found themselves in the same sections of the required courses in history, philosophy, religion, and physical education. Professors Frank O'Malley and Robert Christen taught most of the English writing classes, and the four honors sections of math were taught by Professors Ky Fan, Richard Otter, Donald Lewis, and R. Catesby Taliaferro. The chair of the department, Dr. Arnold Ross, told me later that he specifically assigned all the students interested in becoming math majors to Taliaferro's section.

"Dr." Taliaferro, as he preferred to be called, was different from all the other professors I encountered in my first year at Notre Dame. Each of the others would start his class with a prayer and then tell us his name; Dr. Taliaferro, on the other hand, would enter the classroom precisely on the hour, march across the room and open a window, return to the desk where he had left his green canvas book bag, and begin the lecture. On the first day he presented the Peano axioms for the positive integers. At the end of the class a student asked, "Would you please tell us your name?" "No," he said, and strode out of the room. We all looked at each other. This teacher was different.

The next class he wanted to use the distributive property he had introduced in the first class. He looked down the roll and called on a student to state that law. The student said, "I don't know." "You don't know," Dr. Taliaferro repeated. "I didn't study," the student said. There was a pause, and Dr. Taliaferro then said, thumping his fist on the desk, "When I teach something in this class, you damn well better know it." We all took note.

By the time of the first exam, we had constructed the positive and negative integers, using the principle of mathematical induction. For the test, we were given double-sized sheets of paper on which to write our answers. The last problem was to show that $(1 + h)^n$ is greater than $1 + nh$ for any positive integer n and any positive h. It was the first time in my career that I faced a problem I did not immediately know how to solve. I handed in a partial solution, and I was surprised (and gratified) to learn that I had done very well, along with a few other students who were the other prospective math majors.

We went on to develop elementary number theory in a very rigorous way, ending with the Euler phi function and the Moebius inversion formula, a substantial accomplishment. The final grades in the course ranged from 100 down to 25. Many of the students placed into the honors math course on the basis of their secondary school rankings and aptitude tests were not prepared for the challenges that Dr. Taliaferro posed in every class. He had a relentless quality and he would not accept anything that was not done up to his standards.

In the second semester, the subject of our course was vector analytic geometry and the rigorous construction of the real number system using the principle of Dedekind cuts. Dr. Taliaferro explained that by the end of his course, we would be prepared to take a serious course in calculus, and he hoped that there would be a separate section for those who were prepared for that challenge. Unfortunately, that didn't happen, and the course, taught to a mixed class using the text from Courant, did not meet his expectations for rigor.

Automatically Dr. Taliaferro became the academic advisor for the Arts and Letters math concentrators. Along with the calculus course, we all took

Dr. Donald Lewis's modern algebra course, along with the first-year graduate students. Several of us had been given the privilege of selecting courses other than those in the standard sophomore curriculum and I was anxious to test myself in upper-level courses. I signed up for a senior philosophy course and graduate courses in history and English. Without consulting with my advisor, I subsequently signed up for two additional courses required for English concentrators. Needless to say, I did not do as well in that first semester of my sophomore year as I had in the year before. My advisor didn't have to berate me for my ambitious folly—he saw immediately that I had learned my lesson and encouraged me as I chose fewer courses, and more wisely, for the second semester, a much more rewarding experience.

In my junior year, I took Dr. Taliaferro's Rational Mechanics, the central course in the Arts and Letters mathematics concentration. It was very serious from the beginning, and highly geometric. Since there was no separate undergraduate linear algebra course taught at Notre Dame at that time, all of the relevant vector and tensor algebra had to be developed within the context of the mechanics course. The same was true for vector analysis and elementary differential geometry. Meticulously practiced diagrams illustrated the lectures. There were a number of detailed problems handed out periodically and graded in a timely way.

Most of the students in the Rational Mechanics class were mathematics majors, but there were some who enrolled as part of the General Program in Liberal Education, the original collection of courses for which Dr. Taliaferro had been recruited to teach by Dr. Otto Bird. At one point, a very highly regarded senior in that program was chastised by Dr. Taliaferro for not knowing how to prove the distributive law for the cross product of vector functions. I can remember discussions of the axis of a gyroscope and a demonstration in class in which a diver on the varsity swimming team was asked to describe how he executed twisting dives. Dr. Taliaferro brought in a heavy cane with a silver top and had the diver swing the cane as he jumped in the air in order to illustrate the physical counter-effect.

Examinations in that course required mastery of the classroom presentations and the ability to apply the principles to new situations. Later on, in graduate school at Berkeley, I appreciated the introduction I had received to the differential geometry of space curves, the subject that led to my thesis research.

We were all impressed when Dr. Taliaferro showed up on campus after a heavy snowfall, having walked several miles in his high-laced leather boots and fur cap. He did not drive a car, we learned, but he never missed class.

During my senior year, I acted as grader for the Rational Mechanics course, giving me an additional insight into its construction. The summer after I graduated, I was asked by the mathematics department chairman, Dr. Arnold Ross, to be an assistant in the Number Theory course in the NSF-sponsored program for secondary school teachers, an offshoot of the interest in improving science education after the shock of Sputnik. While I was there at Notre Dame during that summer, I realized I had graduated in more than one way, since more than one of my former professors now treated me as a novice teacher. I recall invitations to 529 North Sunnyside Avenue in South Bend, Indiana, where Dr. Taliaferro would prepare gourmet meals and serve specially chosen port. I was impressed by his collection of classical mathematical texts from the eighteenth and nineteenth centuries and also by his appreciation of fine furniture and silverware. He pointed out the portrait of his grandmother that had been in his family for generations. We did not talk about politics (where he was reputed to be a Royalist) or religion (where he was known to have converted to the Roman Catholic church while studying in France after taking a philosophy degree in Virginia).

The following summer, I was again a teaching assistant in Dr. Ross's summer program. Once again, I had very pleasant visits and conversations with Dr. Taliaferro, then and at a number of times during my graduate career.

After he retired from teaching at Notre Dame, Dr. Taliaferro lived on the second floor of a house that he and his sister had purchased in Rocky Hill, New Jersey, not far from Princeton. I spent several summers at the Institute for Advanced Study and visiting relatives nearby, and I would always stop and see Dr. Taliaferro on those visits. Frequently I would take him to dinner at one of his favorite restaurants on Nassau Street. It was sad to see the development of his Parkinson's Syndrome, especially as his involuntary sweeping motions intensified. Ultimately, he was unable to sit upright in a chair. I learned that he put himself through a regimen that I considered exhausting, taking medications that would permit him to have one or two working hours a day. He would move from an almost catatonic state up to a productive working time at a large table covered with papers and books. Then the disease would progress to a period of increasing agitation when he could no longer work productively. Ultimately it would suddenly drop back to catatonia, and as he took his medications, the whole cycle would start over again. I felt privileged to be with him during these periods of incapacity interspersed with normal activity. I was able to locate for some volumes he requested from the Princeton University library, and to listen to his ideas as he worked on topics in general relativity for his Rational Mechanics text. At one point I volunteered to show his initial chapters on Greek astronomy to my former colleague at Brown University, Prof. Otto Neugebauer, then at the IAS. Dr. Taliaferro was a bit apprehensive since the History of Mathematics faculty at Brown was legendary for its critical appraisals. He didn't have to be worried, as Prof. Neugebauer said to me that the chapters were "all right," high praise from that quarter.

I was very sorry to hear about Dr. Taliaferro's demise. It was gratifying to see the number of former students who came to the funeral services. In his will, he left books and silver to several of his former students. He specifically provided for the publication of his text on Rational Mechanics, and a few of us worked to bring the book to publication. Most recently the careful efforts of the late Prof. Fred Crosson, Dr. Taliaferro's longtime colleague in the General Program, and Prof. Alex Hahn in the Notre Dame mathematics department, made it possible to put the manuscript in final form. I felt proud that my notes from the Rational Mechanics course, and the corresponding set of notes the following year from Prof. Robert Burckel at Kansas State University, served to help shape the ending material on Special Relativity. Unfortunately, the general relativity topics on which Dr. Taliaferro was working at the end of his life could not be located among his papers. I look forward to the publication of this book. It was a great privilege to be a student of Dr. R. Catesby Taliaferro.

Thomas Banchoff

Preface

Catesby Taliaferro was already advanced in years when I first became aware of him in the late 1960s. We were both in the Mathematics Department of the University of Notre Dame, Dr. Taliaferro as a senior professor and I as a fresh graduate student. Clad in what was then "professorial casual"—a dark tweed sports coat buttoned up over a flannel shirt—Professor Taliaferro would walk slowly and unsteadily through the long corridors of the department (still housed in the old Computer Center at the time), his hands trembling from the onset of Parkinson's disease.

I did learn a few basics about Catesby's background at the time: That he had translated several major classical works of mathematics from Greek and Latin into English, including the first translation of the famous *Almagest* of Claudius Ptolemy into English in the early 1950s. It seemed remarkable to me at the time—and still does—that this historic volume did not see its English light of day until then. Much later, after I had developed a more serious interest in the history of mathematics and science, I would learn that Catesby had also translated Euclid's *Elements* and the *Conics* of Apollonius, and that he had written two books of his own, one on number systems, and another about the concept of matter in Leibniz and Descartes. Given his background and qualifications, I was not surprised to learn that he was called to participate in the development of the Great Books program at the University of Chicago and that he was instrumental in setting up the Great Books curriculum at St. John's in Annapolis. When he came to Notre Dame in 1952 he was ideally positioned to contribute to the university's own Great Books Program. Under the title "Program of Liberal Studies," this program thrives still.

Catesby's impact on Notre Dame and its students grew substantially when he moved to the Mathematics Department in the mid-fifties and introduced a course on rational mechanics at a level of rigor and abstraction that went far beyond the scope of any program with a focus on original sources. Catesby taught the course at Notre Dame from the late 50s to the early 70s, developing it into the Rational Mechanics that is the subject of this book. Readers will see for themselves how it unfolds its topic by building the important historical aspects, in particular the work of Ptolemy, Huygens, Kepler, Galileo, Newton, and Euler, into a course of mathematics and physics that includes inner product spaces, differential geometry, and relativity theory. Catesby's course was very influential. It aroused a deeper interest in mathematics in many very gifted Notre Dame undergraduates (Tom Banchoff and Bob Burckel among them) and enticed them to pursue graduate studies in higher mathematics.

Catesby retired from Notre Dame in 1972, moving back to Rocky Hill, New Jersey, and fought valiantly against the advancing symptoms of his Parkinson condition. He passed away in 1989 at the age of 80. His memory lives on at Notre Dame. Former students, colleagues, and friends created a memorial fund in 1989 to endow a yearly Catesby Taliaferro Essay Competition. Essays by second-year honors mathematics majors on aspects of the history and philosophy of mathematics contend for monetary prizes awarded by a faculty selection committee. The competition is testament to Catesby's devotion to his students and the intellectual dimension and breadth that his courses brought to Notre Dame's mathematics curriculum. Catesby's contributions are best summarized by his former student Fred Rickey (later a professor of mathematics at Bowling Green University and at West Point) with the reflection "I have long felt that Catesby was one of the unrecognized people that made Notre Dame a great university."

Soon after I began to direct Notre Dame's Kaneb Center for Learning and Teaching in the fall of the academic year 2002, Professor Frederick Crosson knocked on the door of the Center. Fred was a highly respected faculty member who had been Dean of Notre Dame's College of Arts and Letters. He had known Catesby well. Fred informed me that Catesby had mentioned to him long ago that he was deeply interested in having the lecture notes of his Rational Mechanics course published, and that he had set aside a good sum of money in his last will to cover the costs. Fred asked if I might be able to assist. This would involve the typesetting of Catesby's faded ditto notes with its many diagrams and figures into an electronic mathematics document. Since I had some administrative capital at the Center and a staff of very capable students, I decided to take this task on. Fred and I agreed that we would approach the University of Notre Dame Press about the publication and that Fred would check into the provisions of Catesby's last testament.

The transcription of the manuscript and its diagrams went forward. By the end of the year 2004 much progress had been made. This continued in the very capable hands of Notre Dame undergraduates, principally Carolyn Blessing, Tony Bendinelli, and Adam Boocher. When the chapter on Einstein's Special Relativity listed in Catesby's Table of Contents was nowhere to be found, Paul Gibson, an undergraduate mathematics major, and Laura Kinneman, a graduate student of physics, were enlisted. They most ably reconstructed this chapter from course notes provided by Tom Banchoff and Bob Burckel. The successful conclusion of the project seemed assured when the University of Notre Dame Press agreed to publish the manuscript. But then came a series of setbacks.

The first was the discovery that the directive in Catesby's Last Testament to underwrite the project had lapsed and that the title to Catesby's lecture notes had passed to a descendant. A second setback concerned Notre Dame Press. The Press had narrowed its focus to the humanities and social sciences, discontinuing its engagement of natural science and mathematics, and did not follow through on its commitment to publish Catesby's lecture notes. Fortunately, all this was resolved. After the publication rights were reacquired, Notre Dame's Colleges of Science and Arts & Letters, as well as its Kaneb Center of Teaching and Learning, agreed to fund the transcription of Catesby's document and the extensive proofreading effort. And Dover Publications stepped in to take on the production of the volume.

The final setback was tragic and beyond repair. It was the sad death of Fred Crosson. In 2008 he had had a severe fall, suffering brain damage that left him weak and confined. In full health, he had been a remarkably warm and spirited person with a probing and wide-ranging intellect who had served as chair of the Program of Liberal Studies, as editor of *The Review of Politics,* and as founding director of Notre Dame's Center for Philosophy of Religion. He was president of the American Catholic Philosophical Association and national president of the Phi Beta Kappa Honors Society. Like his colleague Catesby, Fred Crosson was a Renaissance man who fought tirelessly for the ideals of liberal education. He died in December 2009. This volume is certainly devoted to Catesby's contributions to Notre Dame, but it is also dedicated to the memory of Fred Crosson. There is no doubt that Fred's friend would have approved.

In conclusion, I would like to thank the many people who have made this volume possible. These include all the Notre Dame students involved in the typesetting of the original manuscript and the Notre Dame graduate students Steven Broad and Bryan Ostdiek for attending to the completion of its production. Warm words of gratitude go to Professor Chris Kolda of the Department of Physics for guiding the transcription effort to its conclusion; to Dean Greg Crawford of the College of Science for much of the funding; and to Alisa Zornig of the Kaneb Center for organizing the effort over the years. Finally, a word of thanks to Tom Banchoff for his unflagging interest in Catesby's Rational Mechanics and to John Grafton for undertaking its publication with Dover Books.

Alexander Hahn

RATIONAL MECHANICS
The Classic Notre Dame Course

Chapter 1

Greek Celestial Mechanics

The science of the motion of bodies began, for obvious reasons, with the science of the motion of celestial bodies: they could be observed at leisure in their movements without apparent interference since they pass slowly through the sky and they could be supposed to move in an unimpeded region without friction as we shall see. It is therefore proper to begin with celestial mechanics; the planets are intellectually the closest to us even if they appear to be farther away than the things we handle every day; and, indeed, it will turn out that the principles of the classical mechanics of Newton came more from the heavens than from the earth.

We start with Greek systems of celestial mechanics because it is from them that the systems of Copernicus, Kepler, and Newton are derived, not from the Babylonian, however interesting it may be. Further the Greek systems of celestial mechanics are rigorously constructed, complete, and run the range of geocentric and heliocentric theories. By the end of the 4th century B.C. both the geocentric and heliocentric theories had been developed with several variations. If the geocentric hypothesis seems the most obvious to the common man, this is not so for the astronomer, and heliocentric theories probably appear just as early as geocentric ones. The first clear statements concerning the nature of the science of celestial mechanics are found in the dialogues of Plato, developing a long previous Pythagorean tradition. The burden of this Pythagorean-Platonic statement is that the motions of the heavenly bodies can be only understood in terms of mathematical models which it is up to the scientist to build. These mathematical models must satisfy at least two conditions: (1) they must fit the appearance as gotten by observations; (2) they must be interesting in themselves so as to reveal further appearances and to suggest further observations which could not have been seen without the suggestions of the structures of the mathematical models themselves or without the presence of an idea which generates these models.

The role of observations is an ambiguous one. Most observations are already the result of a theory, perhaps quite rudimentary, embodied in the very instruments used to make them. Although the science of mechanics would be sterile without experiments, it is in no sense the slave of them nor just an abstraction from observations as the development of this science in its historical perspective will amply prove. There is a dialectic between theory and appearance and between theory and theory which is the thread of discovery and life in this and perhaps all science and knowledge. In this sense the second condition above is the more important and the more subtle. Because of it, there are times when one must look away from the observations and the world of appearance, as Plato points out in a usually misunderstood passage of Book VII of the Republic where he says one must look away from the stars and concentrate on the mathematical models themselves. Again and again this has been done in the history of mechanics as notably in the case of Galileo as we shall see. The more spectacular manifestations of condition (2) will be seen in the great crises and discoveries of the science as with Kepler, Hamilton, Einstein, and de Broglie.

We shall not go into the system of Eudoxus of Cnidos (about 370 B.C.), since it has little direct influence as far as details are concerned on the subsequent developments of the main body of Western Astronomy and mechanics, although it was a brilliant mathematical achievement and gave a first impetus to the construction of astronomical models. We shall develop first the geocentric model which had its roots in the work of the scientists associated with the Platonic Academy about the same time, which was continued by Hipparchus of Rhodes (150 B.C.), and set down in its most detailed form by Ptolemy of Alexandria (150 A.D.). In fact, this is the only complete Greek theory to come down to us intact in literary form; the rest of the literature is lost. We shall then pass to the theory of Aristarchus of Samos (24.5 B.C.).

A   The System of Hypparchus and Ptolemy

A.1   The Appearances

To understand the development of this system, we shall set out the simpler appearances of the planets and stars as they could be observed by one watching nights over a long period of time. He would need only instruments to measure angles - this was all the Greeks possessed - and simple timepieces such as water clocks or the movements of the stars themselves. Modern watches and pendulum clocks depend on Newtonian mechanics.

First, we regard the heavens as a sphere (the celestial sphere) and ourselves at the center. The stars appear to be fixed on this sphere which moves about a fixed axis one end of which is roughly at the North Star or North Pole for those observing in the northern hemisphere of the earth. That is, the North Star seems to stay unmoved and the stars near it to revolve about it in circles of which it is the center, moving from east to west. As these circles get larger for stars further from the North Star, they dip beneath the horizon, the apparent plane surface of the earth which projects as a circle on the celestial sphere. The time it takes a star to revolve once about the North Star is called a true sidereal day; it is very nearly what we call a day of 24 hours.

If now we image great circles through the North Pole, the meridians, then there is one through the point immediately overhead, the zenith point. The time of the passage of a star from this meridian back to this meridian is the true sidereal day. The great circle at right angles to the axis of the celestial sphere is called the celestial equator. Since, as we shall see, in the course of a year new parts of the celestial sphere gradually appear at night so that we can compare the two parts cut off by the plane of the horizon; we shall find the parts apparently equal so that the plane of the horizon seems to cut the celestial sphere exactly in half.

Thus in the figure we have the celestial sphere with its two poles N and S, E the center of the earth, P the point of the observer on earth's surface, TU the celestial equator, and PR the projection of the tangent plane of the horizon cutting the celestial sphere in R. Since as we have said PR appears to bisect the sphere, therefore R and S' seem to coincide. This leads us to assume that PE, the radius of the earth, is negligible with respect to the radius of the celestial sphere, ES'. This means that from this point of view the stars are infinitely far away com-

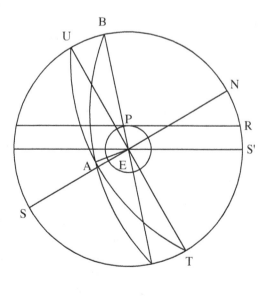

Fig. 1

pared to the radius of the earth.

This also seems to justify the assumption that the earth is indeed the center of the celestial sphere.

We have tacitly assumed here that the earth is roughly a sphere itself, a fact which has hardly ever been doubted since the 5th century B.C,. in Greece, contrary to several modern myths started by ignorant historians. The reasons for this assumption are well known: - (1) the disappearance of ships below the horizon, symmetrically in every direction, (2) the circular shadow of the earth on the moon at eclipses of the moon; (3) the progressive appearance and disappearance of stars for travelers on the earth.

If now we observe the motion of the sun through the heavens, it appears to revolve as the stars once in a day, from east to west. But at the same time it appears to have another much slower motion compounded with this one; it appears to move from west to east on a great circle inclined to the equator at about 23°; and to move very slowly in this way so that it completes this cycle in what is defined to be a solar year. This great circle, AB in the figure, is called the ecliptic, and the points where it crosses the equator, the equinoctial points. The time from the passage of the sun at an equinoctial point back to the same equinoctial point is defined as the solar or equinoctial year. Hence the sun as it appears to revolve about in its daily motion from east to west, appears also to move slowly backwards, allowing different stars gradually to appear at night so that, over a period of a year, there is a full cycle of such stars. That is, a star appearing overhead at midnight at a certain time of the year, appears farther and farther west of the meridian through our zenith until at the end of a year it is back at that same meridian at midnight. It will be found, moreover, that the sun does not move at constant angular speed along the ecliptic, but at one part always more slowly and in the opposite part faster.

The question immediately arises what we mean by slower and faster. We have to take some motion as a standard for constant angular speed. The Greeks assumed that the daily motion of the fixed stars was a constant angular motion and this choice is consistent with the later theory of Newton. Hence we shall see that the sun's yearly motion along the ecliptic is irregular with respect to the assumed constant daily motion of the fixed stars.

But there are other stars which appear to move back along the ecliptic more or less, the moon and planets (or wandering stars) Mercury, Venus, Mars, Jupiter, and Saturn. These were all that were visible to the Greeks. All the other stars were called fixed stars since they keep their positions with respect to each other, but each of these planets in addition to the daily motion appears to move back from west to east along the ecliptic, weaving slightly

Fig. 2

from side to side, each in its own way and each with its own speed. Leaving aside the moon, we can divide the five planets into two sets.

Venus and Mercury are restricted in their motions with respect to the sun: Venus is never more than 48° from the sun on either side, and Mercury never more than about 27°. Hence these stars move backwards along the ecliptic with the sun, crossing first to the east of the sun and then to the west and back again. When, in such a cycle, the planet reaches its farthest angular point from the sun, it is said to be at its greatest eastern or western elongation from the sun. These greatest elongations are different for different parts of the ecliptic in such a way that, when the sun is at a certain point of the ecliptic, the sum of the greatest eastern and western elongations of the planet for that point is greater than the sum for any other point; and, when the sun is exactly opposite on the ecliptic, the sum of the greatest eastern and western elongations is the least. The sums decrease or increase symmetrically on either side of these points.

On the other hand, Mars, Jupiter, and Saturn can be found at any angular distance from the sun on the ecliptic. But they appear tied to the sun in a more subtle way. Let us take Mars for example. It takes Mars almost two years to make a revolution about the ecliptic while it takes the sun only one by definition. If we observe Mars continually, we notice that as the sun approaches Mars on the ecliptic, Mars appears to move faster and faster away from the sun from west to east; but after the sun has passed it, Mars appears to slow up, and, at a certain point, nearly opposite the sun, it stops (a station point) and then appears to move in the opposite direction from east to west (retrogradation), and then to stop again at a point symmetrically situated on the other side of its opposition to the sun, finally continuing its west-east motion. To be more exact, it actually describes a loop. This pattern is repeated for each of the other planets, Jupiter and Saturn, but with different tempo.

It is to be remembered also that each of the planets weaves back and forth on either side of the ecliptic in what are called its latitudinal variations. We shall, however, ignore these variations at first in our search for a theory, and consider only the projection of these motions on the ecliptic, that is, the projection of the planet on the ecliptic by a great circle through the poles of the ecliptic.

The problem of Greek mechanics, as so eloquently presented by Plato in many of his dialogues, was to discover some mathematical law which would explain these strange irregular appearances. And Plato went further; since the world of appearance is ambiguous, there would probably be many different ways of explaining them. We shall see at least three variations on the same theme which appeared in the Platonic Academy itself and several others which were inspired by it in the course of historical tradition.

We have already noticed certain cycles in the planetary movements, that is, movements which return the planet again and again to the same point relative to the "fixed" stars; and, moreover, each planet, except the sun, seems to combine in its motion several cycles. Thus Venus moves from west to east along the ecliptic at an average angular speed of the sun, returning again and again to the same point of the ecliptic in its cycle of longitude or first irregularity; but, at the same time, it moves back and forth across the sun from greatest western elongation to greatest eastern elongation in another cycle, its second or heliacal irregularity. We could, of course, suppose this motion to belong to the sun relative to Venus; but this view will not work since Mercury also has such a cycle but of a different period, and we could not move the sun with respect to both. Furthermore, we should then have to move all the fixed stars with the sun. We have, therefore, Venus' two cycles combined together in such a way that we cannot immediately separate them; one overlaps the other. To find the relation between them, we note, say, the position of Venus on the ecliptic at a greatest western elongation and wait until Venus reaches that same point of the ecliptic again, and again at a greatest western elongation, so that we can say there are so many cycles of second or heliacal irregularity to so many cycles of longitude.[1] We may look in vain for such an exact return and, indeed, we have to settle for less, for a very approximate return to the same position. Thus Ptolemy used 5 cycles of second irregularity to $\left(8 - \frac{1}{160}\right)$ cycles of longitude or 8 solar years less a little more than 2 days. If we did not demand that both cycles end together or very nearly so, we would have one impinging on the other in such a way that, what appeared to be the end of one cycle, might not be at all, but only appear so because of the influence of the uncompleted irregularity of the other.

We may wonder whether we are right in assuming that the return to the point of greatest elongation marks properly the period of the second irregularity. We shall see later in the development of the mathematical theory that

---

[1]Longitude in Greek astronomy is with respect to the ecliptic, not with respect to the equation as in modern tables.

this is not a bad choice. We shall be forced to recognize an irregularity in the longitudinal cycle of the planet, just as there is none in the case of the sun which, as we have said, does not appear to move at a constant angular speed for all points of the ecliptic. But the fact that we choose the final point of greatest elongation at the same point of the ecliptic as the first will guarantee the consistency of our theory and practice.

In a similar way, we must find a relation between the two cycles of the planets of the second set, Mars, for instance. Starting with Mars at a station point, we could count the cycles of longitude and the cycles of second irregularity until Mars reached a corresponding station point at the same point of the ecliptic. We would find, according to Ptolemy, that in $79 + \frac{193}{(60)(365)}$ solar years, Mars accomplishes $42 + \frac{190}{(60)(360)}$, cycles of longitude and 37 cycles of the second or heliacal irregularity. In this case, too, we have to settle for an approximation. Later on, we shall find other points of the heliacal cycle chosen for reference, for reasons we cannot understand at this stage of the game. The other two planets follow the same pattern.

Since here the very numbers will be extremely important in the theoretical development of the science of mechanics, we sum up all these results in a table.

Table 1

Planet	Cycles of Heliacal Irregularity	Cycles of Longitude	Solar Years
Mercury	145	$46 + \frac{1}{360}$	$46 + \frac{31}{(30)(365)}$
Venus	5	$8 - \frac{5}{(4)(360)}$	$8 - \frac{26}{(20)(365)}$
Sun		1	1
Mars	37	$42 + \frac{190}{(60)(360)}$	$79 + \frac{193}{(60)(365)}$
Jupiter	65	$6 - \frac{29}{(6)(360)}$	$71 - \frac{147}{(30)(365)}$
Saturn	57	$2 + \frac{103}{(60)(360)}$	$59 + \frac{7}{(4)(365)}$

It will be noticed that, in the case of the planets of the second set, the sum of the first two columns of figures is always equal to the third. This is an important property of the numbers as we shall see.

A further appearance, so slow that it takes centuries to perceive, was found probably first by Hipparchus and reported in his treatise *On the Precession of the Tropic and Equinoctial Points*. The precession of the equinoxes is the slow rotation of the fixed stars, from west to east, about the poles of the ecliptic, in the same direction as the sun's motion. Ptolemy computed it to be about 1° in a century; modern computations fix it at about 1.39° in a century. Hence the fixed stars also have a double motion: (1) the daily motion from east to west about the poles of the Equator, (2) the precessional motion from west to east about the poles of the ecliptic.

It is also observed for all the planets except the sun that the time from their apparent mean speed through maximum to mean is greater than from mean through minimum to mean. The sum appears to affect the opposite.

A.2   The Theory and Its Mathematical Instruments

Given the cyclic character of the appearances of the sun and planets and the simple formulation of constant angular speed on a circle about the center of that circle, it is easy to understand why the following principle was assumed for astronomical investigations:

A. **Every apparent celestial motion must be explained in terms of the constant angular speed of the planet on a circle about the center of that circle, which center in turn can move at constant angular speed on another circle about its center; and this can be continued to any required degree of complication.**

We shall call this the Neo-aristotelian principle of inertia for celestial motions, because it is a generalization of a more restrictive principle taken over by Aristotle from Eudoxus' system. In Eudoxus' system, all motions had to be explained in terms of a sphere whirling at constant speed about an axis through its center. This sphere in turn could carry the poles of another sphere concentric with it, this second sphere whirling at constant speed about the axis. This could be continued as far as desired. These concentric spheres had the earth as their common center. This principle was adopted by Aristotle who assumed it to be the universal law of motion for the moon and all bodies beyond, and he built a whole cosmological system about it. These details are mentioned here for the later comprehension of the spirit of Copernicus' reconstruction of Aristarchus' heliocentric theory.

But Principle (A) is itself too inflexible and Ptolemy, or someone before him, enunciated a broader principle which is at the basis of the Ptolemaic system in its final form.

B. Every apparent celestial motion must be explained in terms of motion of the planet on a circle at constant angular speed about some one point, not necessarily the center of that circle, and the center of this circle may in turn move on a circle at a constant angular speed about some point, not necessarily the center of the second circle. This method can be continued to any required degree of complication.

And a supplementary general principle stated by Ptolemy in his work *On the Hypothesis*:

C. Every planet must be considered in isolation from every other. No planet has any influence on any other; each moves in accordance with its own peculiar being and will. - This is an explicit denial of the universal law of gravitation later assumed by Kepler under the influence of Nicholas Cusanus.

We now turn to the basic mathematical instruments to implement and apply this principle, which we may call the Ptolemaic law of inertia for celestial mechanics. These mathematical instruments are very simple and were used quite early by Greek mathematicians, certainly by those of the 4th century B.C.

The simplest is the motion of an eccentric circle where the planet moves at constant angular speed about the center of the circle for an observer off center. Suppose C the center of the circle on which the planet moves at constant angular speed about C, that is, equal arcs in equal times, or equal angles in equal times. Let E be the eye of the observer, the earth for the geocentric theory.

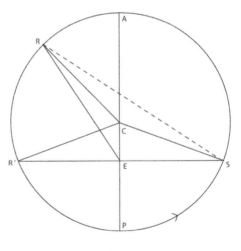

Fig. 3

Then A is the point farthest from the earth called the apogee, and P the nearest called the perigee.

It is intuitively clear that, since the planet moves at constant angular speed about C, it will take more time to describe SAR' than to describe $R'PS$. But from the point of view of E, it describes an angle of 180° in both cases and, since the observer is judging angles and angular speeds only, there being no perceptible difference in size to the naked eye, the planet appears to move more slowly in SAR' than in $R'PS$. As we shall now prove, it will appear to move slowest at A, fastest at P, and at R' and S it will appear to have the speed it has about C. This last is called the mean speed.

We shall first prove these things by the geometrical method of the Greeks and then by modern analysis.

By elementary theorems,

$$RE > R'E \text{ and } R'E = ES;$$

therefore $\angle ESR > \angle ERS$, while $\angle CRS = \angle CSR$.

Therefore $\angle CRE < \angle CSE$. But $\angle CSE = \angle CR'E$.

Hence $\angle CR'E > \angle CRE$ and likewise $\angle CR'E$ is larger than any corresponding angle between R' and P. The same theorem obviously holds on the other side S.

But

$$\angle ACR = \angle AER + \angle CRE$$

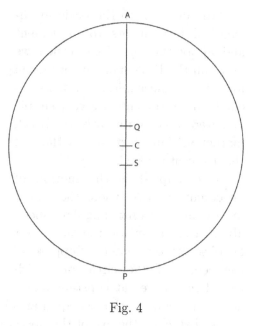

where R is any point on the circle including A, R', P, and S. Now $\angle ACR$ represents the angle of motion about C and $\angle AER$ the angle of observed motion taking place in the same period of time as seen from E. The difference between the two increases from zero at A to its greatest amount at R'; it then decreases from R' to reach zero again at P. Therefore the angular speed about C from A to R' is larger than the apparent speed about E since the differences accumulate in its favor. But at R' they begin to decrease. Assuming the continuity of

Fig. 4

motion, the apparent angular speed about E at R' is the same as the constant angular speed about C.

Again, if we follow the same argument, considering the planet's motion from S on towards A, the apparent speed at S is equal to the constant speed about C, but decreases from S to A since the angle of difference CRE decreases until it is zero at A. Hence the apparent angular speed is slowest at A. But beginning from A, the difference to be added to the angle about E to make it equal to that about C, increases to R' where, as we have seen, the two speeds are again equal. By the same argument, we can prove the apparent motion is at a maximum at P.

By analysis, if θ is angle ACR, then

$$\theta = Kt, \qquad \frac{d\theta}{dt} = K$$

where K is a constant, the t represents time measured from the initial point of θ. Then letting $\delta = \angle CRE$ and $\varphi = \angle CER, CE = e$ and $CR = a$, we have

$$\theta = \varphi + \delta.$$

But

$$e \sin \varphi = a \sin \delta,$$

$$e \cos \varphi \cdot \frac{d\varphi}{dt} = a \cos \delta \cdot \frac{d\delta}{dt},$$

$$e \cos \varphi \cdot \frac{d\varphi}{dt} = a \cos(\theta - \varphi) \cdot \frac{d(\theta - \varphi)}{dt} \qquad (\alpha)$$

$$e \cos \varphi \cdot \frac{d\varphi}{dt} = a \cos(\theta - \varphi) \left(K - \frac{d\varphi}{dt} \right)$$

$$\frac{d\varphi}{dt} = \frac{Ka\cos(\theta - \varphi)}{e \cos \varphi + a \cos(\theta - \varphi)} = \frac{Ka}{a + \dfrac{e \cos \varphi}{\cos(\theta - \varphi)}} \qquad (\beta)$$

From (α) we see that, when $\varphi = 90°$ or $270°$, $\dfrac{d\varphi}{dt} = \dfrac{d\theta}{dt} = K$. From (β), since

$$\left| \frac{\cos \varphi}{\cos(\theta - \varphi)} \right| \le 1,$$

we see, by inspection, $\dfrac{d\varphi}{dt}$ is a minimum at $\varphi = \theta = 0$ where $\dfrac{\cos \varphi}{\cos(\theta - \varphi)} = 1$; and at $\varphi - \theta = 180°$, a maximum where $\dfrac{\cos \varphi}{\cos(\theta - \varphi)} = -1$.

An extension of this problem was noticed and solved by Kepler. The results were important, as we shall see, in the Keplerian revolution in celestial mechanics. Suppose C is center of the circle a planet moves on, Q the point about which the planet has constant angular speed, and S a reference point such that $QC = CS$. It can be proved that

$$\frac{\text{linear speed at } A}{\text{linear speed at } P} = \frac{SP}{SA},$$

the proof of which is left to the reader.

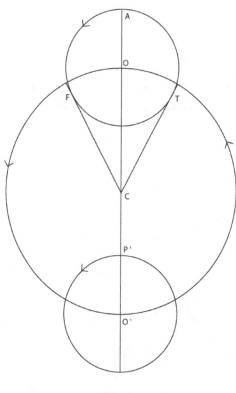

Fig. 5

The second mathematical instrument is the epicycle, whose discovery and use is attributed to Apollonius of Perga although it was probably found and used much earlier.

Suppose the center O of a small circle called the epicycle moves on a larger circle called the deferent at a constant angular speed about the deferent's center C. The planet is placed on the epicycle moving at constant angular speed about the epicycles's center O. There are two possibilities; (1) the motion of planet on epicycle is in same direction, say counterclockwise, as motion of epicycle's center on deferent. (2) The motion of planet on epicycle is clockwise while motion of epicycle's center on deferent is counterclockwise.

Let us consider the first case. Intuitively we can see that, if we suppose the observer at center of deferent C, then, when the planet is at apogee A, its motion on the epicycle is in the same sense and direction as the motion of the epicycle's center O on the deferent, whereas anywhere else on the arc TAT', T and T' being the points of tangency of epicycle for lines from C, only part of the velocity of the planet on the epicycle is added, and on $T'PT$, indeed, subtracts. Therefore the angular speed of planet about C will appear for C to be greatest at apogee A. By the same argument, the angular speed will appear least at perigee P.

By the same argument, since, at T and T', the motion of the epicycle is in the direction of the line to C, the observer, it will not affect the angular speed about C. Therefore when the planet is at T and T', it will appear, relative to C, to have the angular speed of 0 about C, the so-called average or mean speed. Again it is to be noticed that **the time from T through A to T' is greater than the time from T' through P to T**, since, because of the constant speed, the angles about O can be used to measure the time. In other words, **the time from mean speed through maximum to mean is greater than the time from mean through minimum to mean.**

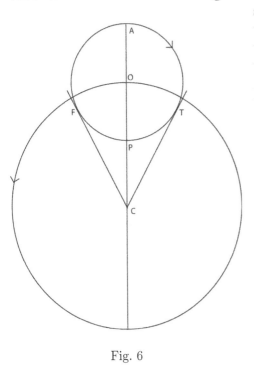

Fig. 6

Let us now consider the second case. Here the reverse holds: the angular speed at the apogee A appears least; that at the perigee P, the greatest. **The time from mean speed through least to mean is greater than the time from mean through greatest to mean.**

Strangely enough, if we take the following special condition of this case, that the angular speed on the epicycle is equal to the angular speed of the epicycles's center on the deferent, then we can show that the planet actually describes a circle with radius equal to deferent's and a center whose distance from C is the radius of epicycle. We are reduced to the case of the eccentric circle.

The proof of this is simple by geometric methods. Consider the epicycle in two positions. In the first, let the planet be at A. Then let O move to O' while the planet moves on the epicycle from A' to R. According to our assumption $\angle A'O'R = \angle ECO'$ and $O'R$ is parallel to CE. Draw RE parallel to $O'C$ meeting OC in E. Then $RE = O'C$ and $RO' = EC$. Therefore the planet describes a circle about E as center with $RE = O'C$ as radius and E at a distance OA from EC. To compare exactly with eccentric circle case we should interchange the notation E and C. This can also be proved, of course, by analytic methods.

These are the main instruments for the Ptolemaic theory. A more complicated one, the moving eccentric, we shall bring in later at the appropriate

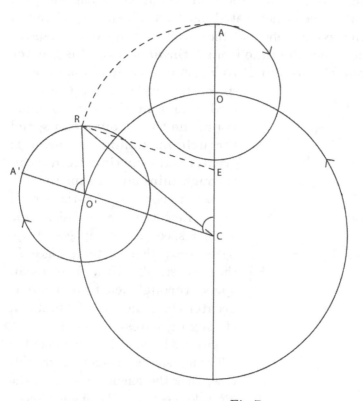

Fig.7

point in the development of the physical theory, and certain refinements on those above will be introduced where they are needed. Let us now begin to apply them to the specific planets. We shall treat first the sun, then Venus, and then Mars, as representatives of their sets, leaving aside the tortured case of Mercury.

A.3   Solar Theory

We have already remarked that the sun moves from west to east on the ecliptic, completing the circuit from equinoctial point back again in what is defined as the solar year, which comes to $365\frac{1}{4}$ solar days, less some small amount. From observations over long periods of time, Ptolemy computed it to be $365 + \frac{14}{60} + \frac{48}{60^2}$ solar days, the same for all years. This means the sun moves back $0°59'8''$

very nearly, each day on the ecliptic. But this is only an average. For, by observation, it is found as we have said that the sun moves faster on the ecliptic in one part and slower in the part opposite; the difference is not great and the change is continuous and symmetrical.

By comparing the shadows thrown by gnomons, it was computed that the sun takes $94\frac{1}{2}$ solar days from the spring equinox (intersection of ecliptic and equator) to the summer tropic (highest point reached by sun on ecliptic at $90°$ from equinox), and $92\frac{1}{2}$ solar days from the summer tropic to the fall equinox.

These solar days are not quite equal obviously; for otherwise there would not be this discrepancy between the two figures. The difference, however, would be too small to be detected by the instruments then used, since Ptolemy computed his instruments were only accurate to within $15'$ of arc. It is easy to find the error resulting in using $94\frac{1}{2}$ and $92\frac{1}{2}$. One need only observe the number of true sidereal days in the two intervals to find the correct ratio of these numbers.

The appearances suggest that one use the eccentric circle to save this difference in apparent speeds, and it will

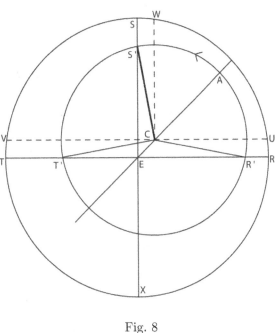

Fig. 8

be found that the time from maximum apparent speed to mean is less than from mean to minimum as for the eccentric circle.

The problem now is to find, from these observations, assuming the sun moves on an eccentric circle, the apogee of the sun's circle and its eccentricity.

Let circle $RSTU$ be the ecliptic on the celestial sphere with center E, the earth. Let R be the projection of the spring equinox on the ecliptic, S of the summer tropic, and T of the fall equinox. The sun's circle lies in the plane of the ecliptic. We have seen the sun takes $94\frac{1}{2}$ days to pass through the angle $RES, 92\frac{1}{2}$ through angle SET or 187 days to pass from the line ER to the line ET. Hence it takes $365\frac{1}{4} - 187 = 178\frac{1}{4}$ days to pass around the rest of the circuit. Therefore the center of the sun's eccentric circle must lie in the semicircle RST, and finally in the quartercircle RES, since that is the quadrant of its slowest motion.

Let C be the center of the sun's eccentric circle $R'AS'T'P'$, to find the angle $R'EC$ and the eccentricity $\frac{EC}{CA}$. Since, by assumption, the sun moves at constant speed on circle $R'S'T'$ about its center C, and, by observation, moves from R' to S' in $94\frac{1}{2}$ units of time and from S' to T' in $92\frac{1}{2}$ to the whole circuit $365\frac{1}{4}$ very nearly, therefore we know

$$\angle R'CS' - \frac{94\frac{1}{2}}{365\frac{1}{4}}(360°) = \alpha, \qquad \angle S'CT' - \frac{92\frac{1}{2}}{365\frac{1}{4}}(360°) = \beta.$$

But, drawing UCV parallel to RET, we have

$$CT' = CR', \qquad \angle CT'E = \angle CR'E = \angle T'CV = \angle UCR';$$

and, drawing WC parallel to ES and CR perpendicular to UCV, we have

$$\angle WCS' = \angle CS'E.$$

And
$$\angle R'CS' = \alpha = 90° + \angle UCR' + \angle WCS'$$
$$\angle S'CT' = \beta = 90° + \angle VCT' - \angle WCS'$$

where α and β are already known. Hence, adding, we have

$$\alpha + \beta = 180° + \angle UCR' + \angle VCT' = 180° + 2\angle UCR',$$

$$\angle UCR' = \frac{1}{2}(\alpha + \beta) - 90° = \angle CR'E;$$

$$\angle CS'E = \angle WCS' = \frac{1}{2}(\alpha - \beta).$$

Now, in triangle $CR'E$,

$$\frac{\sin CER'}{\sin CR'E} = \frac{CR'}{CE} \qquad \text{or} \qquad \frac{\sin CER'}{\sin\left\{\frac{1}{2}(\alpha + \beta) - 90°\right\}} = \frac{CA}{CE};$$

and, in triangle $CS'E$,

$$\frac{\sin CS'E}{\sin CES'} = \frac{CE}{CS'} \qquad \text{or} \qquad \frac{\sin\frac{1}{2}(\alpha - \beta)}{\sin CES'} = \frac{CE}{CA}.$$

Multiplying these two equations together term by term, we have

$$\frac{\sin CER'}{\sin CES'} \cdot \frac{\sin \frac{1}{2}(\alpha - \beta)}{-\cos \frac{1}{2}(\alpha + \beta)} = 1$$

$$\frac{\sin CER'}{\sin CES'} = \frac{\sin CER'}{\cos CER'} = \tan CER' = -\frac{\sin \frac{1}{2}(\alpha - \beta)}{\cos \frac{1}{2}(\alpha + \beta)} = \gamma$$

and $\cos CER' = \frac{1}{\sqrt{1+\gamma^2}}$, so that

$$\frac{CE}{CA} = \frac{\sin \frac{1}{2}(\alpha - \beta)}{\cos CER'} = \sin \frac{1}{2}(\alpha - \beta) \cdot (1 + \gamma^2)^{\frac{1}{4}}.$$

The results of the computation are, by Ptolemy's figures,

$$\angle CER' = 65°30' \quad \text{and} \quad \frac{CE}{CA} = \frac{2 + \frac{29}{60} + \frac{30}{60^2}}{60} = 0.0415.$$

This simple theory of the sun's motion with its one irregularity remained practically unchallenged for nearly fifteen hundred years and many observations through Copernicus and even Tycho Brahe. It was finally challenged by Kepler for reasons of pure theory, because of a new idea, and then destroyed by Kepler's use of Tycho Brahe's observations.

Another fatal but perfectly logical step in the Ptolemaic theory was the substituting for the apparent sun, whose motion on the ecliptic is irregular, a fictitious sun moving at constant speed about the earth. This is called the mean sun. Since we now have a mathematical structure of the motion, we can always reduce an observation made with respect to the apparent sun to an observation made with respect to the mean sun. The mean sun and apparent sun coincide at the apogee and perigee. That such a step should be made in this theory follows from the spirit of the theory itself: the principal aim is to build a mathematical model which follows the appearances. Of the two conditions which a physical theory must fulfill, mentioned at the beginning of this chapter, it is the first one which is emphasized in the system of Hipparchus and Ptolemy; little or no place is found for the second where the theory anticipates the appearances and presents a revolutionary vision of the world.

A.4   The Theory of Venus

We turn now to the theory of the simpler planet of the first set. As we have said, if one has observations of Venus over centuries, one finds that the sum of the greatest eastern and western elongations with respect to the sun reach a maximum for the sun at one point of the ecliptic and a minimum for the point exactly opposite. These maxima and minima will be judged with respect to the mean sun in accord with the practice of this system which treats each planet separately precluding any connection between them.

It is found that the maximum is the sum of a greatest western and eastern elongation each of $\left(47 + \frac{1}{2} + \frac{1}{30}\right)^{\circ}$, and the minimum for the point opposite is a sum of two elongations each $\left(47 + \frac{1}{4}\right)^{\circ}$. To take care of the cycle of Venus' swings back and forth on either side of the sun, we use the epicycle on a deferent as in the first case, i.e., the motion on the epicycle is in the same sense as that of the motion of the epicycle's center on the deferent, since the time from greatest western elongation through the maximum speed to greatest eastern elongation is the greater. To explain the fact that the greatest elongations vary for different parts of the ecliptic, we place the epicycle's center on an eccentric circle. The epicycle's center moves on the eccentric circle about its center at the angular speed of the mean sun while the planet moves about the epicycle's center 5 times in $8 - \frac{26}{(20)(365)}$ solar years in accordance with Table 1.

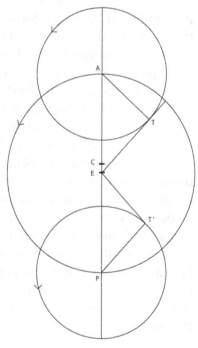

The line from the earth, cutting in half the angle between the two greatest elongations of the maximum or minimum, thus passes through the apogee and perigee of Venus' eccentric. It is remarkable that Ptolemy found in his time that the apogee of Venus' eccentric was only $10°30'$ west of the apogee of the sun's eccentric.

From these observations of the greatest elongations, we can now compute the eccentricity of Venus' deferent and the ratio of the radius of its epicycle to the radius of its eccentric.

Let C be the center of the eccentric deferent, E the earth. Let A be the position of the epicycle's center when at the apogee of the eccentric, that is, when the sum of the greatest elongations is least, the mean sun

Fig. 9

being on the line ECA and Venus at its greatest elongation at T. Likewise P is the eccentric's perigee where the sun of the greatest elongations is a maximum, and Venus is in its greatest elongation at T'. Then, by observation,

$$\angle AET = \left(44 + \tfrac{4}{3}\right)^{\circ} = \alpha,$$

$$\angle PET' = \left(47 + \tfrac{1}{3}\right)^{\circ} = \beta;$$

and angles ATE and $PT'E$ are right. Hence

$$AE = \frac{AT}{\sin(\alpha)}, \qquad PE = \frac{PT'}{\sin(\beta)} = \frac{AT}{\sin(\beta)}$$

or

$$AE + PE = 2AC = AT\left(\frac{1}{\sin(\alpha)} + \frac{1}{\sin(\beta)}\right),$$

$$\frac{AT}{AC} = 2\left[\frac{\sin(\alpha)\cdot\sin(\beta)}{\sin(\alpha) + \sin(\beta)}\right]$$

and again

$$AE - PE = 2CE = AT\left(\frac{1}{\sin\alpha} - \frac{1}{\sin\beta}\right)$$

$$\frac{CE}{AT} = \frac{1}{2}\left(\frac{1}{\sin\alpha} - \frac{1}{\sin\beta}\right)$$

$$\frac{CE}{AC} = \frac{\sin\alpha\cdot\sin\beta}{(\sin\alpha + \sin\beta)}\left(\frac{1}{\sin\alpha} - \frac{1}{\sin\beta}\right)$$

$$= \frac{\sin\beta - \sin\alpha}{\sin\alpha + \sin\beta}.$$

Letting $AC = 60$,

$$\frac{\text{epicycle's radius}}{\text{eccentric's radius}} \approx \frac{43 + \tfrac{1}{10}}{60} \approx 0.7183$$

$$\text{eccentricity of eccentric} = \frac{CE}{AC} \approx \frac{1 + \tfrac{1}{4}}{60} \approx 0.0208.$$

This gives Venus an extremely large epicycle, almost as large as the deferent, and a very small eccentricity. With the instruments at hand, it was not possible, in a geocentric theory, to assign Venus' position with respect to the earth and the sun, relative to linear distance. In the heliocentric theory of Aristarchus and Copernicus, this is no longer the case as we shall see; we need no new instruments and no new observations to assign immediately the relative distances of the earth and Venus from the sun; we need only the results we have just gotten.

Since, in the geocentric theory, there is no way of placing Venus in linear distance with respect to the sun, Ptolemy for very weak reasons of symmetry, placed Mercury and Venus on their epicycles, between the earth and the sun in this order, and the second set of planets beyond the sun. But Heraclides of Pontus in the Platonic Academy (350 B.C.) already seems to have placed them about the sun, that is, with the sun approximately at A and P, the center of the epicycle, and with Mercury's smaller epicycle concentric with Venus'.

In much the same way, but with added complications which are of no direct interest to us here, we find the eccentricity of Mercury and the ratio of its epicycle's radius to its eccentric's radius:

$$\frac{\text{epicycle's radius}}{\text{eccentric's radius}} \approx \frac{22\frac{1}{2}}{60} = \frac{3}{8} = 0.375$$

$$\text{eccentricity of eccentric} \approx \frac{12}{60} = \frac{1}{5} = 0.2$$

although this last ratio does not have the same simple meaning as with Venus.

When observations are made or old ones reconsidered to check this theory, it is found that the observed positions of Venus fit well with those predicted for positions of the mean sun about the apogee and perigee of Venus' eccentric; but, when the mean sun is a quadrant's distance from this apogee or perigee, that is, the line from the earth to the mean sun is perpendicular to the line of centers, the line through the earth and the center of Venus' eccentric and so through the apogee and perigee of this eccentric, then the observations are off by several degrees, but less so as the mean sun approaches the apogee or perigee.

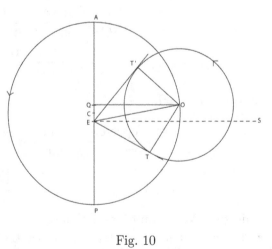

Fig. 10

To investigate this problem, Ptolemy considered a greatest western and a greatest eastern elongation of Venus for the mean sun 90° east of the eccentric's perigee. Both observations were made by Ptolemy himself, the first in the year 18 of Hadrian, Pharmouthi[2] 2-3 (140 A.D. historical way) and the second in the year 3 of Antonine, Pharmouthi 4-5, 6 years later. In the first, Venus was sighted in its greatest western elongation $\left(43 + \frac{1}{2} + \frac{1}{12}\right)^\circ$ from the mean sun; in the second, it was sighted in its greatest eastern elongation $48\frac{1}{3}^\circ$ from the mean sun.

In the figure ES is the line from the earth to the mean sun for both observations, according to which

$$\angle TES = \left(43 + \frac{1}{2} + \frac{1}{12}\right)^\circ = \alpha, \qquad \angle T'ES = 48\frac{1}{3}^\circ = \beta.$$

Draw the line from O, the center of the epicycle, perpendicular to the line of centers AP and parallel to ES, the line from earth to mean sun. According to the theory so far developed, Q should coincide with C, the eccentric's center, but these observations will show it does not. For

$$\angle TEO = \frac{1}{2}(\alpha + \beta),$$

$$\angle SEO = \angle EOQ = \frac{1}{2}(\alpha + \beta) - \alpha = \frac{1}{2}(\beta - \alpha).$$

But

$$EO^2 = (OT^2)\left[\sin \frac{1}{2}(\alpha + \beta)\right]^{-2} = QE^2 + OQ^2,$$

$$OQ^2 = (QE^2)\cot^2 \frac{1}{2}(\beta - \alpha);$$

$$(OT^2)\left[\sin \frac{1}{2}(\alpha + \beta)\right]^{-2} = QE^2 + (QE)^2 \cot^2 \frac{1}{2}(\beta - \alpha);$$

$$\frac{QE^2}{a^2} = \frac{OT^2}{a^2} \frac{1}{\left[1 + \cot^2 \frac{1}{2}(\beta - \alpha)\right]\left[\sin \frac{1}{2}(\alpha + \beta)\right]^2} = \frac{OT^2}{a^2} \frac{\sin^2 \frac{1}{2}(\beta - \alpha)}{\sin^2 \frac{1}{2}(\beta + \alpha)},$$

where a is eccentric's radius. We have already from above [p.19]

$$\frac{OT^2}{a^2} = \left(\frac{431}{600}\right)^2.$$

Computing, we find

$$\frac{QE}{a} \approx 2\frac{CE}{a}$$

---

[2] An Egyptian month.

and, indeed, Ptolemy considered it to be exactly double the eccentricity of the eccentric within the error of his observations. There are we assume

$$QE = 2CE.$$

Hence it seems that the epicycle's center O must move at constant angular speed, not about C the eccentric's center, but about another point Q, at twice the distance of C from E in the same sense. For it is obvious that the fact that O moves regularly about Q instead of C will affect the positions of Venus not at all for O at A and P, and very little for positions of O near A and P, and that it will affect the positions of Venus to the greatest extent when O is at the quadrants as it should.

Q is called the **equant point** and seems to be an original contribution of Ptolemy himself to the geocentric theory since he refers to no one before him and uses only his own observations made for this very purpose. Such a point will be used by Ptolemy and his followers for every planet except the sun. True to the positivistic spirit of his methods, Ptolemy does not ask why such points exist; he is content to use them to save the appearances. And we see now why Ptolemy adopted Principle B instead of the stricter Principle A for his construction of the world.

The equant point will have a great history. Copernicus will amend the heliocentric theory of Aristarchus in order to eliminate the equant points, and will consider the possibility of this elimination the greatest justification for his heliocentric theory, proclaiming Principle A as the only proper foundation for the science of astronomy. Kepler, in a youthful vision, will divine in the appearance of the equant point the presence of a new principle of mechanics which will make possible the classical system of Newton and Leibniz.

A.5   The Theory of Mars

The theory of Mars is precisely the same as that of Jupiter and Saturn, except for a difference in numbers. The problems here become more complicated, but it is on this planet that the Keplerian revolution will depend.

In the first place, we have no greatest elongations from the sun to indicate sharply the relation of the epicycle's motion to the possible eccentricity of the deferent. We do have station points and retrogradations, and a change in the angular distances between the two station points bounding the retrogradations for different positions on the ecliptic would indicate the deferent is an eccentric circle. But the measurements here are very difficult since it is hard to know exactly when the star is at a station point. In order, therefore, to have some precise way of identifying the sighting of the center of Mars' epicycle, Ptolemy made the assumption that once, when Mars, the earth, and the mean sun were in a straight line in that order in what is called an acronychal opposition (since Mars is on meridian at midnight), then the center of Mars' epicycle was also on that line[3]

We shall show that, since, in the case of all the planets of the second set, the number of cycles of heliacal irregularity plus the corresponding number of cycles of longitude is always equal to the corresponding number of solar years, therefore, if at one acronychal opposition, the center of Mars' epicycle was on the line of Mars, earth, and mean sun, then at every acronychal opposition it will be so. The reason for the assumption is not difficult to see: - when Mars is in acronychal opposition, it is in its retrogradation about halfway between the two station points. Since the planet is in its apparent slowest motion when it is at its position nearest the earth (for it is moving ,as we have seen, on its epicycle counter-clockwise as the epicycle's cen-

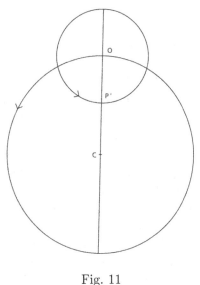

Fig. 11

ter moves counterclockwise in the deferent), therefore at some point nearly halfway between the two station points it will cross the line between the earth and the center of the epicycle. But we have said that Mars is in its retrogradation when it is more or less opposite the sun on the ecliptic and in the maximum retrogration when just about opposite the sun. To be precise, the assumption is made that Mars is at the maximum retrogradation exactly when it is in acronychal opposition with the mean sun. One could ask why the mean sun is chosen instead of the apparent sun, as Kepler will. But Kepler had a new viewpoint and a new leitmotiv to guide him; the question makes little or no sense in the system we are now completing.

It is intuitively clear that the planet's appearing to retrograde around P' to an observer at C will depend on the ratio of the epicycle's radius to the deferents and also to the ratio of the angular speed of planet on epicycle to the

---

[3] "Acronychal opposition" and "straight line" here refer to angles about ecliptic C giving possible latitudinal variations of Mars. Those without latitudinal variation of Mars become very important later on.

angular speed of epicycle's center on deferent. Thus the larger the epicycle's radius and the greater the angular speed of the planet on the epicycle, the greater the chance of the planet's appearing to retrograde around P' to an observer at C and greater the arc of the epicycle over which it will appear to retrograde. But Apollonius of Perga (about 225 B.C.) proved a beautiful theorem showing the exact relation of these conditions for retrogradation. But we shall put off proving this remarkable theorem until we shall have found the radius of Mars' epicycle.

We turn to prove now that, if the center of Mars epicycle (or that of any planet of the same set) at any one acronychal opposition lies on the line of Mars, the earth and the mean sun, then it will at every acronychal opposition. Let E be the observer, C the center of the deferent, and Q the equant point which may or may not be different from C but which we suppose to be on the line CE. Let also C, M, E, and S where S is the mean sun be the acronychal opposition in which it is assumed that all four points are in the same straight line. Let O', M', S' be the position the center of Mars' eccentric, of Mars, and of the mean sun in the next acronychal opposition. We have to show that O' lies on the straight line $M'ES'$.

We first remark that M and M' are called the apparent perigees of Mars to distinguish them from the regular perigees N and N' where Mars would be nearest the equant point, not the earth. The cycles of heliacal irregularity are to be measured with respect to N, the regular perigee, not with respect of M, the apparent perigee. Thus, if Mars were at N instead of M for position O of its eccentric it would have finished one cycle of heliacal irregularity for O' if it were at

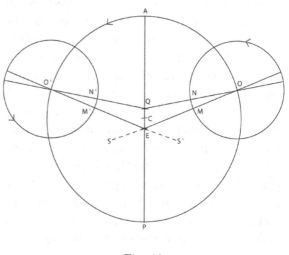

Fig. 12

N' at that time. It will obviously make no difference which perigee one uses if one counts the number of cycles of heliacal irregularity for Mars at a certain point of the ecliptic back to that same point, and the ratios of our table hold for either way of counting.

Now since the number of cycles of heliacal irregularity plus the number

of cycles of longitude always equals the corresponding number of solar years and, using radians, $2r\pi + \angle OQO'$ = cycles of longitude, $2m\pi - \angle NOM - \angle N'OM'$ = cycles of heliacal irregularity, $2n\pi + \angle SES'$ = number of solar years, therefore

$$2r\pi + \angle OQO' + 2m\pi - \angle NOM - N'O'M' = 2n\pi + \angle SES'$$
$$2\pi(r+m) + \angle OQO' - (\angle NOM + \angle N'O'M') = 2n\pi + \angle SES'. \quad (\alpha)$$

Now it is easy to see for the positions in the figure that, if O' is on the line $M'ES'$, then

$$\angle OQO' = \angle SES' + \angle NOM + \angle N'O'M',$$

since the exterior angle of a triangle equals the sum of the opposite interior angles. It then follows that

$$2\pi(r+m) = 2n\pi, \qquad r + m = n.$$

Further, if O' were not on the line $M'ES'$, then the equation (α) could not hold. For

$$\angle OQO' = \angle NOM + \angle N'O'M' + \angle SES'$$

could not equal any multiple of 2π radians since the differences between $\angle AQO'$ and $\angle N'O'M'$ for O' on $M'ES'$ and any other possible position of O' consistent with the fixed position of M' could not be as much as 2π radians, the other angles remaining constant. Hence (α) could not hold. The reader can convince himself that the same holds for N' in any other quadrant.

With this established, we can now turn to Ptolemy's classical method of computing the eccentricity of Mars' equant point and the direction of the line of centers of the eccentric. To do this, Ptolemy uses three astronychal oppositions of Mars on which occasions we can assume, as we have just shown, the center of Mars' epicycle is on the line with Mars, the earth, and the mean sun. These three oppositions were observed by Ptolemy himself; we shall not give them in detail but simply show the method of their use.

Let $POAO'O''$ by Mars' eccentric or deferent with center C, M be Mars at the first acronychal opposition, M' at the second, and M'' at the third. Let E be the earth and Q be the equant point about which O moves at constant angular speed. We shall assume Q lies on the line CE by analogy with the equant in the case of Venus and because ultimately it will enable us to construct a satisfactory theory. For the same reasons, we shall begin by assuming $EC = CQ$.

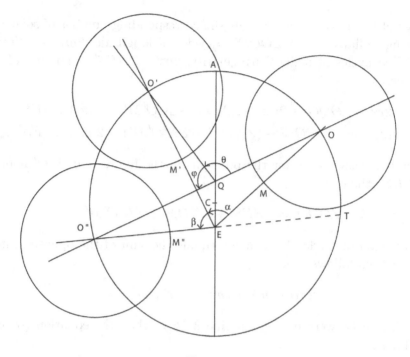

Fig. 13

From the observations we know the angles

$$\angle MEM' = 67°50' = \alpha, \quad \angle M'EM'' = 93°44' = \beta.$$

Again, since the time from the first observation to the second was 4 Egyptian years (of 365 days), 69 days, and 20 hours, therefore we know angle OQO'; for from Table 1 we know that Mars makes $42 + \frac{190}{(60)(360)}$ cycles of longitude in $79 + \frac{193}{(60)(365)}$ solar years. We can therefore compute how far it moves about Q in the given time. Likewise we can compute angle $O'QO''$ since the time from the second observation the third was 4 years, 96 days, 1 hour. Hence

$$\angle OQO' = 81°44' = \theta, \quad \angle O'QO'' = 95°28' = \varphi.$$

We are to find the ratio of QE to CO and the angle OEA.

Ptolemy found he could only solve this problem by a method of successive approximations. We assume for the first approximation that Q and C coincide. This is a small error which will be corrected to any desired degree by successive substitutions. Thus, having found $\frac{QE}{CO}$ assumed equal to $\frac{CE}{CO}$, we then take C as the midpoint and find what change this entails. We shall only outline this

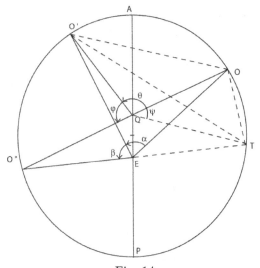

Fig. 14

long and tedious method, and include only this much of a discussion because of its enormous importance for the subsequent developments in Kepler.

We draw the figure again with $Q = C$. Continue $O''E$ to T; join $OT, O'T, QT$. Then, in triangle $O'TE$,

$$\frac{O'T}{ET} = \frac{\sin(\pi - \beta)}{\sin(\beta - \frac{\varphi}{2})} = \frac{\sin\beta}{\sin(\beta - \frac{\varphi}{2})}$$

and, in triangle OET,

$$\frac{OT}{ET} = \frac{\sin(\pi - \alpha - \beta)}{\sin(\alpha + \beta - \frac{\theta+\varphi}{2})} = \frac{\sin(\alpha + \beta)}{\sin(\alpha + \beta - \frac{\theta+\varphi}{2})}$$

Therefore

$$\frac{O'T}{OT} = \frac{\sin\beta}{\sin(\beta - \frac{\varphi}{2})} \cdot \frac{\sin(\alpha + \beta - \frac{\theta+\varphi}{2})}{\sin(\alpha + \beta)} = \lambda.$$

Again, in triangle OTO', since

$$\angle TO'O = \frac{\psi}{2}$$

where we call angle OQT the angle ψ,

$$\frac{O'O}{OT} = \frac{\sin\frac{\theta}{2}}{\sin\frac{\psi}{2}}$$

and $OT = 2a \sin \frac{\psi}{2}$ where a is radius of the eccentric.

Now, by Law of Cosines in triangle OTO', we have

$$O'O^2 = O'T^2 + OT^2 - 2(O'T)(OT) \cos \frac{\theta}{2},$$

$$\frac{O'O}{OT} = \left(\lambda^2 - 2\lambda \cos \frac{\theta}{2} + 1 \right)^{\frac{1}{2}},$$

and, hence

$$\frac{\sin \frac{\theta}{2}}{\sin \frac{\psi}{2}} = \left(\lambda^2 - 2\lambda \cos \frac{\theta}{2} + 1 \right)^{\frac{1}{2}},$$

$$\sin \frac{\psi}{2} = \frac{\sin \frac{\theta}{2}}{\left(\lambda^2 - 2\lambda \cos \frac{\theta}{2} + 1 \right)^{\frac{1}{2}}},$$

Computed by Ptolemy, ψ is $21°41'$.

Hence, by substitution

$$ET = \frac{\sin \left(\alpha + \beta - \frac{\theta + \varphi}{2} \right)}{\sin(\alpha + \beta)} OT = \frac{\sin(\alpha + \beta - \frac{\theta+\varphi}{2})}{\sin(\alpha + \beta)} \left(\sin \frac{\psi}{2} \right) (2a)$$

and

$$O''T = 2a \sin \frac{2\pi - (\theta + \varphi + \psi)}{2} = 2a \sin \frac{\theta + \varphi + \psi}{2}$$

and

$$O''T - ET = O''E = 2a \left[\sin \frac{\varphi + \theta + \psi}{2} - \frac{\sin(\alpha + \beta - \frac{\theta+\varphi}{2})}{\sin(\alpha + \beta)} \cdot \sin \frac{\psi}{2} \right]$$

But, by Euclid III, 35, and II, 5,

$$(ET)(O''E) = (AE)(EP),$$
$$(QE)^2 + (AE)(EP) = a^2.$$

Therefore, substituting in this last equation, we get

$$QE^2 = a^2 - 4a^2 \left[\frac{\sin(\alpha + \beta - \frac{\theta+\varphi}{2})}{\sin(\alpha + \beta)} \cdot \sin \frac{\psi}{2} \right]$$

$$\times \left[\sin \frac{\theta + \varphi + \psi}{2} - \frac{\sin(\alpha + \beta - \frac{\theta+\varphi}{2})}{\sin(\alpha + \beta)} \cdot \sin \frac{\psi}{2} \right]$$

From this, Ptolemy computed

$$\frac{QE}{a} = \frac{13 + \frac{4}{60}}{60} \approx 0.2186$$

Using what we have found, it is not difficult to compute angle ABO or angle AQO. Ptolemy actually computed

$$\angle AQO \approx 36°31'.$$

But these results follow from the premise that C and Q coincide and therefore only furnish a first approximation. Following Ptolemy, we now take C as the midpoint of QE and from this compute what the observed angles α and β would be with these measurements of $\frac{QE}{a}$ and $\angle AQO$. It turns out that

$$\alpha' = 68°55', \qquad \beta' = 92°21'.$$

If now again we use the same formulas as before to find $\frac{QE}{a}$ and $\angle AQO$, we have

$$\frac{QE}{a} = \frac{11 + \frac{50}{60}}{60} \qquad \text{and} \qquad \angle AQO = 42°45'.$$

We again take C as the midpoint of this new QE and compute what α and β would have to be, finding $\alpha'' = 68°46'$, $\beta'' = 92°36'$.

Using the same formulas, we get

$$\frac{QE}{a} \approx \frac{12}{60}$$

and

$$\angle AQO \approx 41°33'.$$

If we use these results to compute the observed angles α and β, we find the original angles actually observed. After a further check from an old observation, these numbers are adopted for the eccentricity and apogee of Mars; they will be the subject of a lengthy investigation by Kepler.

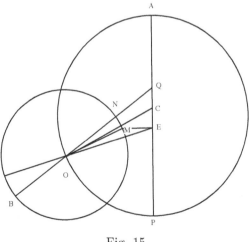

Fig. 15

We are now able to find the ratio of the radius of Mars' epicycle to the radius of Mars' eccentric. To this end, Mars was observed 3 days after the third acronychal opposition and appeared to have moved $0°58'$ westward since it is in retrogradation.

Letting E, C, and Q represent, as usual, the earth, the eccentric's center, and the equant point respectively, we have Mars at M sighted $53°54'$ west of the perigee P. Again from Table 1, we can compute that it has added, in the 3 days, $1°32'$ in the cycle of longitude and very nearly $1°21'$ in heliacal irregularity. If we add these to the positions at the third opposition above where we could have computed the angles AQO'' and $BO''M''$ in figure 13, we have here

$$\angle AQO = 137°11' = \gamma, \qquad \angle BOM = 172°46' = \delta.,$$

and from observation,

$$\angle PEM = 53°54' = \varepsilon.$$

We wish to find $\frac{OM}{OC}$.

Now, in triangle QOC, since $\frac{QC}{OC} = \mu = \frac{1}{10}$,

$$\mu = \frac{\sin QOC}{\sin(\pi - \gamma)}, \qquad \sin QOC = \mu \sin \gamma.$$

And, in triangle COE,

$$OE^2 = CE^2 + OC^2 - 2(CE)(OC)\cos(OCE)$$

$$\frac{OE^2}{OC^2} = \mu^2 + 1 + 2\mu \cos OCE.$$

But

$$\angle OCE = \angle QOC + 180° - \gamma$$

and hence is known. Therefore $\frac{OE}{OC}$ is a known constant σ.
And again, in triangle COE,

$$\frac{OE}{CE} = \frac{\sin OCE}{\sin COE}, \qquad \sin COE = \frac{OE}{CE}\sin OCE.$$

But

$$\frac{CE}{OC} = \mu, \qquad \frac{OE}{OC} \cdot \frac{OC}{CE} = \frac{\sigma}{\mu} = \frac{OE}{CE}$$

and therefore is known. And $\sin OCE$ is known; therefore also $\sin COE$. But

$$\angle OEP = \angle COE + \angle OCE$$

and is known also. Therefore

$$\angle MEO = \varepsilon - \angle OEP$$

and is known. Further

$$\angle MOE - \delta - \angle BOE = \delta - (180° - \angle QOE)$$
$$= \delta + \angle QOE - 180°$$

which is known. And, in triangle OME,

$$\frac{OM}{OE} = \frac{\sin MEO}{\sin MOE},$$

and so this ratio can be computed. We already have $\frac{OE}{OC}$; so we have $\frac{OM}{OC}$. Ptolemy computed

$$\frac{OM}{OC} = \frac{39 + \frac{1}{2}}{60}.$$

In precisely the same way, we can find the eccentricity, apogees, and ratios of radii for the planets Jupiter and Saturn. We place these results conveniently in

TABLE 2

	Eccentricity of Eccentric	Ratio of epicycle's radius to Eccentric's
Mercury	$\frac{12}{60}$	$\frac{22\frac{1}{2}}{60}$
Venus	$\frac{1\frac{1}{4}}{60}$	$\frac{43 + \frac{1}{10}}{60}$
Sun	$\frac{2\frac{1}{2}}{60}$	
Mars	$\frac{6}{60}$	$\frac{39\frac{1}{2}}{60}$
Jupiter	$\frac{2}{60}$	$\frac{11\frac{1}{2}}{60}$
Saturn	$\frac{3}{60}$	$\frac{6\frac{1}{2}}{60}$

It should be noted that the eccentricities given are those of the center of the eccentric circle or deferent. To have the eccentricity of the equant point, one doubles the eccentricity given in the table except for the case of Mercury.

A.6   Apollonius' Theorem on Retrogradation and Stations

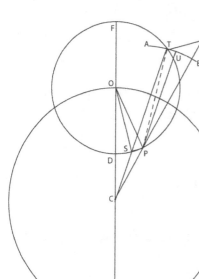

Fig. 16

The question still remains whether according to the theory developed above Mars and the other planets should actually appear to stop and retrograde as they do in their observed movements. The answer to this lies in an elegant theorem of Apollonius of Perga, the great geometer and authority on conic sections.

For the sake of simplicity, we suppose the epicycle's center to move regularly about the deferent's center which is also the point of observation. Correction can be made subsequently for the equant point and eccentricity.

Working analytically, let us suppose that the motion is such that at point S on the epicycle, the planet appears to stop for an observer at C. Take P any other point on the epicycle; for our purposes let it be between S and the point on the tangent from C to the epicycle. For it is obvious that all points beyond the point of tangency will not be points of retrogradation.

Let CS again intersect the epicycle at T. Draw CP, TV parallel to SP, and PU parallel to CST, intersecting TV at U. Then

$$\frac{\angle TPU}{\angle UPV} > \frac{TU}{UV}$$

because, drawing an arc AB through U with center P,

$$\frac{\angle TPU}{\angle UPV} = \frac{\text{sector } PAU}{\text{sector } PUV} > \frac{\text{triangle } PTU}{\text{triangle } PUV} = \frac{TU}{UV}.$$

But

$$\frac{TU}{UV} = \frac{CP}{PU} = \frac{CS}{ST}.$$

Hence, since

$$\angle SOP = 2\angle STP = 2\angle TPU,$$

therefore

$$\frac{\angle SOP}{\angle UPV} > \frac{CS}{\frac{1}{2}ST}.$$

Now $\frac{SOP}{\Delta t}$, where Δt is interval of time from S to P, represents angular speed of planet about center of epicycle, ω_2; and

$$\frac{\angle UPV}{\Delta t} = \frac{\angle SCP}{\Delta t}$$

represents the average angular speed, in going from S to P, about C, which results from the motion of planet on the epicycle about its center independently of the motion of epicycle's center about C.

Since

$$\frac{\angle SOP/\Delta t}{\angle UPV/\Delta t} > \frac{CS}{\frac{1}{2}ST},$$

therefore, if

$$\frac{2CS}{ST} = \frac{\omega_2}{\omega_1}$$

where ω_1 is angular speed of epicycle's center about C, then

$$\frac{\angle UPV}{\Delta t} < \omega_1.$$

But $\frac{\angle UPV}{\Delta t}$ is the angular speed about C, caused by the motion on the epicycle alone. Since it is less than ω_1 in the opposite sense, therefore ω_1 prevails and the planet appears not to retrograde at any point P after S. It should be remarked that, although $\frac{\angle UPV}{\delta t}$ is only an average, intuitively it is clear it is greater the nearer P is to S, and so is monotonic decreasing as P moves from S.

In the same way, if we take any point P' between D and S, we can show that the opposite effect prevails, and that the planet appears to retrograde. Hence the necessary and sufficient condition that S be a station point is that

$$\frac{CS}{\frac{1}{2}ST} = \frac{\omega_2}{\omega_1}.$$

Since $\frac{CS}{ST}$ decreases as S approaches D, with ST reaching a maximum DF and CS a minimum CD, but increases without limit as S approaches the point of tangency of the line from C, therefore the necessary and sufficient condition that the planet appear to retrograde is that

$$\frac{CD}{DO} = \frac{CO - DO}{DO} < \frac{\omega_2}{\omega_1}.$$

Let us make this test for Mars, assuming that the equant point, deferent's center, and observer coincide. Then

$$\frac{60 - 39\frac{1}{2}}{60} = \frac{43}{120} < \frac{37}{42 + \frac{19}{(6)(360)}}$$

and the inequality is more than sufficient to take care of the corrections for the fact that the equant point and the deferent's center are different from each other and the observer. We shall not stop to make these corrections by means of which we could predict the station points.

A.7 Latitudinal Variations

We pause to give a very rapid account of the variations of the planets (See p.5) off the ecliptic. The variation in latitude from the ecliptic for each planet appears to have two cycles, one tied to the deferent and one to the epicycle. Thus these variations will be explained (1) by inclining the deferents to the plane of ecliptic about the ecliptic's center, the earth, and (2) by inclining the epicycles to the planes of their deferents. The deferents or eccentrics are inclined about the center of the ecliptic because the planets appear to be exactly in the plane of the ecliptic only when the epicycle's center is at a quadrant's distance from the eccentric's farthest point north or south of the ecliptic and the planet on the epicycle is at quadrant's distance from the epicycle's apparent apogee or perigee. Therefore the epicycle is inclined about its diameter perpendicular to the line through its apparent apogee and perigee.

In the case of Mars, Jupiter, and Saturn, when the epicycle's center is at the northernmost limit of its eccentric, the planet appears farthest north (farther north than the eccentric's northernmost point) when it is at the epicycle's apparent perigee; and the planet appears farthest south of the ecliptic when the epicycle's center is at the southernmost limit of its eccentric and the planet again at the apparent perigee of the epicycle. For these three planets, therefore, the epicycles are tilted to account for this. For Mars, the eccentric's northernmost limit is very near its apogee.

But in the case of Venus and Mercury, the phenomena are more complicated. When the epicycle's center is at the eccentric's apogee or perigee, the latitudinal variation of the planet from the ecliptic is the same for the planet at the epicycle's apparent perigee as at the epicycle's apparent apogee. Hence, when the epicycle's center is at the eccentric's apogee or perigee, the line through the epicycle's apparent perigee and apogee lies in the plane of the eccentric. But when the planet is at the greatest elongations, and the epicycle's center at the eccentric's apogee or perigee, then the latitudes of the greatest eastern differ the most from those of the greatest western elongations and from the plane of the eccentric, Venus' greatest eastern elongation being north at the eccentric's apogee and south at its perigee. On the other hand, when the epicycle's center is at the intersections of the ecliptic and the eccentric (the nodes), that is, a quadrant's distance from the eccentric's apogee or perigee, then near the greatest elongations the planet is in the plane of the ecliptic, but the latitude for the epicycle's apparent apogee will differ the most from the latitude for the epicycle's apparent perigee. Thus the Ptolemaic Theory makes the plane of the epicycle shift so that the epicycle's diameter through the apparent apogee is in the plane of the eccentric at the eccentric's apogee and perigee with the diameter perpendicular to this at its greatest inclination to the eccentric's plane; precisely the opposite takes place at the nodes.

Because of its importance for us later, let us consider the treatment of Mars's latitudinal variations in some detail.

Let the plane of the figure be perpendicular to the plane of the ecliptic intersecting it in OO', and intersecting the plane of Mars' eccentric in AP where A is eccentric's apogee and P its perigee. E is the earth. Further let the plane of the figure intersect the inclined planes of the epicycles in lines RAS and TPU where R and T are the apogees of epicycle and S and U its perigees.

The assumption is made that the inclination of the plane of the epicycle to the eccentric's remains the same; that is,

$$\angle EPU = \angle EAS$$

Ptolemy observed the angles of latitudinal variations when Mars was at the epicycle's true perigee in the acronychal opposition, finding

$$\angle OES = 4 - \frac{1}{3}^{\circ}, \qquad \angle O'EU = 7^{\circ}.$$

where $\angle OES$ is north of the ecliptic and $\angle O'EU$ is south. Since the angles AES and PEU are very

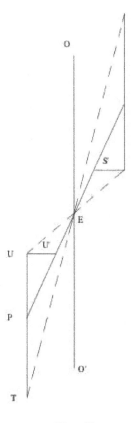

Fig. 17

small, Ptolemy considered the arcs through U and S
as straight lines perpendicular to the lines of sight from E and proportioned
to the sines of the angles. Then

$$\frac{\sin AES}{\sin PEU} \approx \frac{SS'/ES'}{UU'/EU'} = \frac{EU'}{ES'} \approx \frac{60 - 39\frac{1}{2} + 6}{60 - 39\frac{1}{2} - 6} \approx \frac{9}{5}$$

and

$$\frac{\angle AES}{\angle PEU} \approx \frac{9}{5},$$

$$\frac{\angle OES = \angle OEA}{\angle OE'U - \angle O'EP} = \frac{4 + \frac{1}{3}^\circ - \angle OEA}{7^\circ - \angle OEA} \approx \frac{5}{9},$$

$$\frac{7^\circ - 4 - \frac{1}{3}^\circ}{4 + \frac{1}{3}^\circ - \angle OEA} \approx \frac{9 - 5}{5},$$

whence

$$\angle OEA = 1^\circ, \qquad \angle AES = 3 + \frac{1}{3}^\circ, \qquad \angle PEU = 6^\circ.$$

We can interpret the situation in terms of Mars' motion on its epicycle
and use the extensive tables worked out for that purpose. we find that corresponding to $\angle AES = 3 + \frac{1}{3}^\circ$, at this distance,

$$\angle EAS = 2 + \frac{1}{4}^\circ,$$

and likewise angle EPU.[4] This, therefore, is the angle of inclination of the
epicycle to the eccentric while the eccentric is inclined at 1° to the ecliptic
plane. We can now compute

$$\angle OER = \angle OEA - \angle AER = 1^\circ - \angle AER.$$

But

$$\frac{\sin AER}{\sin \left(2\frac{1}{4}^\circ - AER\right)} = \frac{AR}{EA} = \frac{39\frac{1}{2}}{60},$$

$$\left(\sin 2\frac{1}{4}^\circ\right) \cot AER - \cos 2\frac{1}{4}^\circ = \frac{132}{79},$$

---

[4]This justifies the assumption that epicycle remains parallel to itself.

$$\angle AER = 0°54',$$

and, therefore, the least angle of northern latitudinal variation

$$\angle OER = 0°4',$$

a result which shall throw Copernicus into error.

A.8 Some Results of Lunar Theory

We cannot in this work follow through the development of lunar theory. This would constitute a history in itself from Ptolemy through Delaunay. But we shall only give some of the grosser results to be used later in a corroboration of the Newtonian theory.

The moon also moves back along the ecliptic from west to east, and the time of its return from a given point back again when disengaged from the other irregularities is about 27.3216 mean solar days which agrees quite well with modern measurements.

The ratio of the moon's distance from the Earth's center to the Earth's radius was found by Ptolemy by the use of the diurnal parallax. Since the moon's distance from the Earth is not so great with respect to the Earth's radius that the Earth can be considered as a point, the position of the moon among the fixed stars will appear different to an observer in a straight line with the Earth's center and the moon, and one who is not. But there is one lunar phenomenon which is independent of the observer, the lunar eclipse, that is, when the moon passes into the shadow cast by the Earth; this appearance is practically the same for all observers both as to magnitude and time. By using such lunar eclipses, one can construct a mathematical theory of the moon for an observer at the center of the Earth. We assume this has been done.

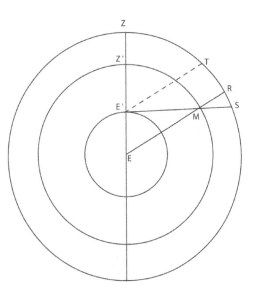

Fig. 18

Let E be the observer at the Earth's center, E' the observer on the Earth's surface, and M the moon sighted on the circle $Z'M$, the meridian through the observer's zenith point Z'. Draw $E'T$ parallel to ER. From the theory constructed from eclipses, we know the direction of the line EMR, that is, the point R of the celestial sphere at which the moon would be sighted by observer E; by actual observation we know the direction of the line $E'MS$. Hence the angle $E'ME$, the angle of parallax, is known on condition that we consider the distance EE' negligible with respect to ER, the radius of the celestial sphere. For then arc RS gives the measure of angle $TE'S$ as if E' were at E. Further we know angle $ZE'S$, and so $\frac{EM}{EE'}$. Ptolemy computed it to be $39 + \frac{3}{4}$ for a given observation which is not very satisfactory from modern measurements which give 60.2 for the mean distance. The Ptolemaic theory of the moon, however, is strange in that it sacrifices the accuracy of linear distances to accuracy for angular distances, and predicts a variation in the moon's distances from 34 radii of the Earth to 65, which is absurd for the grossest observation as Copernicus will point out. Luckily, the ratio for the moon's mean distance, in Ptolemaic Theory, comes out to be 59 which agrees well with modern measurements. Using the same method of parallaxes but with an improved theory of the moon's movements, Copernicus achieved an accurate result and without luck.

Finally, we add Eratosthenes' (about 260 B.C.) method of measuring the radius and circumference of the Earth. Taking two cities having almost the same meridian, Alexandria and Syene in Egypt, he compared the angle of the sun indicated by its shadow cast by a gnomon at Alexandria at noon of the day on which a gnomon at Syene cast no shadow.

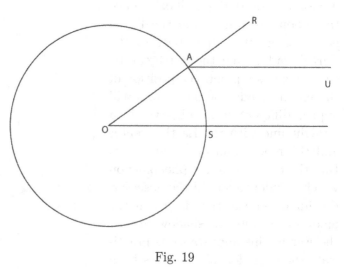

Fig. 19

Assuming that the lines to the sun from both cities are parallel because of the sun's great distance, then from the shadow cast at Alexandria, he knew angle RAU or angle AOS at the center of the Earth. He then measured the distance from Alexandria to Syene (this was difficult) and hence knew arc AS. Therefore he could find the radius and circumference of the Earth. By some happy coincidence, which historians have not been able to account for, Eratosthenes came up with the astonishing result of 3925 miles for the Earth's radius, the modern measurement is 3950 miles.

B The Heliocentric Theories of Aristarchus and Copernicus

B.1 Transition to the Theory of Aristarchus

In Book XII of the Almagest, Ptolemy reports an alternative theory for the planets of the second set, Mars, Jupiter, and Saturn, which he says was used by Apollonius of Perga and other mathematicians. This is the theory of the moving eccentric which we shall proceed to describe, showing that it is equivalent to the epicyclic theory of Mars already demonstrated and that it is a bridge from the geocentric to the heliocentric theory of Aristarchus of Samos (265 B.C.). We shall assume, for this demonstration, that the equant point, eccentric's center, and observer are identical. Suppose a circle about the Earth E as center with radius of Mars' epicycle in the plane of the ecliptic (we are eliminating again considerations of latitude); and suppose

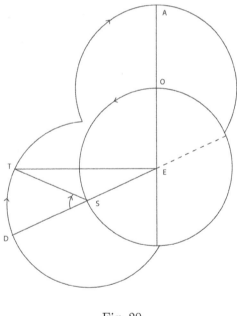

Fig. 20

a circle with radius equal to that Mars' eccentric has its center on this first circle moving from west to east about E at the angular speed of the mean sun while the planet moves on this second circle at the angular speed of the heliacal irregularity from east to west. Thus in the figure O moves to S at speed of mean sun while the planet moves from $D = A$ to T at the speed of heliacal irregularity. We shall show that this gives the same appearance as the theory of Ptolemy given in 1.A.5.

We note that

$\angle AED$ = angle of mean sun's motion,

$\angle DST$ = angle of heliacal irregularity,

$\angle AET$ = angle of apparent distance between two positions of Mars.

We see that
$$\angle AET = \angle AED - \angle SET.$$
But, by Table 1,
$$\angle AED = \text{(angle of longitudinal cycle)} + \angle DST.$$
Therefore
$$\angle AET = \text{(angle of longitudinal cycle)} + \angle DST - \angle SET$$
$$= \text{(angle of longitudinal cycle)} + \angle STE.$$

Let us compare this with the theory developed previously. Here in Figure 21,

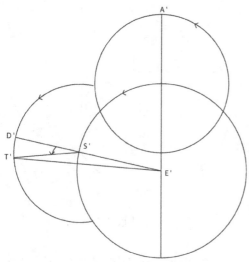

$\angle A'E'D'$ =angle of longitudinal cycle,

$\angle D'S'T'$ =angle of heliacal irregularity,

$\angle A'E'T'$ =observed angle.

Now, from Table 1, $\angle A'E'D' + \angle D'S'T'$ = angle of mean sun = $\angle AED$. We wish to show that the observed angles in both theories are equal:
$$\angle AET = \angle A'E'T'.$$

Fig. 21

But
$$\Delta TES \cong \Delta T'E'S'$$
since

$$S'E' = TS, \qquad SE = T'S', \qquad \angle D'S'T' = \angle DST.$$

Hence

$$\angle S'E'T' = \angle STE.$$

And

$$\angle A'E'T' = \angle A'E'D' + \angle S'E'T'$$

or

$$\angle A'E'T' = \text{(angle of long. cycle)} + \angle S'E'T' = \angle AET.$$

It is easy to see that the relative linear distances are also preserved.

In this new theory of the moving eccentric, the mean sun must lie always on the line DSE, between D and E when Mars is at its apogee A, $T = D = A$ and $S = O$; and on DE produced when Mars is at its perigee P in acronychal opposition PES, that is, $T = P$, as in Fig. 22. For we have already seen this in the epicyclic theory. But here the line DSE moves in the sense of and at the rate of the mean sun about E. Hence the mean sun always lies on it. It is therefore consistent with the theory and the appearances to place the mean sun at O, S, and $\bar{S}$ respectively in Figs. 20 and 22 so that the mean sun is precisely the center of the moving eccentric of Mars. If for each of the three planets Mars, Jupiter, and Saturn we were to keep the same circle EOS and, of course, the same center S, varying the radius of the eccentric and the angular speed of the planet on the eccentric as required in each, then we have achieved a remarkably unified theory for the planets of the second set which now all move on concentric circles with the

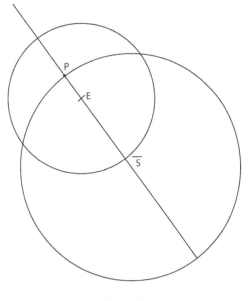

Fig. 22

mean sun as center, the mean sun in turn moving about the fixed Earth, with the epicycles of the previous theory all reduced to this circle of the mean sun about the Earth. The same appearances are exactly preserved.

It is not difficult to fit the planets of the first set, Venus and Mercury, into this system. The eccentrics of Mars, Jupiter, and Saturn are so large that the Earth and mean sun always lie within them, that is, these planets always lie outside the circle EOS, and hence can be seen at any angle from the mean sun. But Mercury and Venus move back and forth across the sun with greatest elongations on either side, and, as we have mentioned, Heraclides of Pontus (350 B.C.) had already placed their epicycles concentric to the sun. We have only to do this here, letting their deferents be identified with the circle OES so that their epicycles, with S as center, pass between E and S.

In this way we have all the planets on concentric circles about the mean sun with the mean sun moving about the fixed Earth. For Venus and Mercury, their epicycles in the Ptolemaic Theory represent their orbits about the mean sun, and their one deferent, the path of the mean sun about the Earth. For Mars, Jupiter, and Saturn, their epicycles in the Ptolemaic Theory represent one and the same circle, the path of the mean sun about the Earth, while their different deferents (or eccentrics) represent their concentric orbits about the mean sun.

Without more ado, we can read off from Tables 1 and 2, the relative distances of the planets and the mean sun from the Earth and their periodic times about the mean sun, but we shall wait for the next section to do so.

This theory was revived by Tycho Brahe more than a generation after Copernicus. But it is not too bold to conjecture that it was held in all its essentials prior to 265 B.C.

B.2   The Heliocentric Theory of Aristarchus

We have so far assumed that the radius of the sphere of the fixed stars is infinite with respect to the radius of the Earth. If we are willing to go a step farther and assume that the radius of the sphere of fixed stars is infinite with respect to the distance of the sun to the Earth, we can immediately pass from the geocentric theory of the moving eccentric to the heliocentric theory for Mars, Jupiter, and Saturn.

This is what Aristarchus assumed according to the report of Archimedes in **The Sand Reckoner**: "But Aristarchus of Sanos brought out a book consisting of some hypotheses, in which the premises lead to the result that the universe is many times greater than that now so called. His hypotheses are that the fixed stars and the sun remain unmoved, that the Earth revolves about the sun in the circumference of a circle, the sun lying in the middle of the orbit, and that the sphere of the fixed stars, situated about the same center as the sun, is so great that the circle in which he supposes the Earth to revolve bears such a proportion to the distance of the fixed stars as the center of the sphere bears to the surface. Now it is easy to see this is impossible... We must however take Aristarchus to mean this: since we conceive the Earth to be, as it were, the center of the universe, the ratio which the Earth bears to what we describe as the universe is the same as the ratio which the sphere containing the circle in which he supposes the Earth to revolve bears to the sphere of the fixed stars."

We shall see how this comes about. Let us consider again the theory of the moving eccentric as shown in Fig. 20 with S, the mean sun, and E the Earth. Let us shift the figure so that S takes the place of E and E moves out on the circle which was the orbit of S. We keep the fundamental triangle TES intact.

Thus in our new theory, A'' corresponds to A in the theory of the moving eccentric (Fig. 20), S'' to S and O, E to E_1'', in its first position, to E_2'' in its second, and T'' to T. Whereas before the observed angle between the two positions of Mars at A and T for an observer at E was angle AET, it is now for the two positions of Mars at A'' and T'' for a moving observer first at E_1'', and then at E_2'', the angle $FE_2''T''$ **if and only if in general, the fixed star F against which Mars is being sighted, is so far away that $\mathbf{FE_2''}$ is parallel to $\mathbf{A''S''}$,** that is, so that the moving observer constantly sees the direction in which it sighted A'' from E_1'' as parallel to itself, as E'' moves around S''.

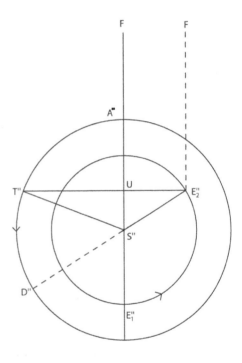

Fig. 23

We shall now show that, under this assumption of Aristarchus, with the Earth and Mars revolving in the same sense about the mean sun S, the Earth at the rate of the mean sun, Mars at the rate of its longitudinal motion, $\angle FE_2''T'' = \angle A''UT'' = \angle AET$ in Fig. 20.

We note that

$$\angle E_1''SE_2'' = \text{the angle of the mean sun},$$
$$\angle A''S''T'' = \text{the angle of Mars' longitudinal motion},$$

Since, as always for the planets of the second set, the angle of longitudinal cycle plus the angle of heliacal irregularity equals the angle of the mean sun,

$\angle E_1'' S'' E_2'' = \angle A'' S'' T'' +$ (angle of heliacal irregularity); but also

$$\angle E_1'' S'' E_2'' = \angle A'' S'' T'' + \angle D'' S'' T'';$$

therefore $\angle D'' S'' T'' =$ angle of heliacal irregularity.

Now, by hypothesis,

$$S'' T'' = ST, S'' E_2'' = SE,$$

and, in Fig. 20,

$$\angle DST = \text{angle of heliacal irregularity},$$

so that $\angle DST = \angle D'' S'' T''$ and we have indeed preserved the fundamental triangle by our new assumptions. Hence the triangles $T'' S'' E_2''$ and TSE are congruent and

$$\angle STE = \angle S'' T'' E_2''.$$

But

$$\begin{aligned} \angle A'' U T'' &= \angle A'' S'' T'' + \angle S'' T'' E_2'' \\ &= \text{(angle of long. cycle)} + \angle S'' T'' E_2'', \end{aligned}$$

and on p.41 we proved

$$\angle AET = \text{(angle of long. cycle)} + \angle STE.$$

Therefore

$$\angle A'' U T'' = \angle AET,$$

and the two theories explain the same appearances.

Therefore, in the heliocentric theory of Aristarchus, if we ignore the eccentricities which are relatively small, the deferent or eccentric of the Ptolemaic geocentric theory represents the true orbit of Mars about the mean sun, while the epicycle represents the orbit of the Earth about the mean sun reflected in our observation of Mars' motion. Hence the ratio of the radius of Mars' deferent to the radius of Mars' epicycle represents the ratio of Mars' distance from the mean sun to the Earth's distance from the mean sun, and the ratio of the number of solar years to the number of cycles of longitude represents the periodic time of Mars about the mean sun, that is, the time it takes Mars to make one revolution about the mean sun with respect to the fixed stars for an observer on the sun. The same laws of conversion hold for Jupiter and Saturn. These are the outer planets since their orbits are outside the Earth's.

This motion of the Earth gives the yearly motion of the sun. For the daily motion, the Earth is rotated on its own axis from west to east.

For Venus and Mercury, the conversion from the geocentric to the heliocentric is simpler. Here the Ptolemaic epicycle represents the orbit of Venus about the mean sun as it swings from greatest eastern to greatest western elongation about it.

Let O be the center of the epicycle and C the center of the deferent with Earth at C. In the Ptolemaic Theory, O rotates about C at the rate of the mean sun; in the Theory of Aristarchus, the Earth C rotates about O at the same rate, O being the mean sun. Since the planet rotates on the epicycle in the same sense as the epicycle's center O rotates about the deferent's center C in the Ptolemaic Theory, therefore, in the Theory of Aristarchus, Venus is revolving more rapidly in its orbit about the mean sun (and in the same sense) than the Earth. According to Table 1, it passes between the Earth and the mean sun, C and O, 5 times in nearly 8 solar years in the same sense, so that it revolves about the mean sun nearly $5 + 8$ times in 8 years, and its periodic time about the mean sun is nearly $\frac{8}{13}$ of a solar year. The ratio of Venus' distances from the mean sun to the Earth's is the ratio of the epicycle's radius to the deferent's radius in the Ptolemaic Theory.

We collect these results together in

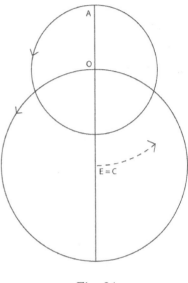

Fig. 24

TABLE 3

Planet	Relative Distances from Mean Sun			Periodic Times about Mean Sun		
Mercury	$22\frac{1}{2}$	0.3750	[0.3871]	$\dfrac{46+\frac{31}{(30)(365)}}{145+46+\frac{1}{360}}$	0.2408	[0.2408]
Venus	$43+\frac{1}{10}$	0.7183	[0.7233]	$\dfrac{8-\frac{26}{(20)(365)}}{13-\frac{5}{(4)(360)}}$	0.6151	[0.6152]
Earth	60	1.000	1	1.		[1.00004]
Mars	$91+\frac{11}{79}$	1.5190	[1.5236]	$\dfrac{79+\frac{193}{(60)(365)}}{42+\frac{190}{(60)(360)}}$	1.8808	[1.8809]
Jupiter	$313+\frac{1}{23}$	5.2174	[5.2028]	$\dfrac{71-\frac{147}{(30)(365)}}{6-\frac{29}{(6)(360)}}$	11.8746	[11.8622]
Saturn	$553+\frac{11}{13}$	9.2313	[9.5388]	$\dfrac{59+\frac{7}{(4)(365)}}{2+\frac{103}{(60)(360)}}$	29.4322	[29.4577]

The first column in each division contains the Ptolemaic figures; the second their decimal equivalents with the Earth's 1; the third column in brackets contains the modern equivalents. But it must be noted that we have given the Ptolemaic figures for the periodic times in terms of a return to the tropic or equinoctial points without correction for the precession. These are not therefore true sidereal returns; but this would affect the fourth place. The modern equivalents are given with these corrections but for the apparent sun. Both are in terms of the tropical or equinoctial year. The modern equivalents for distances are the mean distances, that is, half the major axis of the ellipse for apparent sun.

It is fairly certain that Aristarchus did not have figures as accurate as Ptolemy's; these are results which could have been deduced from his theory at the time of Ptolemy. That no one continued the heliocentric tradition after the 3rd century B.C., as far as we know, until the time of Nicolas Cusanus and Copernicus, nearly seventeen hundred years later, is probably due to the unfortunate ascendancy of the Aristotelian physics; it is a strange historical fact in any case.

For, astronomically, from the point of view of system, the heliocentric is obviously superior. Many appearances which are accidental in the geocentric theory become necessary in the heliocentric theory or at least have a deeper meaning.

Thus the fact that there are two sets of planets in the Ptolemaic geocentric theory follows from the fundamental geometry of the heliocentric theory. The fact that all planets in the Ptolemaic theory move on their epicycles in the same sense as their epicycles move on their deferents has little meaning there, but immediately translates into the planets' revolving all in the same sense about the mean sun in the heliocentric theory in such a way that the angular speed of the smaller orbit is greater than that of the greater orbit. Again the geocentric theory of Ptolemy, it seems quite arbitrary that Mars, Jupiter, and Saturn should retrograde at the acronychal oppositions, but this directly follows from the structure of the heliocen-

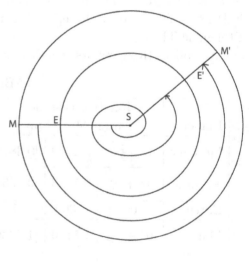

Fig. 25

tric system. And Ptolemy's assumption that at the acronychal oppositions the center of Mars' epicycle lies on the straight line through Mars, the Earth, and the mean sun, having a loose connection only with the rest of his theory, becomes an integral part of the heliocentric system.

For, in the latter, the mean sun is the center of the Earth's orbit which represents the Ptolemaic epicycle.

B.3 The Copernican Emendations

The heliocentric theory has been developed above without regard to the eccentricities of the deferents and equants, and it is very probably that Aristarchus paid no heed to them, and almost certainly did not know of the existence of the equants.

But, when Copernicus revived the Aristarchean Theory in 1535-40 A.D., he was naturally forced to bring these irregularities into the heliocentric system. In the case of Mars, Jupiter, and Saturn, since their deferents represent truly their orbits about the sun, therefore the eccentricities of deferent and equant in the geocentric theory would belong to their orbits about the mean sun in the heliocentric theory. But Venus' deferent in the geocentric theory is the Earth's orbit about the mean sun in the heliocentric theory; therefore it would seem that these irregularities belong not to Venus' orbit but to the Earth's. Strangely enough Copernicus ignored this distinction be-

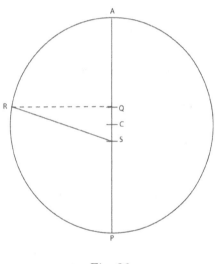

Fig. 26

tween the planets and incorporated the eccentricities into the orbits of all the planets leaving the Earth's orbit a circle with the mean sun as center. If he had incorporated the eccentricities of Venus' deferent into the Earth's orbit, this would have given the Earth an equant point and therefore in the geocentric theory an equant for the sun's motion about the Earth. But this was not found in the Ptolemaic Theory in its traditional treatment as we have seen above and the appearances had never presented any need for it. This would also give Venus an orbit about the mean sun without eccentricities since the epicycle is Venus' orbit in the heliocentric theory. Or it could be that the eccentricities of Venus' deferent represent a combination of the eccentricities of the Earth's orbit and Venus'. This problem, however, will be raised by Kepler in a striking fashion and for revolutionary reasons.

Let us repeat; Copernicus will incorporate the eccentricity of the Ptolemaic Theory into the orbits of the planets about the mean sun. If we consider the case of Mars, then Mars moves on its deferent with the mean sun replacing the Earth. Hence in Fig. 26, S is the mean sun, C the center of Mar's orbit, and Q the equant point about which Mars moves at constant angular speed. A is the aphelion and P the perihelion. The ratios of the Ptolemaic Theory are exactly preserved, that is

$$\frac{QC}{AC} = \frac{6}{60} = \frac{CS}{AC},$$

and Mars revolves about Q at the angular speed of nearly $\frac{42}{79}$ rev. per solar year. We shall call this the Ptolemaic form of the Copernican Theory.

For Copernicus the heliocentric theory has a greater mission than the superior cohesion and deductive power we have spelled out in the last section. At the beginning of the 16th century, two intellectual movements confronted each other; on the one hand a Neo-Aristotelian renaissance and on the other, a new Platonian advanced by Cardinal Nicolas Cusanus with the revival of Greek mathematical texts. Pico della Mirandola, at the court of Julius II, tried to marry the two in his philosophical treatises, and Raphael, in a series of paintings, expounded their opposition. Copernicus representing faithfully Aristotle, held that each science has its separate and distinct subject matter with its own proper set of axioms, and that it should be separate and distinct subject matter with its own proper set of axioms, and that it should be separate from every other science. The science of astronomy, according to this view, must be based strictly on Principle A; to introduce equant points is to mix the science of astronomy with the science of motions not celestial. For Aristotle himself, there were two distinct sciences of motion, the celestial mechanics built on the principle of concentric spheres, a materialization of the mathematical system of Eudoxus, and the science of motions proper to the sublunar sphere about the Earth. This distinction is kept by Copernicus and even by Galileo nearly a hundred years later; but the motions of the sublunar sphere will have to be extended to the neighborhood of every planet, since the position of the Earth and its space are no longer unique as in the geocentric theories of Aristotle and Ptolemy. Just where the regions dominated by the one or the other science end and begin is no longer very clear and not explained by Copernicus except in his affirming there must be gravity towards the center of each planet while any influence from one planet to another is excluded as in the Ptolemaic Theory.

In the matter of the Earth's daily rotation, Ptolemy had argued, using Aristotle's principles, that objects projected upwards would fall far to the west as the Earth moved beneath them. Copernicus replied that any object on the Earth participates in the rotating motion proper to the Earth and preserves it with the addition of a movement upwards. William of Ockham had indeed, two hundred and fifty years before, enunciated something like the classical law of inertia of Newton, but it is doubtful Copernicus knew of it since even Ockhams' followers neglected it. But exactly the argument of Copernicus providing an addition of the uniform circular motion of the Earth with the upward motion of the projectile to account for its return to the position it left on the Earth appeared a generation after Ockham in the work of Nicolas of Oresme; and it would seem Copernicus was here following a tradition. But even if Copernicus in these matters followed the nominalist mechanics of Paris, yet, in a more general sense, he advanced the Aristotelian notion of a science.

In the light of the Neo-Aristotelian world picture, Copernicus saw the heliocentric theory as one in which the equant points could be eliminated. He proceeded in this way. Consider the Ptolemaic form of the Copernican Theory in Fig. 27. We shall now use an epicycle to give the irregularity expressed by the equant point Q taking C' as the center of the deferent but $C'S$ as $\frac{3}{4}$ the distance QS in the Ptolemaic form. This change is compensated for by the addition of an epicycle with radius equal to $\frac{1}{4}$ of QS in such a way that the planet on the epicycle is at the proper distance AS when the epicycle's center is at the eccentric's aphelion O, hence below O; and opposite Q at R' when the epicycle's center O' is at the quadrant; and below O'' at P so that SP is the same length

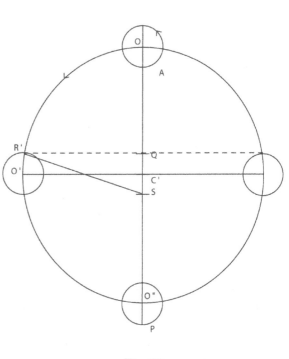

Fig. 27

as in the Ptolemaic form. Hence it is clear that the planet must move on the epicycle in the same sense as the epicycle's center moves on the eccentric and at the same angular speed. This preserves the same distances from S at aphelion and perihelion; and keeps the planet at the proper angular distances at these points and at the quadrants, as it should. But the distance $R'S$ is greater than RS in the Ptolemaic form. Whether this difference is an observable one remains to be seen; it is very small in its effect on angular distances for an observer at S. If we compute the angles ASR and ASR' in the two theories, we have, in Fig. 26.

$$\tan ASR = \frac{RQ}{QS} \approx \frac{\sqrt{60^2 - 6^2}}{12} \approx 4.9763$$
$$\angle ASR \approx 78°38'30'',$$

while in Fig. 27

$$\tan ASR' = \frac{R'Q}{QS} \approx \frac{60}{12} = 5.0000$$
$$\angle ASR' \approx 78°41'31''.$$

This difference of 3' is far below the possibility of error claimed by Ptolemy. It will turn out, however, that as far as the appearances are concerned, the Ptolemaic form is nearer the truth than the Copernican; but, at this point of development, one cannot decide between them.

In addition to eliminating the equant points which because of its salutary effects in lunar theory led Copernicus to believe he had truly restored the science of astronomy, he also adapted the precession of the equinoxes to the system of Aristarchus. Since, in the system the Earth rotates on its own axis for the daily motion and about the ecliptic for the yearly motion, its axis remains almost invariant in direction with respect to the fixed stars. In the geocentric theory, the precession appears as a slow motion of the sphere of the fixed stars from west to east about the poles of the ecliptic which means, in the heliocentric theory, that the axis of the Earth's rotation appears to revolve about the poles of the ecliptic in circles a little more than 23° from them from west to east. Interpreted further, this means that the Earth in revolving about the mean sun is a circle in the plane of the ecliptic varies the tilting of its axis of daily rotation in such a way that the Earth returns to the equinoctial point before it has made a whole revolution about the sun.

Using Greek, Arabic, and observations of his own, Copernicus thought to find an irregularity in the precessional cycle which has not been confirmed since.

In the matter of latitudinal variations, we shall only record Copernicus' treatment of the outer planets as essential to our exposition; and, in particular, we choose Mars. Since, in the heliocentric theory, Mars' Ptolemaic epicycle is the Earth's orbit, therefore Copernicus adapts the Ptolemaic Theory of latitudes in such a way as to make the plane of the epicycle parallel to the plane of the ecliptic. He uses Ptolemy's observations, confirmed by his own or contemporary ones, that at acronychal oppositions, when the center of the epicycle, Ptolemaic style, is near the eccentric's apogee or, Copernican style, when Mars is at its aphelion, and the Earth, in both styles, is at its point nearest Mars, then the latitudinal angle is nearly $4 + \frac{1}{3}^\circ$ north of the ecliptic; but, when Mars is at its perihelion and the Earth at its nearest point, then the angle of latitude is nearly 7° south of the ecliptic. Let ESE' be in the plane

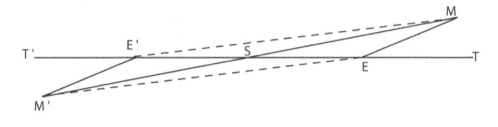

Fig. 28

of the ecliptic with S the mean sun, E the Earth in acronychal opposition with Mars at M at its aphelion, while E' and M' are in acronychal opposition with Mars at its perihelion. Then $M'SM$ is a line in Mars' eccentric and the plane of the figure is the plane perpendicular to the ecliptic. Then

$$\angle MET = 4°30',$$
$$\angle M'E'T' = 6°50',$$

which are the Copernican values replacing the Ptolemaic values of $4 + \frac{1}{3}^\circ$ and 7° respectively. Mars is nearer the Earth in its perihelion than in its aphelion. We have, from the Ptolemaic theory,

$$\frac{E'S}{SM'} = \frac{39\frac{1}{2}}{60 - 6}, \frac{ES}{SM} = \frac{39\frac{1}{2}}{60 + 6}$$

and Copernicus, from new observations of his own, modifies these values slightly so that

$$\frac{E'S}{SM'} = \frac{6580}{10,000 - 1460}, \frac{ES}{SM} = \frac{6580}{10,000 + 1460}.$$

If now we calculate, from these assumptions, angles $SE'M$ and SEM', the latitudinal angles at the conjunctions of the sun and Mars (when Earth, Sun, and Mars are on the same straight line in that order) which were not easily observed, we shall get angles greater than those gotten by Ptolemy from his theory. Deferring to the results of the Ptolemaic Theory, Copernicus introduced a slight oscillation of the plane of Mars' eccentric to and from the ecliptic which later gave Kepler the occasion to write in his **Astronomia Nova**, Chap.13: *"Copernicus divitiarum suarum ipse ignarus Ptolemaeum sibi exprimendum omnino sumpsit, non rerum naturam, ad quam tamen omnium proxime accesserat. - Copernicus, ignorant of his own riches, tried expressing Ptolemy to himself, not the nature of things to which he of all persons had approached most nearly."*

In addition to these theoretical emendations to the theory of Aristarchus, Copernicus discovered the very slow shift from west to east about the ecliptic of the line of apsides of the planetary orbits over and above the shift of precession and which he judged to be larger than we now do. Again his measurement of the sun's eccentricity gave a result smaller than Ptolemy's $\frac{1}{31}$ approximately instead of $\frac{1}{24}$.

Chapter 2

The Keplerian Revolution

The Copernican Theory, as we have seen, revived the Greek heliocentric theory of Aristarchus, absorbing at the same time the more sophisticated irregularities of Ptolemy, and, in doing so, it tried to reestablish a very special notion of a science in general and of the science of astronomy in particular.

The other intellectual position we have spoken of, opposing the spirit of both the Ptolemaic and the Copernican notions of science, considered mathematics not simply as a tool, but as holding a privileged place for the direction of all the disciplines of knowing including theology, and as being a sort of mirror of the universe. In this view, the very construction of a science with its heuristic techniques and its analogies with all the other domains of knowledge is far more important than the axiomatic structure and apartness. Here, instead of the neat deduction from sets of axioms hierarchically arranged, we have the dialogue of the enquiring mind in the search for a deeper unity of all knowing. Especially in its eyes, it is not sufficient for an astronomical theory to account for the appearances; it must anticipate them and furnish a unity and rationals far beyond the appearances themselves.

Cusanus, the philosophical leader of this movement, had, more than a century before, created the intellectual climate for its development with these leading themes. (1) The predominance of unity over being. (2) The importance of the notions of function and relation as opposed to the notions of substance and accident: we know things in their relations to everything else, not in their essences. (3) These relations are best known through number and proportion. (4) Human knowledge is constructive and conjectural; the proposition is created by the human mind just as God creates the universe, so that abstraction is not its fundamental operation.

From these general themes followed more particular ones: (1) There is no absolute rest in the sensible world: hence it is absurd to build a system in which the Earth is absolutely at rest. (2) The planets and stars influence each other and the Earth, and the notion of absolutely separate planets with laws of motion based on such absolute separation is false. And finally it is suggested in the **Idiota de Staticis Experimentis** that (3) all things could be related by weighing with a balance, and, although it is not stated explicitly that the planets too will be weighed, it is certainly implicit in the spirit of the treatise; and Kepler later will speak indeed of translating Archimedes' balance into the heavens from which Aristotle and his successors had thought to banish it forever.

A   The Youthful Manifesto

In 1596, at the age of twenty-five, Kepler, a student in theology and pupil of Mästlin, published his first work called for short the **Mysterium Cosmologicum**, in which he laid down a series of conjectures based on the presumption that the structure of the world must be the most beautiful possible, exhibiting the harmonies of geometry, number, and music and reflecting the unity of its Creator.

In view of this presumption, Kepler sought to find a rationale for the distances of the planets from the sun and a relation between their distances and their periodic times about the sun. Further, using the notions of interactions, attractions, and repulsions between bodies on the Earth, he sought to find a relation of force between the sun as center and the planets revolving about it.

For the distances, Kepler, having six planets and the five regular solids of Euclid, tried to fit the distances of the planets to the spheres inscribing and circumscribing the regular solids in a certain order. Thus, except for Jupiter, the following order is fairly successful if one takes the radius of the circumscribed sphere as the shortest distance of the planet from the sun and the inscribed sphere as the greatest distance of the next planet below:

> a cube between Saturn and Jupiter;
> a tetrahedron between Jupiter and Mars;
> a dodecahedron between Mars and Earth;
> an icosahedron between Earth and Venus;
> an octahedron between Venus and Mercury.

For the distances as functions of the times, Kepler was not successful at this time, although he first tried the relation

$$\frac{r_1}{r_2} = \frac{T_1}{\frac{T_1+T_2}{2}}$$

where r_1 is the mean distance of the inner planet of the two, r_2 of the next planet, and T_1 and T_2, their periodic time respectively. In a later edition, he tried

$$\frac{r_1}{r_2} = \frac{T_1}{\sqrt{T_1 \cdot T_2}}.$$

He was only to succeed in this particular matter at the end of his career. Since the periodic times are greater for planets at greater distances, this suggested that the power of the sun to rotate planets about it is weaker the farther the planet is from it, and for the first time the analogy of illumination and the motor force of the sun is considered.

But these questions were too general and the solutions were not forthcoming. But the questions were of a new kind and they served to direct astronomical research in a new direction, and his very failure in answering them forced Kepler to question and criticize, in terms of his own fundamental insights, the Copernican results.

Moreover, two conjectures consequent upon these general hypotheses were more adapted for opening the way for concrete investigation and for analyzing away the assumptions of Greek celestial mechanics as traditionally practiced. First, if the sun is the center of force, then one cannot with impunity make assumptions with respect to the mean sun instead of the apparent sun. The distances of the planets were computed by Copernicus with respect to the mean sun as the center of the world. But the apparent sun is in the center of the world in Kepler's Theory since it is the center of force. And, in this regard, Kepler will remember later that Ptolemy assumed, for the planets of the second set, that the centers of their epicycles are

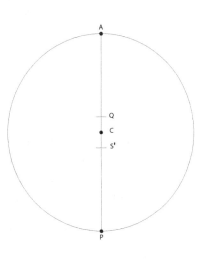

Fig. 29

on a line with the planet and the Earth, at acronychal oppositions with respect to the mean sun.

Second, – and this was immediately the most fruitful consequence,– when one considers the planets of the second set in the amended Ptolemaic form of the Copernican Theory, because of the existence of the equant point, the planet is moving more slowly when it is farther from the apparent sun S' at the aphelion A, than when it is nearer the apparent sun at the perihelion P. The existence of the equant point can now be explained as exhibiting the force of the sun on each individual planet of the second set since the planet moves more slowly in angular speed and also in linear speed the farther it is from the sun. The epicycle represents the motion of the Earth about mean sun, and no equant point had ever been found for this orbit.

But the Keplerian Theory immediately presents a difficulty. If the equant represents the variation of the force of the sun with the distance of the planet from the sun, then every planet which has an eccentricity and hence a variation of distance from the sun must have an equant point. Thus the Earth must have an equant point. In the case of Venus, the problem is more complicated: the deferent of the Ptolemaic Theory represents the path of the Earth about the sun. It would seem then that the eccentricities of Venus belong properly to the Earth with respect to the sun so that Venus would have no equant point and the Earth would.

Kepler predicted that there must be an equant for the Earth and for Venus. To establish his theory, he needed the observations of Tycho Brahe in Prague; these were the most recent and made with superior instruments. And so he joined Tycho Brahe in 1600.

B   The Commentaries on Mars

Since Mars has the greatest eccentricity except for Mercury, and Brahe fortunately was observing Mars most particularly, Kepler began the test of his theories on this planet. The work was to take nine years, resulting in the publication of what many have considered to be the greatest single work in physical theory, **Astronomia Nova**, or **The Commentaries on Mars.**

In the introduction to this work, we find stated:

> *Therefore the true doctrine of gravity rests on these axioms.*

> *Every corporeal substance, in so far as it is corporeal, is by nature fitted to remain at rest in every place where it is placed alone beyond the orbit of the force of a like body.*

> *Gravity is a corporeal affect between like bodies for their union or conjunction (and magnetic force is of this sort) so that the Earth pulls a stone much more than the stone attracts the Earth.*

Heavy things (even if we place the Earth in the center of the world) are not carried to the center of the world as to the center of the world, but to the center of a round body of the same kind, that is, the Earth. And so wherever the Earth is carried by its animate power, always to it will heavy things be borne.

This statement of a universal law of attraction, which brings in magnetism as an example under the influence of William Gilbert's **DeMagnete** published in 1600, completely denies the Aristotelian Theory of proper place according to which only bodies below the lunar sphere seek the center of the universe, the Earth. But from stating such a general and revolutionary theory of universal attraction, it is a far cry to the ability to formulate it precisely in kinematic and mathematical form. It was a long road that Kepler

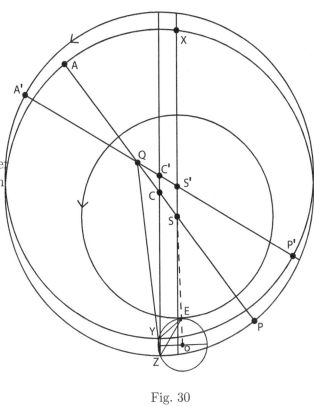

Fig. 30

had yet to travel to apply it to the problems of Mars' orbit. Nor was he ever to solve completely the problem of the motion of two bodies about each other as an example of this mutual attraction.

We now proceed with an outline of the long dialectic of discovery by which he applied the theses of the **Mysterium Cosmographicum** to the motions of Mars.

As he had already indicated in his youthful work, in view of his theory that the planet should have its slowest motion at the aphelion with respect to the apparent sun and not with respect to the mean sun, as in Copernicus' treatment where the mean sun replaces the Earth of the Ptolemaic Theory,

Kepler saw that one must construct the line of apsides, SCQ, through the apparent sun and not through the mean sun. This means one must consider the acronychal oppositions with respect to the apparent sun and not with respect to the mean, so that Mars is seen on its orbit with respect to the apparent sun without the interference of the Earth's motion at such oppositions.

For this reason, Kepler set out to compute the position of greatest angular difference between the two theories so that one could check by observation.

Let S' be apparent sun, and S the mean sun or center of Earth's orbit and the center about which it revolves at constant speed.

Let Q be the equant point of Mars, C the center of eccentric path with respect to mean sun and C' with respect to apparent sun. If the two eccentrics are drawn with C and C' as centers, then the greatest difference in the path, will be YZ on the line CC' (or in the opposite direction). Kepler has to take the intersection of QZ with the eccentric about C' instead of Y to consider the planet in the two theories at the same time. The question is at what position of the Earth will YZ be seen in its greatest angular distance and the answer is when the Earth is at E, the point where a circle through YZ is tangent to the Earth's orbit. Such a circle will have its center on the perpendicular bisector of YZ and the line from S through the point of tangency of the two circles.

Then all angles on this little circle subtended by YZ are equal while the angles subtended by any point different from E on the circle of the Earth's orbit are less.

Kepler computed this angle YFZ larger than the corresponding angle on the other side of S' because the Earth is nearer YZ, to be approximately $1°3'32''$, an angle which could be certainly observed if the rest of the theory is correct.

Since Tycho Brahe had questioned the eccentricities of Mars as established by Ptolemy and Copernicus, notably the bisection of the equant's eccentricity by that of the deferent's, therefore to test the eccentricities of Mars, Kepler first devised a new method of approximation using four acronychal oppositions instead of Ptolemy's three. Secondly, he reformed the theory of planetary latitudes in the light of his new principles and established the eccentricities in a new way.

First, in the method of four acronychal oppositions, three were used by Kepler to find the aphelion and eccentricity according to the general method of Ptolemy in Chapter 1, p.25, but with respect to the apparent sun according to the theory of Kepler. Hence let S' be sun, C eccentric's center, and Q equant. By observation, we know, from the acronychal oppositions, angles $M_1S'M_4$, $M_1S'M_2$, $M_2S'M_3$, and $M_3S'M_4$. For at the acronychal oppositions, Mars, Earth, and mean sun (or apparent sun) are on the same straight line except for latitudinal variations. Hence observers on Earth and sun see Mars longitudinally against the same fixed star. But in the heliocentric theory, the fixed stars are infinitely distant with respect to the distance of the Earth from the sun. Therefore the observed angles between two oppositions of Mars as seen from the Earth and the sun are the same. - From the mean longitudinal motion, we have angles $M_1QM_4, M_1QM_2, M_2QM_3, M_3QM_4$. We have the approximate position of $AQS'P$ from the Ptolemaic method and also the approximate ratios of QC, CS', CA.

Kepler then explicitly assumed all through the subsequent adjustment that M_1, M_2, M_3, and M_4 all lie on a circle (from an ancient metaphysical prejudice, he says) and the Q, C are on the same line with S' in agreement with his theory concerning the dynamical meaning of the equant point. He then determined angles $M_2M_1M_4$ and $M_2M_3M_4$. If their sum is not $180°$, then the four positions of Mars are not on a circle, and a change is made in angle $M_1S'A$ until their sum is $180°$ so that the four positions are on the circle. With this established, the position of C is computed, and if it turns out to be not on

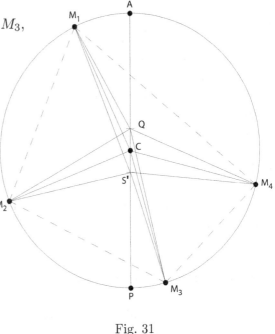

Fig. 31

QS', a shift is made of angle M_1QA (and consequently of angle $M_1S'A$) until M_1, M_2, M_3, M_4 fall on one circle and C falls on $QS'A$. Kepler reports he computed in this way seventy times.

The results, for several sets of four oppositions, were the following proportions:

$$CA = 100,000, \quad S'Q = 18,564 \quad \text{and} \quad CQ = 7,232.$$

This contradicts Ptolemy's bisection and is similar to the results of Tycho Brahe, with no great difference if referred to the mean sun. This result, so accurate for the longitudes of Mars, was called the vicarious hypothesis.

Secondly, an attack was made on this problem from the point of view of the latitudinal variations. Here again Kepler applied his new ideas to effect a reform in the Ptolemaic and Copernican theory of the latitudes. Since the planet's motion is subject only to the force of the sun because of its relative great size, therefore the plane of the planet's motion should be an immovable one through the apparent sun. With this hypothesis, it is now possible to find the true inclination of the planet's plane to that of the ecliptic, and eliminate the Copernican oscillation.

Let M_1, M_2, M_3 be positions of Mars in its plane, S' sun, and E Earth moving in the plane of ecliptic. To find the line of intersection of the two planes, that is, the line of nodes, we have only to observe an acronychal opposition of Mars where Mars has no latitudinal variation at M_1. To find the true angle of inclination, we have only to observe the latitude of Mars

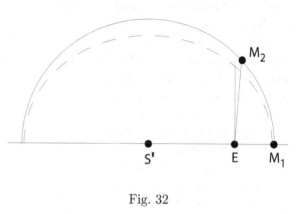

Fig. 32

when the Earth is on the line of nodes through the apparent sun and Mars is at a quadrant's distance from it at M_2. In this way, Kepler computed the angle of inclination to be $1°50'$.

We now consider two acronychal oppositions of Mars with the apparent sun, one when Mars is at its northern latitudinal limit and the other at its southern. At least this is the ideal situation, although it was not perfectly realized in the actual observations of Kepler who made the proper corrections. Hence let M_1 be Mars at its northern-most limit, M_2 at its southern-most, M_1M_2 Mars' plane, E_1 and E_2 the Earth in the ecliptic's plane, S' the apparent sun, and C the center of the sun's orbit. Then the observed angles are $O_1E_1M_1$ and $O_2E_2M_2$, and the angles $E_1S'M_1$ and $M_2S'E_2$ are computed from the latitudinal theory of Mars just given. Taking as given E_1S' and E_2S' from the theory of the sun and Earth as established by Ptolemy and corrected by Copernicus, M_1S' and M_2S' can be computed, and then

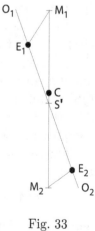

Fig. 33

$CS' = M_1S' - \frac{1}{2}(M_1S' + M_2S') = \frac{1}{2}(M_1S' - M_2S')$. But here the results give the bisection of the eccentricity demanded by Ptolemy. And the same contradictions result if the oppositions are considered with respect to the mean sun.

The same bisection of the eccentricity of the equant is required for Mars at its aphelion or perihelion when not in acronychal opposition.

Since this theory follows primarily from two assumptions: (1) the path of the planet is a perfect circle; and (2) the planet moves at constant angular speed about one definite point on the line of apsides, therefore it would seem that either one or both of these assumptions is false.

But before resolving this contradiction, it was natural for Kepler to decide to use the conjecture of his youthful work that the Earth must have an equant point to test his theory and, if found true, to clear up any error in the theory of Mars which might follow from an error in the theory of the Earth and sun. For we used the ratio of E_1S' to E_2S' from this theory. To investigate the path of the Earth without introducing any hypothesis about

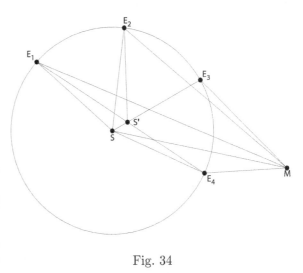

Fig. 34

Mars' distances from S and S', Kepler found this ingenious method. We assume only that Mars returns to the same point of the ecliptic in its longitudinal motion in equal times (with correction for precession) and that we know the equant point of the Earth's path about the sun, that is, the mean sun. For the procedure used by Ptolemy and Copernicus for establishing the eccentricity of the sun or Earth really determined the equant point, the center of uniform angular motion; they then assumed it to be the center of the eccentric. We also assume the accuracy of the vicarious hypothesis for angular positions of Mars as attested by the observations of Tycho Brahe, which have also been used to make more precise, within the framework of the Ptolemaic and Copernican theories, the numbers of the solar theory.

With these assumptions, Kepler took four positions of the Earth with Mars at the same longitudinal point of its path, that is, he took four observations of Mars from the Earth at three intervals of time equal to the periodic time of Mars abut the sun. Let these four positions be E_1, E_2, E_3, E_4 one of which falls on the line of apsides SS'; let S be the mean sun supposedly the equant point of the Earth (for Ptolemy and Copernicus also the center of its orbit), S' the apparent sun, and M the position of Mars, the same for all four positions of the Earth.

Now angle E_1SM is known from the vicarious hypothesis and also the direction among the fixed stars of SM. But the direction of E_1M is observed, so that angle SME_1 is known and the ratio of E_1S to SM. Further angles $E_1SE_2, E_2SE_3, E_3SE_4$ are equal and known from the uniform motion of the sun so that the ratios of E_1S, E_2S, E_3S, E_4S to SM can be computed. The results of the computation are

$$\left.\begin{array}{l} SE_1 = 66774 \\ SE_2 = 67467 \\ SE_3 = 67794 \\ SE_4 = 67478 \end{array}\right\} \text{ to } SM = 100,000.$$

But, if S is not only the equant, but also the circle's center, they should be equal. Not only are they not equal, but SE_3 is the longest at the sun's perigee, and the others shorter the farther they are from the perigee as would be expected from the other planets. S is not the center of the circle, and the center of the circle lies in the direction of S'.

Using these figures and others for different sets of 4 positions, Kepler computed the eccentricity of the equant to be 3600 to 100,000 or 0.036 which agreed with Brahe's rectification of Ptolemy's 0.042, but, contrary to all his predecessors, found that the center of the eccentric bisects this eccentricity of the equant, as predicted in the **Mysterium Cosmologicum.** The dynamic theory of Kepler is thus corroborated in an essential point: the equant seems to indicate the influence of the sun on the planet. Again, while the Ptolemaic and Copernican theories could give no good reason for the position of the eccentric's center on the line between the sun and the equant point, the Keplerian theory of solar influence demands it.

The reform of solar theory effected by Kepler still does not remove the contradiction between the vicarious hypothesis and the theory of latitudinal variations. It is for this reason that Kepler now turned to the question of the law of force realized in the astronomy of his predecessors by the equant point. Although he did not have the formal calculus by which we were able easily to compute that the linear speed of the planet at the aphelion and perihelion in the Ptolemaic form of the Copernican theory is inversely as the distance from the real sun [p. 12], he was able to arrive at the same conclusion by a laborious and ingenious method we shall not repeat here. He was also able to show that the same law did not hold exactly in other points of orbit. It was proper, therefore, to drop the geometric picture of the equant point altogether for the mathematically formulated physical law that the influence of the sun varies inversely as the distance from it as a first attempt at a formulation of a celestial dynamics.

Further Kepler used two physical analogies to explain this law. (1) The motor virtue of the sun whirls the planet about it like light. For by geometrical optics, it was argued that the illumination of light varies inversely as the square of the distance, since the number of lines of light from S on the surfaces of two concentric spheres about S would be the same, but the areas would vary as the squares of the distances. If, however, we considered only the circle as representing the path of the planet along which the planet is being moved by the motor virtue of the sun, then, as with light, the number of lines of force or of light are the same, but the lengths of the paths on which they are acting are as the distances. Hence the densities of the lines of light for the respective arcs are inversely as the distances. (2) As Gilbert had already suggested, the Earth is a magnet and perhaps also the sun, and it might be possible to explain in this way if only the precise mathematical formula could be found, the fact that the planets do not move in concentric circles about the sun, but are now nearer now farther away.

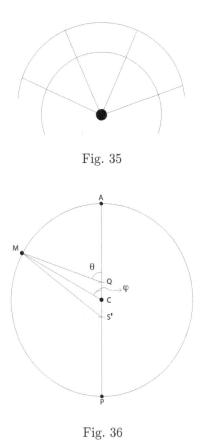

Fig. 35

Fig. 36

The equant point's purpose was to predict the angles of the planet given the times. If the distance law which now replaces the equant, is to have any efficacy, it too must allow us to predict the angles from the times. But this presents a difficult problem of integration. In modern notation, if φ is about center of eccentric C, and r distance of Mars from apparent sun S', radius of eccentric and $ae = QC = CS'$, assuming bisection[1], then $a\frac{d\varphi}{dt} = \frac{e}{r}$ where c is

---

[1] It should be remarked that the distance law is used with the results of the theory of latitudinal variation rather than those of the vicarious hypothesis as far as the position of the eccentric is concerned. For the former theory deals more directly with distances from the sun while the latter deals rather with angular positions.

a constant, and $r = (a^2 + a^2 e^2 + 2a^2 e \cos \varphi)^{\frac{1}{2}}$, so that

$$t = \frac{a}{c} \int_0^\varphi r d\varphi.$$

Since the problem of integration immediately bought to mind areas, it occurred to Kepler that it would be interesting to see how close an approximation to this integration the corresponding areas swept out about S' by the planet M would be. They are obviously not the same.

Kepler considered the difference geometrically in this way. For geometrical reasons, we shall consider the integration with respect to $(ad\varphi)$ instead of $d\varphi$, that is, we shall consider equal increments $(a\Delta\varphi)$ instead of $\Delta\varphi$. This will only change the results by a constant a. Let us suppose, in the manner of Archimedes, who is here the source of Kepler's inspiration, that the semicircle AM_2P is unrolled as a straight line that is equal to $a\pi$, and that a rectangle with length $a\pi$ and width $AC = a$ is constructed whose area is twice the area of the semicircle. Then the areas swept out by M_1, M_2, etc. according to the distance law will be given by the figure with base AM_2P and widths AS', M_1S', etc.

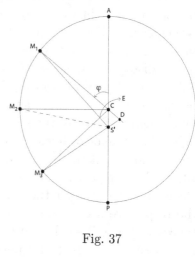

Fig. 37

corresponding to lengths along AM_2P equal to a $\Delta\varphi_1, a\Delta\varphi_2$, etc. Therefore the area in the figure bounded by the straight lines AM_2P, PS', and AS', and the curved line of the S', is the geometrical representation of the integral of the distance law

$$t = \frac{a}{c} \int_0^\varphi r d\varphi, \frac{1}{c} \text{ a constant of proportionality.}$$

Let us now effect a comparison of this distance law with the areal law. The integral representing the areal law is

$$t = \frac{1}{C'} \int_0^\varphi a(r') d\varphi,$$

where $r' = DM$, etc. in Fig. 37, where D is the foot of the normal from S' on CM_1 produced etc. That is

$$r' = a + ae \cos \varphi$$

and

$$\int_0^{\varphi} a(r')d\varphi = \int_0^{\varphi} a^2 d\varphi + \int_0^{\varphi} ae(a\cos\varphi)d\varphi,$$
$$= 2[\text{sector } ACM_1 + (ae)(a\sin\varphi)],$$
$$= 2[\text{sector } ACM_1 + \text{triangle } S'M_1C].$$

[2]

If we now turn to a geometrical representation in the manner of Fig. 38, we again take abscissas $(a\Delta\varphi)$ along AP and ordinates r' perpendicular to them. Now $r' = r$ at A and P, $r' < r$ at all other points with greatest differences at the quadrants; at $\varphi = \frac{\pi}{2}$ and $\frac{3\pi}{2}$, $r' = a$. Hence the area represented by this integral, which is the area swept out by r about S', is also the area in Fig. 38 bounded by the straight lines AP, AS', PC and the curved line $S'DCES'$.

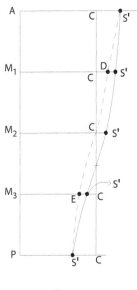

Fig. 38

In using the areas represented in Fig. 38 for the distance law and the areal law, to measure the times of the planet, we must remember that each total area represents the same half periodic time of the planet about which there is not doubt. In the analytic representation by the integrals, this is taken care of by the multiplication by constants $\frac{1}{c}$ and $\frac{1}{c'}$. But Kepler's Fig. 38 shows the distribution of the times for each law. Now the greatest difference in the two laws occurs around the quadrants. For the sequel, it is important to notice that, in passing from the positions M_1 to M_2, the area M_1DCEM_3 of the areal law is a smaller part of its total area than the area $M_1S'S'M_3$ of the distance law is of its total area. Therefore, by the areal law, the planet takes less time in passing from M_1 to M_3 than it does by the distance law.

The distance law was only an approximation in the first place, and the areal law is only an approximation to it. But the strange thing about all of this is that, as it will turn out, the two assumptions of the perfect circle and the areal law cancel each other so that, while it was assumed the first is true and the second false, it turns out that the first is false and the second true. Kepler called this the miracle. We shall now see how this came about.

---

[2]Kepler, of course, did not prove this, in this way, but by geometrical symmetry. For every position M above M_2, in Fig. 37, there is a corresponding position of M below M_2 whose angular distance from P equals φ so that r' in the latter position is shorter than CA by the same amount that it is longer than CA in the former position. Thus, in the figure, triangles CDS' and CES' are congruent.

Since we have developed above a method by which we can construct an accurate table of the distances of the Earth from the sun with respect to the distance of Mars at a given point of the orbit, we can in this way find also the distances of Mars from the sun for different positions and in this way pin down more accurately the position of Mars' aphelion.

Suppose now we compute with Kepler the time it takes Mars to move from $\varphi = 45°$ to $\varphi = 135°$ according to the areal law and according to the vicarious hypothesis of unequal eccentricities. Since the latter is quite accurate for longitudinal angles, this is a fair test of the accuracy of the areal law, and through it of the unmanageable distance law.

Beginning with the areal law, the most accurate figures on this are, for Kepler, $CS' = 9264$ to the 100,000 of the eccentric's radius. Thus given a and ae and $\cos \varphi$, we find

$$r = S'M_1 = \left[a^2 + a^2 e^2 + 2a^2 e \cos \varphi\right]^{\frac{1}{2}}$$

and hence angle δ from

$$\frac{\sin \delta}{ae} = \frac{\sin \varphi}{r}.$$

We therefore can compute angles $AS'M_1$ and $AS'M_3$ for $\varphi = 45°$ and $135°$. We have

$$\angle AS'M_1 = 41°28'54'',$$
$$\angle AS'M_3 = 130°59'25''.$$

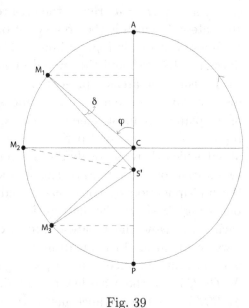

Fig. 39

For the times, we have to find the ratios of the areas swept out. We shall in fact reduce the areas to angles of regular motion in this way. The periodic time, the area of the circle, will correspond to the complete revolution of 360°, and the other corresponding areas swept out by r about S' will correspond to so many degrees as they are parts of the total area of the circle. To compute the number of degrees corresponding to area $AS'M_1$, we have only to compute triangle $CS'M_1$, since sector ACM_1 obviously corresponds to 45° of regular motion. To compute triangle $CS'M_1$, we first compute triangle $CS'M_2$.

$$\frac{\Delta CS'M_2}{\pi a^2} = \frac{(9264)a}{2\pi a^2} = \frac{x}{360°},$$

where $a = 100,000$ and x is the number of degrees of regular motion measured by $\triangle CS'M_2$. By computation, $x = 19,108''$ of regular motion. Then

$$\frac{\triangle CS'M_1}{\triangle CS'M_2} = \frac{\sin 45°}{\sin 90°} = \frac{y}{19,108''} = \frac{1}{\sqrt{2}},$$

since the triangles are on the same base and hence as their heights. Then y is the angle of regular motion measured by triangle $CS'M_1$. By computations, $y = 13,512'' = 3°45'12''$. The same obviously holds for triangle $CS'M_3$. Therefore

$$\text{area } AS'M_1 = 45° + 3°45'12'' = 48°45'12''$$
$$\text{area } AS'M_3 = 135° + 3°45'12'' = 138°45'12''.$$

Now computing according to the vicarious hypothesis, $CS' = 11,000$, $CQ = 7,000$, $a = 100,000$. We want to find the movement of Mars from the same length of time by this hypothesis. Therefore we compare the movement of the same angles of regular motion here measured about the equant point Q. Corresponding to the angles of the areal law, we take

$$\angle AQM_1 = 48°45'12'',$$
$$\angle AQM_2 = 138°45'12'',$$

and we compute the corresponding angles $AS'M_1$ and $AS'M_3$ by successive applications of the law of sines and cosines:

$$\angle AS'M_1 = 41°20'30'',$$
$$\angle AS'M_3 = 131°7'26''.$$

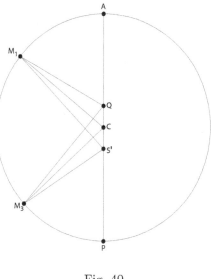

Fig. 40

We have now shown that, in the same time, Mars moves by the areal law, $(130°59'25'' - 41°28'54'') = 89°30'31''$, but, by the vicarious hypothesis, $(131°7'26'' - 41°20'30'') = 89°46'53''$. Since the vicarious hypothesis gives the angular positions most accurately, the areal law represents the planet as moving too slowly by $0°16'22''$ which is an observable angle in Tycho Brahe's observations. We can now assert that the areal law represents the planet as moving too slowly in the quadrants. The areal law, however, was only brought into consideration as a computable substitute for the distance law.

But the distance law represents the planet as moving even more slowly about the quadrants. And so we are forced to admit that the areal law is nearer the true law of force under the distance law. This latter, in spite of its plausibility, we must now abandon.[3]

But if the areal law as judged by the vicarious hypothesis is to be the true law, the path of the planet must be drawn in at the quadrants so that, as suggested by Fig. 38, the area swept out from M_1 and M_3 should be even less, relatively to the whole. It is now appropriate to scrutinize the distances of Mars from the sun at and near the quadrant. From the method used for finding the Earth's equant point with respect to the sun, we were able to build an improved table of the distances of Mars from the sun and the position of its aphelion. With several ingenious methods of refining this table, Kepler finally came to the conclusion that the distance of M from S' is not $S'M = r$, but MCD the projection of $S'M$ on MC. Hence

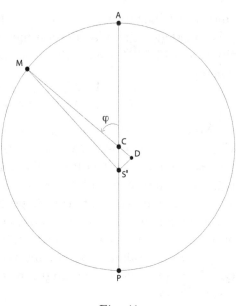

Fig. 41

$$S'M = r' = a + ae\cos\varphi.$$

instead of $r = \left(a^2 + a^2e^2 + 2a^2e\cos\varphi\right)^{\frac{1}{2}}$. Kepler than saw this makes Mars move on an ellipse with S' at one of the foci, and with the original circle as the major auxiliary circle of the ellipse.

For in an ellipse, we have

$$\frac{x^2}{a^2} + \frac{y^2}{a^2(1 - e^2)} = 1,$$

---

[3]If we should use the equant point and bisection of the eccentricity as in the Ptolemaic form of the Copernican theory and compare it with the vicarious hypothesis, the error is almost as much but in the opposite direction.

where $CA = a, x = CT, y = MT, CS' = ae$. If $S'M = r'$, we have

$$(r')^2 = (ae + x)^2 + y^2$$
$$= (ae + x)^2 + a^2(1 - e^2)\left(1 - \frac{x^2}{d^2}\right)$$
$$= (a + ex)^2 = (a + ae\cos\varphi)^2$$

where

$$\varphi = \angle TCM'.$$

Then

$$a\int_0^\varphi r'd\varphi = a^2\varphi + a^2 e\sin\varphi$$

which, as we have seen above, is the area of the section $S'M'A$ of the major auxiliary circle for the point M' on the circle which corresponds to the point M on the ellipse. But

$$S'MA = \frac{b}{a}(S'M'A), \quad b = a\left(1 - e^2\right)^{\frac{1}{2}},$$

since

$$\frac{b}{a} = \frac{\text{section } TMA \text{ of ellipse}}{\text{section } TM'A \text{ of circle}} = \frac{\Delta S'MC}{\Delta S'M'C},$$

so that the areal law holding for the circle holds also for the ellipse with a different constant of proportionality.

This corrects the error of the areal law with respect to the vicarious hypothesis, since, at $\varphi = 45°$, M is seen from S' nearer to A by angle $M'S'M$, and, at $\varphi = 135°$, M is seen farther from A by the corresponding angle so that M will traverse a greater angle from S' as demanded by the vicarious hypothesis. In fact, when $\varphi = 45°$ angle $M'CM = 0°7'26''$ and angle $M'S'M$ a little less. At $\varphi = 135°$, angle $M'S'M$ is a little more, so that

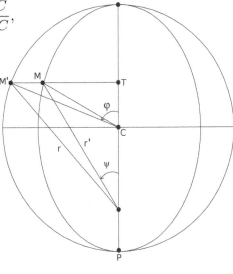

Fig. 42

the total difference is about $0°15'$, while the difference above between the vicarious hypothesis and the areal law was $0°16'$.

Considering these equations and Fig. 42 again, we have

$$r' = a + ae\cos\varphi$$

and, letting $\psi = \angle TS'M$,

$$\cos \psi = \frac{ae + x}{r'} = \frac{ae + a\cos\varphi}{a + ae\cos\varphi},$$

$$\cos \psi(1 + e\cos\varphi) = e + \cos\varphi,$$

$$(1 - e\cos\psi)(1 + e\cos\varphi) = 1 - e^2.$$

Now, since the times are proportional to the areas of the ellipse swept out about S', we have with b as semi-minor axis of the ellipse,

$$kt = \Delta S'MT + \text{sect. } TMA = \Delta S'MC + (\Delta CMT + \text{sect. } TMA)$$

$$= \frac{1}{2}bae\sin\varphi + \frac{b}{a}\left(\frac{a^2\varphi}{2}\right)$$

$$= \frac{1}{2}ab(\varphi + e\sin\varphi),$$

so that

$$\frac{2k}{ab}t = \varphi + e\sin\varphi,$$

or

$$nt = \varphi + e\sin\varphi,$$

where k is the areal speed, φ the eccentric anomaly, the ψ the true anomaly. This is the famous Kepler equation. The problem of solving it for φ in terms of given t, n, and e has been discussed for a long time. The first method of solution was given by Newton, and we shall give another in a later chapter.

Although Kepler began his inquiry with the intention of establishing that the path of Mars must be symmetrical with a line of apsides through the apparent sun rather than through the mean sun, it is only at the end of the treatise that he finally establishes, by means of the distances of Mars from the apparent sun and the mean sun, the correctness of his conjecture in this matter. Meanwhile the dialectic which began from this point has evolved far more important things. The **Commentaries on Mars** have established three major laws: -

1. From the point of view of Copernicus, the Earth has an equant point distinct from the center of its eccentric circle with respect to the apparent sun as predicted in the **Mysterium Cosmologicum.** Because of its small eccentricity, the circle with the eccentric's center bisecting the equant's eccentricity is sufficient to save the appearances. But this law is to be modified in view of the next two.

2. Mars moves on an ellipse with the apparent sun at a focus.

3. Mars moves on its ellipse in such a way that the areas it sweeps out about the apparent sun are proportional to the times.

4. The Earth is assumed to move about the apparent sun with the same laws as Mars.

Although Kepler had succeeded in revolutionizing astronomy in view of his general principle that the sun is the center of force, yet he had not succeeded in deducing the path of the planet from a clear mathematical formulation of this notion; nor had he explained the shift in the line of apsides, nor the number and distances of the planets, nor the relation of their periodic times to their mean distances from the sun. And the lower planets, Venus and Mercury, still had to be absorbed into the Keplerian system. These and other tasks were to occupy the rest of his life.

C   The Magnetic Theory of Planetary Motion

It was a part of the Keplerian theory from the very beginning that the matter of celestial bodies was essentially the same as that of Earthly bodies. On this was based his theory of universal attraction. In accord with this notion, Kepler had correctly interpreted the appearance of the novas of 1572 and 1607 as exploding stars and not as a meteor below the lunar sphere as the Aristotelians wanted to do for obvious reasons. The revelations of Galileo's telescope in 1609 were not revelations for Kepler but only a confirmation of a previous theoretical position which had already been confirmed by the theory of Mars.

Thus, in the same spirit, he sought to give a mechanical and mathematical explanation of the elliptical path of Mars in terms of magnetism. This theory, which has had little success, was started in **The Commentaries on Mars** and finished in books IV and V of **The Epitome of the Copernican Astronomy**, published in 1620 and 1621.

As we have already reported, the planet is, according to Kepler, whirled about the sun by the rays of motor virtue emanating from the rotating sun. But the planet moves more slowly than the rotating rays because of its inertia or resistance to this motor virtue of the sun, an inertia which is proportional to the volume and density of the planet. But this would only explain the planet's motion in a circle about the sun as center. Why does the planet

move on an ellipse with the sun at the focus? At this point Kepler introduced
his theory of the planet as a magnet.

For the planet as a magnet is composed of magnetic threads which are
all parallel and which are on one side solipetal (attracted to the sun) and on
the other are solifugal (repulsive to the sun). They are always directed to the
same part of the heavens except for a slight inflection caused by the sun's own
power, an inflection which is restored according to the cycle of the planet's
revolution. Roughly, at the aphelion and perihelion the magnetic threads are
perpendicular to the rays from the sun and neither attract nor repel, but, as
they move from the aphelion, present more and more directly to these rays
the solipetal side; and as they leave the perihelion, present more and more
the solifugal side. This explains the successive approaches and withdrawals of
the planet to the sun.

Let us see the ex-
act mathematical for-
mulation of this the-
ory.   In the figure,
the little circles are
exaggerated represen-
tations of the plane-
tary globe and the ar-
rows represent the di-
rection and sense of
the magnetic threads
for different positions
of the planet, the ar-
row head being the
solipetal side.

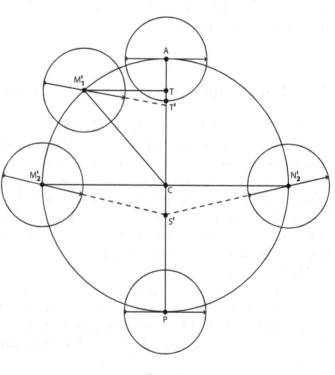

It is assumed by
Kepler that the threads
are deflected from their
perpendicular positions
at A and P in such a
way that they point to
the sun at the quad-
rants M_2' and N_2'. To

Fig. 43

this end, two assumptions are made:

1. The sines of the angles of deflection are proportional to the sines of the

corresponding angles of S'. That is,

$$\frac{\sin TM_2'T'}{\sin S'M_2'C} = \frac{\sin AS'M_1'}{\sin AS'M_2'}.$$

This will give no deflection at A and P, and a deflection directly to S' at the quadrants.

2. The attraction of the planet to (or repulsion from) the sun is proportional to the cosine of the angle that the ray from planet's center to the sun makes with the thread in the planet. This is positive from A to P and negative from P to A, denoting the measure of approach in the first case and of withdrawal in the second.

From (1) we get, since

$$\frac{M_1'C}{CS'} = \frac{\sin AS'M_1'}{\sin S'M_1'C},$$
$$\frac{M_2'C}{CS} = \frac{\sin AS'M_2'}{\sin S'M_2'C},$$

and hence

$$\frac{\sin AS'M_1'}{\sin AS'M_2'} = \frac{\sin S'M_1'C}{\sin S'M_2'C},$$

that

$$\frac{\sin TM_1'T'}{\sin S'M_2'C} = \frac{\sin S'M_1'C}{\sin S'M_2'C},$$

and

$$\begin{aligned}
\sin TM_1'T' &= \sin S'M_1'C, \\
\angle TM_1'T' &= \angle S'M_1'C, \\
\angle S'M_1'T' &= \angle CM_1'T;
\end{aligned}$$

and therefore

$$\cos S'M_1'T' = \cos CM_1'T = \sin \varphi.$$

The deductions from (2) are, however, somewhat perplexing. It is assumed that the planet would move on a circle if it were not attracted and repelled by the magnetic properties of the planet and the sun. In other words, Kepler is still bemused by the ancient law of inertia that celestial bodies move on circles, but here with the restriction that they do so only independently of the magnetic forces acting.

Now we know that the measure of the approach of the planet to the sun at each point of the path is $K \sin \varphi$ assuming that the planet remains on the circle. But, of course, as the result of this attraction, it does not remain on a circle, with a consequent change in the angle made by the sun's ray with the planet's thread. Kepler recognized this problem, but judged that, for his purposes, the resulting disturbance was negligible. First, therefore, we find the constant of proportionality K, by an integration. We know that in going from A to M_2' at the quadrant, the planet's distance from S', the sun, changes by ae'. Hence

$$\int_0^{\frac{\pi}{2}} K \sin \varphi d\varphi = K = ae',$$

and likewise, the result of the summation of the attraction at each point for the planet's passage from A to M_1' at angle φ is

$$\int_0^{\varphi} ae' \sin \varphi d\varphi = ae'(1 - \cos \varphi).$$

Kepler did not effect these integrations as easily as we do; he used a theorem of Pappus and the harsh method of indivisibles in the manner of Archimedes when proceeding heuristically.

Now the planet at A was distant from S' by $ae' + a$; at angle φ, it should be nearer to S' by $ae'(1 - \cos \varphi)$. Therefore the distance of the planet from the sun S' is

$$r = ae' + a - ae'(1 - \cos \varphi) = a + ae' \cos \varphi,$$

and this gives precisely the ellipse as shown above, p.68

This method is certainly not entirely satisfactory but we give it here to show the spirit which constantly animated the Keplerian search for an explanation, quite other than what is usually reported. It could be further extended to explain the precession of the line of apsides of the planets, that is, if one could properly explain that the position of the threads in the planet are not quite restored at the aphelion and perihelion.

D The Contributions of Terrestrial Mechanics

The Newtonian synthesis, which will form the base of classical mechanics, is largely inspired by the dynamic celestial mechanics of Kepler as we have expounded it from the **Commentaries on Mars**. But for a satisfactory mathematical formulation of force and mass, Newton will use two contributions of terrestrial mechanics; The theory of falling bodies of Galileo and the theories of percussion of Beeckman, Descartes, Huygens, and Leibniz.

We have seen Kepler enunciated a special case of the classical law of inertia that any particle when unaffected by any other moves in a straight line at a constant velocity forever; namely, that any body when placed at rest anywhere, remains at rest if unaffected by any other. In this respect, Kepler betrayed his Cusanean heritage; absolute rest has no meaning in this world. In assuming the sun, the center of force, as absolutely at rest, he betrayed his own statement of universal attraction and repulsion, and his own theory of the new star should have made him suspicious of an absolutely placed starry sphere, however distant from the center. But such are perhaps exigencies of applications of theory. In a way, the classical law of inertia is already implicit (except for its absolute frame of reference) in the theory of Cusanus that nothing is absolutely at rest and that motion belongs inherently in all things, a theory which harks back to Plato's **Sophist** where motion is a form and is opposed to the essential Aristotelian notion that motion has to be explained. The theory that even in a vacuum all things on the surface of the Earth would have their proper velocities was advanced in the Sixth Century A.D. by Philoponus of Alexandria in a treatise widely read in the late Middle Ages and in the Renaissance.

First, let us turn to the problem of the percussion of bodies, Beeckman in 1615 or 1618 postulated a vacuum with indivisible bodies or atoms so that all apparent motions would have to be explained in terms of the impacts of these atoms on each other in a vacuum. To this end, he assumed:

1. The classical law of inertia.

2. The classical law of the conservation of momentum in one straight line,

$$m_1v_1 + m_2v_2 = m_1v_1' + m_2v_2'$$

   where v_1 and v_2 are the velocities before impact, and v_1' and v_2' after impact. The measure of m is not clear.

3. After impact, these perfectly hard atoms move off together at the same velocity.

This last law seems strange when we consider our usual experience with bodies which bounce off each other. But this law, of course, is to hold for the **elementary** particles. It has, however, stranger consequences; it will be shown later that a universe subject to it runs down. Beeckman, aware of this, could only keep things moving by a special divine intervention. This notion of perfectly hard bodies lasts through much of the Seventeenth Century and is held notably by Wallis. Today, we would rather speak of them as perfectly soft bodies.

Descartes, who was in direct communication with Beeckman, assumed the physical world to be completely reducible to the continuum of three dimensional Euclidean geometry with relative motion between its parts. A perfectly hard body is a three dimensional volume which has no relatively moving parts, and is not essentially hard. Descartes' laws of percussion derive from the classical law of inertia, now stated for the first time in a general form, and from the law of the conservation of the absolute quantity of motion. This last law states that the $\sum_i m_i |v_i|$, where m_i is the volume and $|v_i|$ is the speed, is conserved. This will give a world where particles have the usual elastic action and which keeps moving of itself. Thus two hard equal volumes, meeting each other with equal but opposite velocities rebound with equal and opposite velocities. It leads, however, to seemingly impossible conclusions, and the seven laws of percussion of Descartes are for the most part false, from the point of view of classical mechanics. There is no reason to consider them here in detail, although the discussions concerning them will lead to the classical laws.

It is Huygens, as early as 1658, who enunciates a set of postulates for the laws of percussion in a vacuum which agree with what will later be incorporated into the classical system. They are independent of any consideration of the celestial mechanics of Kepler, and, describing in the eyes of Huygens the fundamental reactions between elementary particles, they would constitute, for him and for Leibniz, the general axioms from which Kepler's mechanics would be deduced. This will only be partly so, in so far as the new concept of energy, seized upon by Leibniz, will develop into a far-reaching principle. The set of laws of Huygens for hard bodies is as follows:

1. The law of inertia.

2. Two bodies, meeting each other with equal speeds but in opposite sense, rebound with the same speeds.

3. The laws must hold for all coordinate systems moving at constant velocity with respect to each other.

4. If $m_1 > m_2, v_2 = 0$, then $v'_1 < v_1$ and $v'_2 = \delta$.

5. If $v_1 = v'_1$, then $v_2 = v'_2$, a special case of the conservation of momentum.

To these is added the axiom that the center of mass of the system cannot rise. This axiom is obviously only meaningful in the special situation at the surface of the Earth. The definition of center of mass is given in a later chapter.

From these hypotheses, Huygens deduced the Principle of the Conservation of Kinetic Energy:

$$m_1 v_1^2 + m_2 v_2^2 = m_1 (v_1')^2 + m(v_2')^2,$$

which was destined to play such an important role in Leibniz' system and in the subsequent development of the science of physics.

The crucial experiment on which depended the checking and justification of all such theories is that of two hard balls (in the sense of Descartes, not of Beeckman) which hang as pendulums, just touching at rest (the double ballistic pendulum). When they are allowed to fall from equal heights, their velocities as they hit at the bottom can be computed by the theory of Galileo which follows; their velocities after impact can likewise be computed by the heights to which they arise; and their relative masses computed by weighing on a scale if one so defines mass. Just such an experiment was carried out by Huygens, Wallis, and Wren in London, in 1661, with the ultimate victory of Huygens' theory. It is to be noted that Leibniz, accepting no hard bodies, theorized on the deformation under impact. We return to this much later.

Secondly, we turn to the theory of falling bodies on the surface of the Earth as advanced by Galileo, which is the other side of the terrestrial mechanics represented in this experiment. Galileo simply assumed that, if one ignores all the phenomena of friction and supposes an empty space at the surface of the Earth restricted to a small area, then the mathematical law of motion in this space is that all bodies, irrespective of size or weight (mass), are constantly accelerated at the same rate in parallel lines perpendicular to the plane tangent to the Earth at that locality. It is well to remember that Galileo had no

Fig. 44

theory of universal attraction; his celestial mechanics was that of Copernicus, not that of Kepler whom he consciously ignored. From this fundamental assumption of a space of constant acceleration, he deduced that for a particle falling vertically or sliding without friction down an inclined plane, from rest, the square of its speed was proportional to the vertical height of fall (this was extended to any smooth curve by Huygens and used in the above experiment). He also deduced that a projectile, in such a mathematical space, given an initial non-vertical velocity, will follow a parabola with a vertical axis. In the notation of the modern calculus which Galileo, of course, did not use, we have for free fall from rest,

$$\frac{d^2y}{dt^2} = g, \qquad \frac{dy}{dt} = gt, \qquad y = \frac{1}{2}gt^2, \qquad y = 0, \qquad t = 0$$

whence

$$t = \frac{1}{g}\frac{dy}{dt} \qquad y = \frac{1}{2g}\left(\frac{dy}{dt}\right)^2.$$

Hence the experiment of the ballistic pendulums, used to justify the laws of Huygens on percussion against those of Descartes, involved also the theory of Galileo. Newton reports on this in the **Principia**, published in 1686, but composed much earlier, in the Scholium to Law III, and adds an account of an ingenious method for eliminating the friction of the air from the motion. If we let one of the pendulums fall from A and return to B, then the arc AB represents

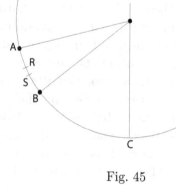

Fig. 45

the action of friction in one oscillation of the pendulum. To judge approximately the action of friction, in the descent from A to C, one takes $\frac{1}{4}AB = RS$ and places RS in the middle of AB so that $AR = SB$. Then for the fall of the pendulum from R, one takes it as falling without friction from S.

E The Newtonian Synthesis

The great achievement of Newton was to put the terrestrial mechanics of percussion and acceleration together with the celestial mechanics of Kepler, adopting the world-view of Kepler which reached him via Beeckman and Horrox and which was ignored by the others, Kepler had not been able to find the precise mathematical formulation of the relation of attraction and repulsion between bodies which lay at the center of his laws. Newton seized on the constant acceleration characterizing the Galilean space at the surface of the Earth as the key to this formulation, although Kepler had already thought in this direction at the very beginning of his career when he interpreted the difference in velocities of each planet with respect to the sun, as symbolizing the action of the sun on the planet. Newton, very early also, as early as 1669, in a manuscript reprinted in **The Correspondence of Isaac Newton**, Vol. I, pp. 297-301, which is reflected in one of the early primitive theorems of the **Principia**, Book I, Sect. II, Prop. 4, set out to generalize the Galilean constant acceleration in a given direction to the general notion of acceleration as change both in speed and direction. But, in these early attempts, he tackles again a special case to simplify the mathematical problem. He supposes a fixed and immovable center of force (as did Kepler) and constrains the planets to move on circles concentric to this center at constant speeds. In other words, he is taking the case where the acceleration consists simply in the change of the direction of the velocity. It is the vectorial character of velocity and acceleration which is here made explicit and Newton uses linesegments with direction and sense in his proofs just as did Kepler for probably the first time in the history of mathematics. Anticipating the full exposition of vector-spaces and tensors and their applications in the ordinary vector algebra of Gibbs in the next chapter, we shall use some of the simple vectorial techniques in expounding these early theorems of Newton to show how he was led to his final axioms of mechanics.

Therefore, in the early theorems, Newton supposes a fixed and immovable center of force and particles moving on circles in one plane concentric to this at constant angular speeds. Generalizing Galileo's theory, it is assumed that, on a given circle at a given distance, the mass of the particle accelerated has no effect on the acceleration imposed.

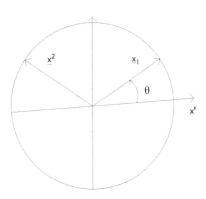

Fig. 46

Let $\underline{x}_1$ be the position-vector of m_1 and $\underline{x}_2$ of m_2. Then

$$\underline{x}_1 = x_1'\underline{e}_1 + x_1^2\underline{e}_2 = r_1 \cos\theta_1\underline{e}_1 + r_1 \sin\theta_1\underline{e}_2, \quad \theta_1 = \omega_1 t,$$

$$\underline{x}_2 = x_2'\underline{e}_1 + x_2^2\underline{e}_2 = r_2 \cos\theta_2\underline{e}_1 + r_2 \sin\theta_2\underline{e}_2, \quad \theta_2 = \omega_2 t,$$

$$\frac{d\theta}{dt} = \omega_1, \frac{d\theta_2}{dt} = \omega_2, \text{the constant angular velocities.}$$

Then the velocity and acceleration vectors are

$$\frac{dx_1}{dt} = (-r_1 \sin \theta_1 \underline{e}_1 + r_1 \cos \theta_1 \underline{e}_2)\omega_1,$$

$$\frac{dx_2}{dt} = (-r_2 \sin \theta_2 \underline{e}_1 + r_2 \cos \theta_2 \underline{e}_2)\omega_2$$

$$\frac{d^2 x_1}{dt^2} = (-r_1 \cos \theta_1 \underline{e}_1 - r_1 \sin \theta_1 \underline{e}_2)\omega_1^2 = -\omega_1^2 \underline{x}_1 = \underline{a}_1$$

$$\frac{d^2 x_2}{dt^2} = -\omega_2^2 \underline{x}_2 = \underline{a}_2.$$

The acceleration-vectors are directed to the center of force as they should be, indicating the direction of the action of the force. If now we consider the magnitudes of the accelerations, we have

$$|\underline{a}_1| = \omega_1^2 r_1, \quad |\underline{a}_2| = \omega_2^2 r_2.$$

Since we have here only the action of the center of force on the particles (or planets) and no action of the particles on the center, we can carry over analogously the procedure of Galileo that all particles irrespective of their size and weight (or mass) are given the same accelerations for a given neighborhood on the surface of the Earth, except that Newton for this general celestial case will have to change it to the assumption that all particles at the **same distance** from the center of force have the same magnitude of acceleration (not direction obviously). With this assumption, in comparing the ratio of the forces exerted, we compare the ratio of the magnitudes of the accelerations. Therefore

$$\frac{|\underline{a}_1|}{|\underline{a}_2|} = \frac{\omega_1^2 r_1}{\omega_2^2 r_2}.$$

Now the periodic times of the particles are respectively $\frac{2\pi}{\omega_1}$ and $\frac{2\pi}{\omega_2}$, the total angle of one revolution divided by the angular speed:

$$P_1 = \frac{2\pi}{\omega_1}, \quad P_2 = \frac{2\pi}{\omega_2}.$$

Kepler's last law stated, we remember, that

$$\frac{P_1}{P_2} = \frac{r_1^{3/2}}{r_2^{3/2}}.$$

If then we assume that the magnitudes of the accelerations on the particles due to the center of force vary inversely as the squares of the distances, as

Kepler even suggested in one of several speculations, then

$$\frac{\omega_1^2 r_1}{\omega_2^2 r_2} = \frac{r_2^2}{r_1^2}\frac{\omega_1}{\omega_2} = \frac{r_2^{3/2}}{r_1^{3/2}} = \frac{P_2}{P_1},$$

and we deduce Kepler's law. This is not, however, too convincing by itself, since we have juggled the variation of the accelerations as function of the distances in a slightly arbitrary manner.

But there is another theorem given by Newton in the **Principia**, obviously an early one also, Prop.I in the same section as the one just quoted. It is another law of Kepler that the planets sweep out equal areas in equal times about the sun as the center of force, and the further law that each planet moves in one plane through the center of force. We prove it by vector methods from the same association of force and acceleration. Let $\underline{x}$ be the position-vector of the planet from the center of force O. Assuming the acceleration vector represents the action of the force and is therefore directed always in the direction of $\underline{x}$,

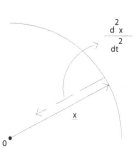

Fig. 47

$$\underline{x} \times \frac{d^2\underline{x}}{dt^2} = 0.$$

Integrating with respect to time,

$$\underline{x} \times \frac{d\underline{x}}{dt} = 2\underline{\kappa}, \ \underline{\kappa} \text{ a constant vector.}$$

But $\underline{x} \times \frac{d\underline{x}}{dt}$ is a vector orthogonal to the plane of $\underline{x}$ and $\frac{d\underline{x}}{dt}$, and, since it is constant, the plane, with the proper assumptions of continuity, is also fixed. But $|\underline{x} \times \frac{d\underline{x}}{dt}|$ is twice the areal speed as we shall show in Chapter 4. It is to be noticed that Newton's deduction shows that the areal law should hold for any central field of force.

So three laws of Kepler can be deduced from the association of force of attraction with the acceleration vector. As we shall see later, with more mathematical technique, we can deduce the remaining law of Kepler that a planet moving freely about the sun moves on an ellipse with the sun, the center of force, at a focus, if the magnitude of the acceleration varies inversely as the square of the distance from the center of force. As has already been remarked by Planck in his **Mechanics**, such a principle of unification cannot be merely conventional.

But Newton did much more. He introduced a relation between the masses of particles and their accelerations so as to bring together the laws of percussion of terrestrial mechanics and the laws of celestial motions, as Kepler had set out to do. This principle states that, for any two particles considered in isolation, with masses m_1 and m_2 and accelerations $\underline{a}_1$ and $\underline{a}_2$,

$$m_1\underline{a}_1 = -m_2\underline{a}_1, \quad m_1\underline{a}_1 + m_2\underline{a}_2 = 0.$$

From this we get, by integration, the conservation of momentum

$$m_1\underline{v}_1 + m_2\underline{v}_2 = \underline{c}; \ \underline{c} \text{ a constant vector.}$$

Let us now formally state the axioms of the classical mechanics as it comes from the hands of Newton, with some precisions added which are only implicit in Newton's exposition.

Axiom 1. The momentum of a particle, that is, its velocity vector multiplied by a number m called its mass remains constant forever if the particle is undisturbed by any force whatsoever, that is, by the presence of any other particle.

Axiom 2. The force imposed on a particle is measured by the derivative of its momentum with respect to time

$$\underline{F} = \frac{d(m\underline{v})}{dt}.$$

Since the mass m of a given particle is assumed constant in classical mechanics, this axiom is usually stated

$$\underline{F} = m\frac{d\underline{v}}{dt} = m\frac{d^2x}{dt^2}.$$

It has often been argued with some reason that this is really a definition of $\underline{F}$ and not properly an axiom. But it will become immediately operative when we assume that $\underline{F}$ is some function of the distances or make some other mathematical assumption about $\underline{F}$ which gives us a differential equation. We notice, however, that m is not yet defined in any meaningful way. That will be taken care of in Axiom 3. Meanwhile, we have our troubles about $\underline{v}$ and $\frac{d^2x}{dt^2}$.

The question immediately arises: $\underline{v}$ and $\frac{d^2x}{dt^2}$ with respect to whom or what? How do we measure the distance and time entering into these expressions? Newton answered the problem by introducing what he called an absolute space and an absolute time which lay behind the sensible motions and sensible distances, and his space was Euclidean 3-space. But since absolute space and time cannot appear to us, we are left with only relative spaces and times as observable. This means we pick a perceptible coordinate system and a perceptible cyclic motion which we hope will approximate to these unperceived absolutes. The only way of judging this approximation is to see if the phenomena then obey the deductions from the Axioms. We can assume, then, that our coordinate system can be taken as absolute, except that we shall later prove any coordinate system moving at constant velocity with respect to it has the same privilege. Such a system was at hand: the so called fixed stars and the rotation of the Earth on its axis. These will suffice for a time, but will later become inadequate, notably in the mechanics of light. There were many contemporary critics of these absolutes of Newton, chief among them Leibniz, but the immediate successes of the theory were too great for people to bother about the strangeness of it, and then Kant, at the end of the Eighteenth Century, gave a seemingly respectable intellectual basis to it.

The most important assumption made implicitly about these measurements of space and time passed unperceived as before. It is that the space measurements of distance and the measurements of time are to be made independently of each other and are the same for all observers, no matter how they are moving relatively to each other. Thus

$$\Delta s = \sqrt{(\Delta x^1)^2 + (\Delta x^2)^2 + (\Delta x^3)^2}, \text{and } \Delta t$$

are independent of each other and the same for all.

Axiom 3. For any two particles with masses m_1 and m_2 in isolation and accelerations $\underline{a}_1$ and $\underline{a}_2$,

$$m_1\underline{a}_1 = -m_2\underline{a}_2, \quad \frac{m_1}{m_2} = \frac{|\underline{a}_2|}{|\underline{a}_1|}$$

where $\underline{a}_1$ is in the direction of the line between the two particles, its sense being determined by whether particle m_1 suffers repulsion or attraction in the presence of m_2. This axiom immediately gives us a method of measuring the relative values of two masses if we can find a situation where the two masses are approximately isolated.

But the notion of mass in this axiom is ambivalent. There are two senses in which m_1 and m_2 appear here: (1) as responding to a gravitational field produced by another particle, and (2) as producing the gravitational field. Thus m_1 responds to a gravitational or accelerative field produced by m_2, and m_2 in turn responds to a gravitational field produced by m_1. We shall distinguish these two uses of gravitational mass by calling the first inertial mass and identifying it with the mass of Axiom 1, and calling the second gravitational mass.[4] It is assumed that a particle m does not produce a gravitational or accelerative field to which it itself responds.

We can find the relation between these two aspects of mass by the assumption of Galileo generalized, that is, that in an isotropic space where the acceleration imposed by a mass on other masses varies as some function of the distance from that mass, the magnitude of the acceleration imposed on different masses is the same at the same distance from the attracting mass. Thus take two isolated masses m_1 and m_2. Then by Ax.1, m_1 and m_2 representing inertial masses,

Fig. 48

$$F_{12} = m_2|\underline{a}_2| = m_2\mu_1\ f(|\underline{x}_{12}|)$$
$$F_{21} = m_1|\underline{a}_1| = m_1\mu_2\ f(|\underline{x}_{21}|)$$

where F_{12} and F_{21} are the magnitudes of forces exerted by gravitational mass of m_1, on inertial mass m_2, and by gravitational mass of m_2 exerted on inertial mass m_1 respectively, and f is the function of the distance. That in the first equation μ_1 is a **constant** representing the gravitational mass of m_1 follows from assumption of Galileo generalized. For if we supposed another inertial mass m_3 replacing m_2, we would have

$$F_{13} = m_3|\underline{a}_2| = m_3\mu_1\ f(|\underline{x}_{13}|) = m_3\mu_1\ f(|\underline{x}_{12}|).$$

Hence μ_1 is a constant properly depending only on the attracting mass and not on the attracted mass. The same argument holds for μ_2. But, by Ax.3,

$$F_{21} = F_{12},\quad m_1\mu_2 f(|\underline{x}_{21}|) = m_2\mu_1 f(|\underline{x}_{12}|)$$
$$m_1\mu_2 = m_2\mu_1,\quad \frac{\mu_1}{m_1} = \frac{\mu_2}{m_2} = \gamma$$

where γ is the universal constant of proportionality between the gravitational and inertial mass of any given particle. It is called the universal constant of gravitation. Therefore

$$F_{12} = m_2 m_1\ \gamma\ f(|\underline{x}_{12}|).$$

In particular, for the inverse square field, we get the usual formula

$$F = \frac{m_1 m_2 \gamma}{|x_{12}|^2}.$$

---

[4]These three distinctions of mass are made notably in this form by Eddington in **The Mathematical Theory of Relativity**, p. 128. There are other usages in the literature.

These are the axioms from which all of the usual classical mechanics, except that depending on the Lagrange equations and Hamilton's Principle, are derived. We shall discuss any further assumptions for the theory of Lagrange when that theory is introduced. The vector entities here introduced will be treated systematically in the next chapter.

Chapter 3

The Mathematical Apparatus

We shall here set out the mathematical language in which the classical science of mechanics is stated without much reference to its historical development. Vectors, in their intuitive form, are already found in Kepler and Newton. We shall here give vectors and tensors in their most abstract form so that they may be developed for specific purposes in their proper places later in the treatise. We shall also give an elementary treatment of the application of the vector calculus to the differential geometry of curves. Further developments from these theories will be given at the place where they arise in mechanics.

A   Vector Spaces and Their Applications

A.1   Groups and Vector Spaces

We remind the reader first of the notions of a group and of a field.

Definition of a Group III.A.1.1. A set of elements a, b, c etc. is a group G if and only if there is an internal law of composition $*$ such that

1. For each a and b in G, there is a c in G such that $a * b = c$.

2. $(a * b) * c = a * (b * c)$;

3. There exists an element e such that for all a in G,

$$a * e = e * a = a;$$

4. For every element a of G, there is an element a^{-1} such that

$$a * a^{-1} = e.$$

If the law of composition is commutative, the group is called Abelian. It is easy to prove from this definition that e is unique; that if $a * c = a * d$, then $c = d$; that a^{-1} is unique and $a^{-1} * a = a * a^{-1}$.

Definition of a Field, III.A.1.2. A set of elements is a field F if and only if there are two internal laws of composition so that it is an Abelian group with respect to the first law (additive) which we will denote as $+$ with its e-element which we shall call 0, and so that it is an Abelian group also with respect to the second law (which we shall denote by $\cdot$), except for the zero element in the first law. There is further a right and left distributive relation between the two laws

$$a \cdot (b + c) = (b + c) \cdot a = (a \cdot b) + (a \cdot c)$$

Definition of a Vector Space, III.A.1.3. Suppose a set of elements $\underline{x}, \underline{y}, \underline{z}$ etc. are members of an additive Abelian group V, and the set of elements a, b, c etc., are members of a field F. They form a vector space if and only if there exists an external law of composition of the elements of F on the elements of V such that

1. $a\underline{x} = \underline{y}$ for all a and $\underline{x}$;

2. $(a + b)\underline{x} = a\underline{x} + b\underline{y}$;

3. $a(\underline{x} + \underline{y}) = (a\underline{x}) + (a\underline{y})$;

4. $a(b\underline{x}) = (ab)\underline{x}$;

5. $1 \cdot \underline{x} = \underline{x}$;

We make no distinction between $a\underline{x}$ and $\underline{x}a$. It immediately follows that

$$(a + 0)\underline{x} = (a\underline{x}) + (0 \cdot \underline{x}) = a\underline{x}$$

and hence

$$0 \cdot \underline{x} = \underline{0}$$

where $\underline{0}$ is the zero-element in V and 0 is the zero-element in F.

In this treatise, we shall only be concerned with F as the field of real numbers or, in the Special Theory of Relativity with F as the field of complex numbers.

Such is the algebraic definition. Let us immediately give

A Geometric Picture. Suppose the elements of V are line-segments of a 3-dimensional space. Notice that we have not yet defined dimension in V; we shall later. Intuitively, they have length, direction, and sense. Two elements x and y of V are said to be equal if they have the same direction, length, and sense. All lines having the same direction are parallel lines according to the definition and axiom of Euclid, that is, two straight lines in the same plane (to be defined algebraically later) which do not meet, if indefinitely extended, are parallel. And, given a point and a line, there is only one straight line through that point parallel to the given line (Playfair's form). By sense, we mean the two ways of moving on the straight line. We compare lengths on the same straight line by an isomorphism of the real number system and the points on the line, and we extend this comparison to all parallel lines. This is implied in our definition of equality, for so-called free vectors. We have no way yet of comparing lengths on lines in different directions and no way of measuring angles.

Our vectors in V can be pictured, therefore, by arrows. By $x+y$ we mean the vector z which is produced as follows. The tail of y is placed at the head of x; the vector z is the vector from the tail of x to the head of y. It is associative, since

$$(x + y) + z = x + (y + z)$$

as indicated in Fig. 49. By ax we mean the vector y such that

$$a = \pm \frac{|y|}{|x|}, \quad ax = y,$$

where $|x|$ means length of vector x, a positive real number. If $a > 0, y$ is in the same sense as x; if $a < 0$, in the opposite sense. The other laws of the external law of composition can be easily verified. Thus

$$a(x + y) = (ax) + (ay)$$

is the relation between two similar triangles as in Fig. 50.

Further

$$x + (-y) = z = x - y, x + (-x) = 0,$$

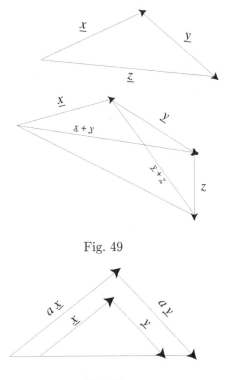

Fig. 49

Fig. 50

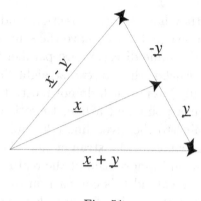

Fig. 51

can be pictured as the addition of $\underline{x}$ and the vector $(-\underline{y})$ which is the vector of same length and direction but opposite in sense to $\underline{y}$. The $\underline{0}$ vector is simply a point, 0 in length, indeterminate in direction and sense.

A.2   The Notion of Dimension

We shall now introduce notions, which shall show us relations implicit in our previous constructions leading to the idea of dimension.

Definition of Independence, III.A.2.1.   A set of vectors $x_1, x_2, \ldots, x_n$ which we shall denote as (x_i) is said to be independent if and only if there exists no sequence $a^1, a^2, \ldots, a^n$ not all zero such that

$$a\underline{x}_1 + \cdots + a^n \underline{x}_n \equiv a^i \underline{x}_i = \underline{0}, \quad i = 1, 2, \ldots, n,$$

where we introduce superscript for the members of F (the scalars) for a reason which will appear later. The Einstein notation $a^i \underline{x}_i$ which indicates summation for repetition of superscript and subscript in the same term will be used from now on.

If this is not so, then the set of vectors is dependent. We can state this in another way. If

$$a^i \underline{x}_i = \underline{0}$$

requires that $a^i = 0$ for all a^i, then $\{\underline{x}_i\}$ is independent and conversely.

Theorem A.2.1. *If some non-empty subset of a set of vectors is dependent, then the whole set is dependent.*

Proof: If of the set $\{\underline{x}_i\}, i = 1, 2, \ldots, n$, we have $\{a_i\}, \ i = 1, 2, \ldots, r, \ r \leq n$ such that

$$a^i \underline{x}_i = \underline{0}, \ a^i \neq 0 \text{ for some } i.$$

Then let $a^{i+1} = \cdots = a^n = 0$, and we have

$$a^i \underline{x}_i = \underline{0}, i = 1, \ldots, n, \text{and not all } a^i = 0.$$

Theorem A.2.2. *Every non-empty subset of a set of independent vectors is independent.*

Proof: For, if there were a dependent subset, by the previous theorem, the whole set would be dependent.

Definition of Basis, III.A.2.2. If every vector $\underline{x}$ of V can be expressed as a linear combination of a set of independent vectors $\{\underline{e}_1, \ldots, \underline{e}_n)$, that is, if there always exist $a^1, \ldots, a^n$ if F such that

$$\underline{x} = a^i \underline{e}_i, i = 1, \ldots, n,$$

for x in V, then the $\{\underline{e}_1\}$, are said to form a basis of V.

Theorem A.2.3. *(Steinitz' Replacement). If all the vectors $\underline{x}$ of a vector space V can be expressed as a linear combination of the set of vectors $\{\underline{y}_1, \ldots, \underline{y}_n\}$ of V, and, if $\{\underline{z}_1, \ldots, \underline{z}_q\}$ is also a set of independent vectors of V, then q of the vectors $\underline{y}_i$ can be replaced by the q vectors $\underline{z}_j$ so that all vectors $\underline{x}$ of V can be expressed as a linear combination of the q vectors $\underline{z}_i$ and the remaining $(n-q)$ vectors $\underline{y}_i$. It will, therefore, result that $q \leq n$.*

Proof: We use finite mathematical induction on the integers beginning with zero.

It is true for $q = 0$. We show if it is true for any $s - 1$ of the q vectors $\underline{z}_i$, then it is true for s of them. Hence we assume for any $\underline{x}$ in V

$$\underline{x} = a^1 \underline{z}_1 + \cdots + a^{s-1} \underline{z}_{s-1} + a^s \underline{y}_s + \cdots + a^n \underline{y}_n \qquad (\alpha)$$

where, for simplicity's sake, we have renumbered the replaced $\underline{y}_i$'s if necessary so as to take the first $s - 1$. But $\underline{z}_s$ is in V; therefore

$$\underline{z}_s = b^1 \underline{z}_1 + \cdots + b^{s-1} \underline{z}_{s-1} + b^s \underline{y}_s + \cdots + \underline{b}^n y_n \qquad (\beta)$$

where not all $b^s, \ldots, b^n$ vanish. For, if they did, $\{\underline{z}_1, \ldots, \underline{z}_s\}$ would be dependent. But every subset of the independent set $\{\underline{z}_1, \ldots, \underline{z}_q\}$ is independent. Suppose $b^s \neq 0$. Then, solving in (β), we have

$$\underline{y}_s = \frac{1}{b^s} \left[\underline{z}_s - \left(b^1 \underline{z}_1 + \cdots + b^{s-1} \underline{z}_{s-1} + b^{s+1} \underline{y}_{s+1} + \cdots + b^n \underline{y}_n \right) \right].$$

Substituting this in (α) for y_s, we see that every $\underline{x}$ in V can be expressed as a linear combination of $\{\underline{z}_1, \ldots, \underline{z}_s, \underline{y}_{s+1}, \ldots, \underline{y}_n\}$, if it can be expressed in terms of $\{\underline{z}_1, \ldots, \underline{z}_{s-1}, \underline{y}_s, \ldots, \underline{y}_n\}$.

It follows immediately that $n \geq q$. It also follows that, given any set of q vectors $\{\underline{z}_1, \ldots, \underline{z}_q\}$ in V where $q > n$, the set is the n $\underline{y}_i$'s, and express the rest of the $\underline{z}_i$'s in terms of them.

It is to be noticed that, in this theorem, the set of $\underline{y}_1$'s is not a basis, since they need not be independent, but they share the property of a basis that every $\underline{x}$ in V can be expressed as a linear combination of them. They are a set of generators of V.

Theorem A.2.4. *Any two bases of a vector space V must have the same number of elements.*

Proof: Suppose two bases $X = \{\underline{x}_1, \ldots, \underline{x}_n\}$ and $Y = \{\underline{y}_1, \ldots, \underline{y}_q\}$. Now, by the previous theorem of Steinitz, considering X as a set of generators and Y as an independent set, we have $q \leq n$. But symmetrically we can consider Y as the set of generators and X as the independent set so that $n \leq q$. Hence $n = q$.

Therefore we can now meaningfully define

Definition of Dimension of V, III.A.2.3.   The number of elements of a basis of V is the dimension of V.

Theorem A.2.5. *Any vector $\underline{x}$ of V can be represented in one way only as a linear combination of basis vectors $\{\underline{e}_1, \ldots, \underline{e}_n\}$ of an n-dimensional vector space.*

Proof: Let $\underline{x}$ be expressible in two ways, so that

$$\underline{x} = a^i \underline{e}_i = b^i \underline{e}_i, i = 1, \ldots, n.$$

Then

$$\left(a^i - b^i\right) \underline{e}_i = 0,$$

and since the $\{\underline{e}_i\}$ is independent,

$$a^i - b^i = 0 \text{ for each } i.$$

Geometric Picture of Independence and Basis in Affine Space.

We first introduce the notion of a position-vector $\underline{x}$ of a given geometric vector-space. Take any point O as the origin from which we consider all vectors drawn. Then any point P in the space is given by the vector $\underline{x}$ from O to P.

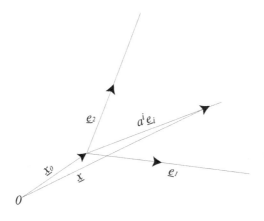

Fig. 52

A plane in n-space through a point P_0 given by $\underline{x}_0$ is an $n-1$ dimensional subspace of the original space. Thus, if $\{\underline{e}_i\}$, $i = 1, \ldots, n$ is a basis of n-space, then any point on a plane through $\underline{x}_0$ will be given by

$$\underline{x} = \underline{x}_0 + a^i \underline{e}_i, \quad i = 1, \ldots, n-1, \quad -\infty < a^i < +\infty.$$

For different planes we have different $\{\underline{e}_i\}$. Thus in 3-space we have

$$\underline{x} = \underline{x}_0 + a^1 \underline{e}_1 + a^2 \underline{e}_2, \qquad \underline{x} - \underline{x}_0 = a^1 \underline{e}_1 + a^2 \underline{e}_2,$$

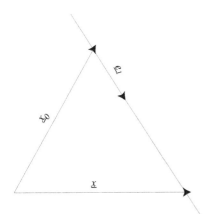

Fig. 53

or since $\underline{x}_0$ is also expressible in terms of $\underline{e}_i$, we have

$$\underline{x} = \left(x_0^1 + a^1\right) \underline{e}_1 + \left(x_0^2 + a^2\right) \underline{e}_2,$$

$$-\infty < a^i < +\infty.$$

In 2-space, the analogue of the plane in 3-space is the straight line

$$\underline{x} = \underline{x}_0 + a^1 \underline{e}_1,$$

$$-\infty < a^1 < +\infty,$$

where $\underline{e}_1$ is in the direction of the line. The fact that the $\{\underline{e}_i\}$ form a basis in 2-space means $\underline{e}_1$ and $\underline{e}_2$ are not collinear; and in 3-space, $\underline{e}_1, \underline{e}_2$, and $\underline{e}_3$ are not coplanar.

On the other hand, a straight line in n-space has the same form

$$\underline{x} = \underline{x}_0 + a^1 \underline{e}_1, \quad -\infty < a^1 < +\infty$$

where $\underline{e}_1$ is in the direction of the line. The straight line is always a one-dimensional subspace of an n-dimensional space, whereas the plane is $(n-1)$-dimensional subspace of an n-space. Another form of this equation which is useful is the following. Suppose P_1, P_2, P points on the same straight line and

$$\overrightarrow{P_1P} = \frac{a^2}{a^1} \overrightarrow{PP_2}.$$

Then

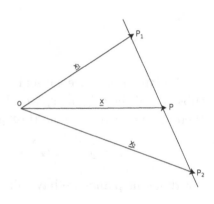

$$\underline{x} = \underline{x}_1 + \overrightarrow{P_1P},$$

$$\underline{x} = \underline{x}_2 - \overrightarrow{PP_2}$$

$$\underline{x} = \underline{x}_1 + \frac{a^2}{a^1}\overrightarrow{PP_2}$$

$$= \underline{x}_1 + \frac{a^2}{a^1}(\underline{x}_2 - \underline{x})$$

$$\underline{x}\left(1 + \frac{a^2}{a^1}\right) = \underline{x}_1 + \frac{a^2}{a^1}(\underline{x}_2)$$

$$\underline{x} = \frac{1}{a^1 + a^2}\left(a^1\underline{x}_1 + a^2\underline{x}_2\right) \quad (\alpha)$$

Fig. 54

It is perhaps instructive to give an example of the use of these concepts in the classic problem of the medians of a triangle which meet in a point 2/3 the distance along median. Since we can compare lengths on any given line only, we can take the midpoint M of AB and L of OB by the equation (α) we have just derived. Further $\overrightarrow{OA}$ and $\overrightarrow{OB}$ which we denote by $\underline{a}$ and $\underline{b}$ are not collinear and are hence independent and form a basis for the plane of

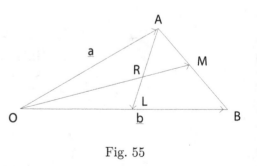

Fig. 55

the triangle. Hence

$$\underline{a} + \overrightarrow{AB} = \underline{b}, \qquad\qquad\qquad \overrightarrow{AB} = \underline{b} - \underline{a}$$

$$\overrightarrow{OM} = \frac{1}{2}(\underline{a} + \underline{b}), \qquad\qquad \overrightarrow{AL} = \frac{1}{2}(-\underline{a} + (\underline{b} - \underline{a}))$$

$$\overrightarrow{OR} = \lambda\overrightarrow{OM} = \frac{\lambda}{2}(\underline{a} + \underline{b}), \qquad \overrightarrow{AR} = \mu\overrightarrow{AL} = \frac{\mu}{2}[-\underline{a} + (\underline{b} - \underline{a})]$$

$$\underline{a} = \overrightarrow{OR} - \overrightarrow{AR} = \frac{\lambda}{2}(\underline{a} + \underline{b}) - \frac{\mu}{2}[-\underline{a} + (\underline{b} - \underline{a})]$$

$$0 = \left(\frac{\lambda}{2} + \mu - 1\right)\underline{a} + \left(\frac{\lambda}{2} - \frac{\mu}{2}\right)\underline{b}.$$

Since $\{\underline{a}, \underline{b}\}$ is an independent set

$$\frac{\lambda}{2} - \frac{\mu}{2} = 0, \qquad \frac{\lambda}{2} + \mu - 1 = 0$$

$$\lambda = \mu, \qquad \frac{3}{2}\mu = 1, \qquad \mu = \frac{2}{3} = \lambda.$$

The same method can be used to complete the proof that the vector from B to the midpoint of OA passes through R.

A.3   Linear Forms and Dual Vector Spaces

We shall now define the linear form f on a vector space V.

Definition III.A.3.1 A linear form f on a vector space V is a function of the members of V such that

1. $f(\underline{x}) = a, \quad a \in F,$ all $\underline{x} \in V$;
2. $f(\underline{x} + \underline{y}) = f(\underline{x}) + f(\underline{y}),$
3. $f(a\underline{x}) = af(\underline{x}).$

If now we make certain notational conventions, we shall find that the linear forms themselves form a vector space V^*. So we write

Definition III.A.3.2 $\left(f^1 + f^2\right)(\underline{x}) = f^1(\underline{x}) + f^2(\underline{x})$

$$(af)(\underline{x}) = af(\underline{x}).$$

Theorem A.3.1. *The linear forms on a vector space V form a vector space V^* over the same F.*

Proof.

1. If f^1 and f^2 are in V^*, so is $(f^1 + f^2)$. For, by the previous definition,

$$(f^1 + f^2)(\underline{x} + \underline{y}) = f^1(\underline{x} + \underline{y}) + f^2(\underline{x} + \underline{y})$$
$$= f^1(\underline{x}) + f^1(\underline{y}) + f^2(\underline{x}) + f^2(\underline{y})$$
$$= (f^1 + f^2)(\underline{x}) + (f^1 + f^2)(\underline{y});$$

and certainly $(f^1 + f^2)(\underline{x})$ is a member of F.

2. Let $f^1, f^2, f^3 \in V^*$, then

$$[(f^1 + f^2) + f^3](\underline{x}) = (f^1 + f^2)(\underline{x}) + f^3(\underline{x})$$
$$= [f^1(\underline{x}) + f^2(\underline{x})] + f^3(\underline{x})$$
$$= f^1(\underline{x}) + [f^2(\underline{x}) + f^3(\underline{x})]$$
$$= [f^1 + (f^2 + f^3)](\underline{x}),$$

and so the law is associative.

3. $[f + (-f)](\underline{x}) = f(\underline{x}) - f(\underline{x}) = 0$. $[f + (-f)]$ is the zero linear form for each f in V^*; we denote it as 0

4. $f^1 + f^2 = f^2 + f^1$. It is an additive Abelian group.

For all $a \in F$, the external laws of composition hold. Thus, by the previous definition

1. $(af)(\underline{x}) = a[f(\underline{x})] = f(a\underline{x}) = f(\underline{y})$

so that $af \in V^*$ if f is, and $a \in F$. The other laws are easily verified. $\qquad\square$

Furthermore we can construct a basis and define independence. If $\{\underline{e}_i\}$, $i = 1, \ldots, n$, is a basis of V, then we can define the dual basis of it in V^*. To this end we define $\{\underline{\varepsilon}^i\}$ in V^* as the linear forms such that

Definition III.A.3.3.

$$\underline{\varepsilon}^i(\underline{e}_j) = \delta^i_j = \begin{cases} 0 & i \neq j \\ 1 & i = j \end{cases}, \qquad i, j = 1, 2, \ldots, n.$$

As before, we shall later explain the reason for the superscript. We are here underlining the linear forms to emphasize their vectorial character and to be able to compare them with the elements of F. Thus we shall often write $\underline{f}$ instead of f.

Theorem A.3.2. *The $\{\varepsilon^i\}$ so defined are a linear form of any $\underline{x} \in V$, and form a basis in V^* such that every $\underline{f} \in V^*$ can be expressed as a linear combination of them over F.*

Proof: First,

$$\varepsilon^i(\underline{x}) = \varepsilon^i(x^j\underline{e}_j) = x^j\varepsilon^i(\underline{e}_j) = x^i, x^i \in F,$$

using the previous definition and the fact that the ε^i are linear forms.

Second, the $\{\varepsilon^i\}$ form an independent set. For, if

$$a_i\varepsilon^i = 0, \quad a_i \in F,$$

then

$$(a_i\varepsilon^i)(\underline{e}_j) = a_i[\varepsilon^i(\underline{e}_j)] = a_j = 0, \text{ for each } j,$$

since $a_i\varepsilon^i$ is the zero-linear form.

Third, any $\underline{f} \in V^*$ can be expressed as a linear combination of $\{\varepsilon^i\}$. For, suppose we are given $\underline{f}$ in V^* such that

$$f(\underline{e}_i) = f_i, \text{ where } f_i \in F, {}^{1}$$

then

$$f(\underline{x}) = f(x^i\underline{e}_i) = x^i f(\underline{e}_i) = x^i f_i.$$

But, also

$$(f_i\varepsilon^i)(\underline{x}) = (f_i\varepsilon^i)(x^j\underline{e}_j) = (f_ix^j)[\varepsilon^i(\underline{e}_j)]$$
$$= x^i f_i$$

Therefore, given $\underline{f}$ in V^* we can always express it as a linear combination of $\{\varepsilon^i\}$ over F. And this, in only one way, since, if

$$f_i^1\varepsilon^i = f_i^2\varepsilon^i,$$

then

$$\left(f_i^1 - f_i^2\right)\varepsilon^i = 0, \quad \left(f_j^1 - f_j^2\right) = 0 \quad \text{ for all } j,$$

the $\{\varepsilon^i\}$ being an independent set.

It follows immediately that the dimensions of V and V^* are the same. V is called the space of contravariant vectors, and V^* of covariant vectors for

---

[1]It should be noted that f^i are different $\underline{f}$'s in V^*, whereas f_i are components of $\underline{f}$ with respect to the basis $\{\varepsilon^i\}$ and are in F. Thus $\underline{x} = x^i\underline{e}_i, x^i \in F, \underline{e}_i \in V, \underline{f} = f_i\varepsilon^i, f_i \in F, \varepsilon^i \in V^*$.

reasons we shall only understand later. It should be noted also that we can quite effectively write

$$\underline{f} \cdot \underline{x} = f_i x^i,$$

which is an example of what will be later defined as inner products. This is an inner product defined on $V \times V^*$.

We can establish another important relation between V and V^*. Let S be any subset of vectors in V and S_0^* the set of all vectors $\underline{f}$ in V^* such that $f(\underline{x}) = \underline{f} \cdot \underline{x} = 0$ for all $\underline{x}$ in S. Then

Theorem A.3.3. S_0^* *is a vector subspace of* V^*.

Proof: If $\underline{f}^1 \cdot \underline{x} = 0$ and $\underline{f}^2 \cdot \underline{x} = 0$ for all $\underline{x}$ in S, then $(\underline{f}^1 + \underline{f}^2) \cdot \underline{x} = 0$. Therefore, if $\underline{f}^1$ and $\underline{f}^2$ are in S_0^*, so also is $\underline{f}^1 + \underline{f}^2$. Likewise if $\underline{f} \cdot \underline{x} = 0$, then $(a\underline{f}) \cdot \underline{x} = 0$, so that, if $\underline{f}$ is in S_0^* so also is $(a\underline{f})$. The rest follows.

Theorem A.3.4. *If* S *is an* r-*dimensional subspace of* V, *then* S_0^* *is an* $(n - r)$-*dimensional subspace of* V^*.

Proof: S_0^* is a subspace of V^* from the previous theorem. Let $\{\underline{e}_1, \ldots, \underline{e}_n\}$ be a basis of V and $\{\underline{e}_1, \ldots, \underline{e}_r\}$ a basis of S; and $\{\underline{\varepsilon}^1, \ldots, \underline{\varepsilon}^n\}$ the dual basis of V^*. Consider further $\{\underline{\varepsilon}^{r+1}, \ldots, \underline{\varepsilon}^n\}$ as a basis of N^*, a subspace of V^*.

We shall show $S_0^* = N^*$.

Since any $\underline{x}$ in S is $\underline{x} = x^i \underline{e}_i, i = 1, 2, \ldots, r$, and $\underline{\varepsilon}^j(x^i \underline{e}_i) = 0$ for $j = r + 1, \ldots, n$; therefore

$$\underline{\varepsilon}^j \in S_0^* \text{ for } j = r + 1, \ldots, n;$$

and so

$$N^* \subseteq S_0^*,$$

to read N^* is a subspace of S_0^*.

On the other hand, if $\underline{f}$ is in S_0^*, then

$$\underline{f} = f_i \underline{\varepsilon}^i, \quad i = 1, \ldots, n \quad (\text{since } \underline{f} \text{ is in } V^*)$$

and

$$\underline{f} \cdot (\underline{e}_j) = f_i \underline{\varepsilon}^i(\underline{e}_j) = f_i = 0, \quad i = 1, 2, \ldots, r.$$

Therefore

$$\underline{f} = f_i \underline{\varepsilon}^i, \quad i = r + 1, \ldots, n$$

and any $\underline{f}$ in S_0* has $n - r$ dimensions. Therefore

$$S_0* \subseteq N^*,$$

whence it follows

$$S_0^* = N^*.$$

These theorems will apply, of course, to the solution of systems of homogeneous linear equations. It should be noted that this theorem could also read: If S^* is an r-dimensional subspace of V^*, then S_0 is an $(n - r)$-dimensional subspace of V, reversing the vector spaces.

A.4 Linear Mappings

We now introduce the notion of linear mapping or transformation of a vector space V into a vector space V'.

Definition III.A.4.1. A linear mapping or transformation A of a vector space V into a vector space V' for some F is a single-valued vector function such that

1. $A(\underline{x}) = \underline{x}'$, $\underline{x} \in V$, $\underline{x}' \in V'$;

2. $A(\underline{x} + \underline{y}) = A(\underline{x}) + A(\underline{y})$, $\underline{x}$ and $\underline{y} \in V$, $A(\underline{x})$, $A(\underline{y}) \in V'$;

3. $A(a\underline{x}) = aA(\underline{x})$, $a \in F$.

This is also called an homomorphism (an endomorphism when $V = V'$) of V to V'. A one-to-one endomorphism is called an automorphism and a one-to-one homomorphism an isomorphism. The difference between a linear mapping and a linear form is that the former maps a vector space into a vector space whereas the latter maps a vector space into a field.

Introducing the proper notations, one can prove, just as in the case of linear forms, that the linear mappings form a vector space over F.

Definition III.A.4.2. If A and B are two linear mappings of V into V' over F, then

$$(A + B)(\underline{x}) = A(\underline{x}) + B(\underline{x})$$
$$(aA)(\underline{x}) = aA(\underline{x}), a \in F.$$

It is easy to show now that the linear mappings of V in V' form an additive Abelian group with external laws from F. For just as in the case of linear forms

$$(A + B)(\underline{x} + \underline{y}) = A(\underline{x} + \underline{y}) + B(\underline{x} + \underline{y}) = A(\underline{x}) + A(\underline{y}) + B(\underline{x}) + B(\underline{y})$$
$$= (A + B)(\underline{x}) + (A + B)(\underline{y})$$
$$(A + B)(a\underline{x}) = A(a\underline{x}) + B(a\underline{x}) = aA(\underline{x}) + aB(\underline{x})$$
$$= a(A + B)(\underline{x})$$

so that

$$A + B = C$$

where C is a linear mapping. Associativity is easy to prove, and, if we take $-A(\underline{x})$ as the inverse of $A(\underline{x})$, we immediately have the zero-mapping.

The external laws from F can be immediately verified. For example, since

$$\begin{aligned}
(aA + aB)(\underline{x}) &= (aA)(\underline{x}) + (aB)(\underline{x}) = aA(\underline{x}) + aB(\underline{x}) \\
&= A(a\underline{x}) + B(a\underline{x}) = (A + B)(a\underline{x}) \\
&= [a(A + B)](\underline{x}),
\end{aligned}$$

we can say

$$(aA + aB) = a(A + B).$$

But we can define another operation which will lead us into much deeper problems. Consider a linear mapping B of V into itself followed by another linear mapping A of V into itself. The resulting mapping is certainly single-valued and from V into V. That it is also linear will be seen after

Definition III.A.4.3. If A and B are two linear mappings of V into V over F, then

$$(A \cdot B)(\underline{x}) = A(B(\underline{x})), \quad \underline{x} \in V.$$

We shall call this the product of A and B.

Theorem A.4.1. *The product $A \cdot B$ is linear.*

Proof.

1. $$\begin{aligned}
(A \cdot B)(\underline{x} + \underline{y}) &= A[B(\underline{x} + \underline{y})] = A[B(\underline{x}) + B(\underline{y})] \\
&= (A \cdot B)(\underline{x}) + (A \cdot B)\underline{y}
\end{aligned}$$

2. $$\begin{aligned}
(A \cdot B)(a\underline{x}) &= A(B(a\underline{x})) = A[aB(\underline{x})] = aA[B(\underline{x})] \\
&= a(A \cdot B)(\underline{x}).
\end{aligned}$$

$\square$

Hence the first two requirements of a multiplicative group are fulfilled:

1. $A \cdot B = C, C$ a linear mapping of V into V,

2. $(A \cdot B) \cdot C = A \cdot (B \cdot C)$.

But the third is here a problem. Obviously the e-element is the identity transformation which we shall call U. To find an X so that, for all A,

$$A \cdot X = X \cdot A = U.$$

First, the inverse will not always exist unless A is restricted to one-to-one mappings, for a single-valued inverse mapping would not then exist. In particular, letting $a = 0$,

$$A(0 \cdot \underline{x}) = A(\underline{0}) = 0 \cdot A(\underline{x}) = \underline{0},$$

which shows that a linear mapping of V into V always maps the null-vector onto itself. If, then, $A(\underline{x}) = \underline{0}$ where $\underline{x} \neq \underline{0}$, we cannot map linearly $\underline{0}$ back to $\underline{x}$. Therefore the inverse A^{-1} does not exist in such a case. Hence we have proved

Theorem A.4.2. *The inverse of A does not exist if $A(\underline{x}) = \underline{0}$ where $\underline{x} \neq \underline{0}$.*

Theorem A.4.3. *If $\{\underline{e}_i\}$ is a basis of V and A is a linear mapping of V into V, then the necessary and sufficient condition that A^{-1} exist is that $\{A(\underline{e}_i)\}$ be an independent set. Then A^{-1} is linear.*

Proof.

1. Let $\underline{x} = x^i \underline{e}_i$. Then $A(\underline{x}) = x^i A(\underline{e}_i)$. If now the $\{A(\underline{e}_i)\}$ is dependent, we can choose x^i not all zero, that is, $\underline{x} \neq \underline{0}$ so that

$$A(\underline{x}) = x^i A(\underline{e}_i) = \underline{0}.$$

Then, by the previous theorem A^{-1} does not exist. Hence if A^{-1} exists, then the $\{A(\underline{e}_i)\}$ is an independent set.

2. If the $\{A(\underline{e}_i)\}$ is independent, it forms a basis in V and $x^i A(\underline{e}_i)$ represents every vector $\underline{x}$ in V exactly once as x^i range over F. The correspondence $x^i A(\underline{e}_i) \to x^i \underline{e}_i$ is single-valued. But this is precisely the definition of the inverse mapping of A, A^{-1}. For $A^{-1}[x^i A(\underline{e}_i)] = (A^{-1}A)(\underline{x}) = \underline{x}$.

   Further A^{-1} is linear, since

   1. $A^{-1}[A(\underline{x}) + A(\underline{y})] = A^{-1}[A(\underline{x} + \underline{y})] = \underline{x} + \underline{y} = (A^{-1}A)(\underline{x}) + (A^{-1}A)(\underline{y})$
      $$= A^{-1}[A(\underline{x})] + A^{-1}[A(\underline{y})].$$
   2. $\quad A^{-1}[A(a\underline{x})] = a\underline{x} = aA^{-1}[A(\underline{x})] = A^{-1}[aA(\underline{x})].$

   $\square$

Again, one could start with $A^{-1}(\underline{x})$ and argue in the same manner to show that A is its inverse. Hence

$$A[(A^{-1})(\underline{x})] = A^{-1}[A(\underline{x})] = U.$$

Theorem A.4.4. *If $\{\underline{e}^i\}$ is a basis on V and $\{A(\underline{e}_i)\}$ is independent and $\{B(\underline{e}_i)\}$ is independent, then $\{(A \cdot B)\underline{e}_i\}$ is independent.*

Proof: Suppose $\{(A \cdot B)(\underline{e}_i)\}$ is dependent, then there exist x^i so that

$$x^i (A \cdot B)(\underline{e}_i) = \underline{0}, \text{ not all } x^i \text{ being zero.}$$

But

$$x^i (A \cdot B)(\underline{e}_i) = A[x^i B(\underline{e}_i)] = \underline{0}.$$

Since, by Theorem A.12,

$$A(\underline{y}) = \underline{0},$$

only if

$$\underline{y} = \underline{0},$$

therefore

$$x^i B(\underline{e}_i) = \underline{0}, \text{ not all } x^i = 0.$$

But this is contradictory since $\{B(\underline{e}_i)\}$ is independent.

Therefore we see that, given a basis $\{\underline{e}_i\}$ of V, the linear mappings A where $\{A(\underline{e}_i)\}$ is independent, form a multiplicative group. We shall see later that it is not Abelian.

A.5   Matrices on F as Representations of Linear Mappings.

We shall now find an analytic representation of the linear mappings of V into V over F. Since $A(\underline{e}_k)$ is a vector in V, and it has a unique representation in terms of the basis $\{\underline{e}_i\}$ of V, (It is understood $i = 1, \ldots, n$ unless otherwise stated), therefore

$$A(\underline{e}_k) = a_k^i \underline{e}_i, \text{ for } k = 1, \ldots, n$$

But

$$\underline{x} = x^i \underline{e}_i$$

and

$$A(\underline{x}) = A(x^k \underline{e}_k) = x^k A(\underline{e}_k) = x^k (a_k^i \underline{e}_i)$$
$$= y^i \underline{e}_i = \underline{y},$$

where

$$y^i = x^k a_k^i \qquad (\alpha)$$

Conversely, given (α), we can retrace our steps to get an A which satisfies the conditions of linearity.

Hence we say the linear mapping A is analytically represented by the matrix $[A]$ for a given basis in V

$$[A] = \begin{pmatrix} a_1^1 & a_2^1 & \cdots & a_n^1 \\ a_1^2 & a_2^2 & \cdots & a_n^2 \\ \cdots & \cdots & \cdots & \cdots \\ a_1^n & a_2^n & \cdots & a_n^n \end{pmatrix}, \quad a_j^i \in F.$$

This is a square $n \times n$ matrix, the number of columns equals the number of rows. For certain considerations, the order is essential as for the elements in a basis, although so far it has not appeared so.

It is evident the same argument can be used for the linear mapping of V_1 into V_2 instead of the vector space into itself. Let

$$y = A(\underline{x}), \quad \underline{x} \in V_1, \quad y \in V_2.$$

Choose a basis $\{\underline{e}_i\}$ in V_1 and $\{\underline{c}_\sigma\}$ in V_2. Then

$$y = y^\sigma \underline{c}_\sigma, \quad \underline{x} = x^i \underline{e}_i, \quad i = 1, \ldots, n; \quad \sigma = 1, \ldots, n.$$

Now

$$A(\underline{x}) = A(x^i \underline{e}_i) = x^i A(\underline{e}_i),$$

where

$$A(\underline{e}_i) = a_i^\sigma \underline{c}_\sigma,$$

since $A(\underline{e})_i$ is a vector in V_2 for every $\underline{e}_i$. Therefore

$$y^\sigma \underline{c}_\sigma = x^i a_i^\sigma \underline{c}_\sigma, \quad y^\sigma = x^i a_1^\sigma.$$

$$[A] = \begin{pmatrix} a_1^1 & a_2^1 & \cdots & a_n^1 \\ \cdots & \cdots & \cdots & \cdots \\ a_1^m & a_2^m & \cdots & a_n^m \end{pmatrix}, a_i^\sigma \in F.$$

Suppose we take another such matrix $B = (b_j^i)$ as representing another linear mapping with respect to the same basis of V into V. Then the result B followed by A would be

$$(A \cdot B)(\underline{x}) = A[B(\underline{x})].$$

But

$$B(\underline{x}) = x^j b^i_j \underline{e}_i = y^i \underline{e}_i = \underline{y},$$

$$A[B(\underline{x})] = \left(x^j b^i_j\right) A(\underline{e}_i) = \left(x^j b^i_j\right)\left(a^k_i \underline{e}_k\right)$$

$$= x^j \left(b^i_j a^k_i\right)\underline{e}_k = x^j \left(a^k_i b^i_j\right)\underline{e}_k$$

$$= x^j \left(a^i_k b^k_j\right)\underline{e}_i,$$

on changing the dummy index i to k. Hence $A \cdot B$ gives the resulting mapping C where

$$c^i_j = a^i_k b^k_j.$$

This directs us to give the usual definition of the multiplication of matrices representing linear mappings

Definition III.A.5.1. The matrix representing C, resulting from the matrix A times the matrix B, consists of terms such that the term in the ith row and jth column is the result of multiplying each term of the ith row of A by the corresponding term of the jth column of B and adding. In other words,

$$c^i_j = a^i_k b^k_j.$$

This product, in general, is not commutative:

$$a^i_k b^k_j \neq b^i_k a^k_j.$$

This definition can, obviously, be extended to any $p \times q$ and $q \times r$ matrices. We can also express

$$cA(\underline{x}) = A(c\underline{x}), c \in F,$$

as a special case of this definition. Let c be represented by the matrix

$$c^i_j = c\delta^i_j.$$

In like manner, from the definition of $A + B$, we are led to a definition of the addition of matrices representing linear mappings. Since

$$(A + B)(\underline{e}_i) = A(\underline{e}_i) + B(\underline{e}_i) = \left(a^j_i \underline{e}_j\right) + \left(b^j_i \underline{e}_j\right)$$

$$= \left(a^j_i + b^j_i\right)\underline{e}_j,$$

we give the usual

Definition III.A.5.2. The $n \times n$ matrix C which is the sum of two $n \times n$ matrices A and B is gotten by adding the corresponding terms of A and B:

$$a_i^j + b_i^j = c_i^j.$$

It follows immediately that the zero-matrix 0 is the matrix every term of which is zero. It is clear also that the unit-matrix U is such that

$$U_j^i = \delta_j^i.$$

When we now turn to the theorem that the linear mapping A has a unique inverse A^{-1} if and only if $\{A(\underline{e}_i)\}$ is an independent set when $\{\underline{e}_i\}$ is a basis of V, we can translate this into the language of matrices:

$$A(\underline{e}_k) = a_k^i \underline{e}_i, \quad i \text{ and } k = 1, \ldots, n.$$

We consider the components or coefficients of the $k\underline{th}$ column $(a_k^1, \ldots, a_k^n)$ as representing $A(\underline{e}_k)$. Such ordered n-tuples of elements of F can be considered as a vector in a vector space of n-dimensions with a fixed basis $\{\underline{e}_i\}$; they are the coefficients of the $\underline{e}_i$. We define addition of two such vectors as the ordered n-tuple resulting from adding the corresponding elements of each; and multiplication by a field element, we define as the vector resulting from multiplying each element of the n-tuple by the field element. We can show then that in terms of $n \times n$ matrices, the condition that the $\{A(\underline{e}_k)\}$ be an independent set becomes the condition that the set of column vectors of the $n \times n$ matrix $[A]$ be an independent set. But we can prove a bit more.

Theorem A.5.1. *If the set of n vectors $\{\underline{e}_i\}$ be independent and the m vectors $\{\underline{b}_k\}$ are such that*

$$\underline{b}_k = a_k^i \underline{e}_i, \quad k = 1, \ldots, m; \quad i = 1, \ldots, n;$$

then the number of independent vectors $\underline{b}_k$ is exactly the number of independent column vectors

$$\underline{a}_k = (a_k^1, a_k^2, \ldots, a_k^n).$$

Proof:

1. If $\lambda^k \underline{a}_k = \underline{0}$, then $\lambda^k a_k^i = 0$, $i = 1, \ldots, n$, $\lambda^k \in F, k = 1, \ldots, n \leq m$. For $\underline{0}$ as an n-tuple is obviously the n-tuple with all elements 0. Then

$$\lambda^k \underline{b}_k = \lambda^k a_k^i \underline{e}_i = \underline{e}_i \left(\lambda^k a_k^i \right) = \underline{0}.$$

2. If $\lambda^k \underline{b}_k = \underline{0}$, then $\underline{e}_i \left(\lambda^k a_k^i \right) = \underline{0}$. But $\{\underline{e}_i\}$ is independent. Hence

$$\lambda^k a_k^i = 0, \quad i = 1, \ldots, n, \quad \text{and} \quad \lambda^k \underline{a}_k = \underline{0}.$$

It is usual to call any linear mapping A which has a unique inverse A^{-1}, non-singular; others are called singular.

We are now in a position to find relations between any two bases $\{\underline{e}_i\}$ and $\{\overline{\underline{e}}_i\}$ of the same vector space V. We have just seen that, if $\{\underline{e}_i\}$ is a basis of V and A a non-singular linear mapping, then $\{A(\underline{e}_k)\}$ is an independent set with n elements, hence it is also a basis of V; we call it $\{\overline{\underline{e}}_i\}$.

Conversely, given any two bases $\{\underline{e}_i\}$ and $\{\overline{\underline{e}}_i\}$ of V, there exists a non-singular linear mapping A between them. For $\overline{\underline{e}}_k$ is in V, hence

$$\overline{\underline{e}}_k = a_k^i \underline{e}_i, \text{ and } k = 1, \ldots, n.$$

The a_k^i form a non-singular matrix $[A]$. So

Theorem A.5.2. *If $\{\underline{e}_i\}$ forms a basis of V, then any non-singular linear mapping A of the $\{\underline{e}_i\}$ also forms a basis in V. And between any two given bases in V, there exists a non-singular linear mapping A.*

We can now compare the matrices representing a linear mapping A of V into V in terms of two bases $\{\underline{e}_i\}$ and $\{\overline{\underline{e}}_i\}$ of V. Let

$$\underline{y} = A(\underline{x}), \quad \overline{\underline{e}}_i = t_i^j \underline{e}_j, \quad \underline{e}_i = (t^{-1})_i^j \overline{\underline{e}}_j.$$

Then

$$y^j \underline{e}_j = A(x^i \underline{e}_i) = x^i A(\underline{e}_i) = x^i a_i^j \underline{e}_j,$$
$$y^j = x^i a_i^j.$$

And likewise,

$$\overline{y}^j \overline{\underline{e}}_j = \overline{x}^i \overline{a}_i^j \overline{\underline{e}}_j, \quad \overline{y}^j = \overline{x}^i \overline{a}_i^j.$$

But also

$$\overline{y}^j \overline{\underline{e}}_j = y^i \underline{e}_i = y^i (t^{-1})_i^j \overline{\underline{e}}_j, \qquad \overline{x}^j \overline{\underline{e}}_j = x^i (t^{-1})_i^j \overline{\underline{e}}_j,$$
$$\overline{y}^j = y^i (t^{-1})_i^j = x^k a_k^i (t^{-1})_i^j,$$

and so also

$$\overline{y}^j = \overline{x}^i \overline{a}_i^j = x^k (t^{-1})_k^i \overline{a}_i^j.$$

Therefore

$$x^k[a_k^i(t^{-1})_i^j - (t^{-1})_k^i \bar{a}_i^j] = 0$$

for every x^k varying independently. Hence

$$(t^{-1})_i^j a_k^i = \bar{a}_i^j(t^{-1})_k^i,$$

Theorem A.5.3. $[(T^{-1})][(A)] = [\bar{A}][T^{-1}]$

$$[\bar{A}] = [T^{-1}][A][T]$$

where $[T]$ is the matrix of the transformation from $\{\underline{e}_i\}$ to $\{\underline{\bar{e}}_i\}$, $[A]$ is the matrix of A in terms of $\{\underline{e}_i\}$, $[\bar{A}]$ the matrix of A in terms of $\{\underline{\bar{e}}_i\}$.

Corollary A.5.4. *If $T = A$, then $[\bar{A}] = [A^{-1}][A][A] = [A]$.*

If now we compare the application of non-singular linear mappings to V and to V^*, we shall see the reason for the use of the subscripts and superscripts. Let

$$\underline{\bar{e}}_i = t_i^j \underline{e}_j.$$

Then

$$\underline{x} = x^i \underline{e}_i = \bar{x}^i \underline{\bar{e}}_i = \bar{x}^i t_i^j \underline{e}_j = \bar{x}^j t_j^i \underline{e}_i$$
$$x^i = \bar{x}^j t_j^i, \bar{x}^i = x^j(t^{-1})_j^i = (t^{-1})_j^i x^j.$$

If we say the basis vectors in V are covariant since they transform by T, then their coefficients (the coordinate vectors) transform by T^{-1} and will be called contravariant. Hence covariant vectors are denoted by a subscript; the contravariant by a superscript.

In V^*, we defined a dual basis $\{\underline{\varepsilon}^i\}$ so that

$$\underline{\varepsilon}^i(\underline{e}_j) = \delta_j^i$$

and likewise

$$\underline{\bar{\varepsilon}}^i(\underline{\bar{e}}_j) = \delta_j^i, \underline{\bar{e}}_j = t_j^k \underline{e}_k.$$

Letting

$$\underline{\bar{\varepsilon}}^i = \underline{\varepsilon}^j s_j^i,$$

we have

$$\underline{\bar{\varepsilon}}^i(\underline{\bar{e}}_j) = \underline{\bar{\varepsilon}}^i t_j^k \underline{e}_k = \underline{\varepsilon}^r s_r^i t_j^k \underline{e}_k$$
$$= s_r^i(\underline{\varepsilon}^r \underline{e}_k)t_j^k = s_r^i t_j^r = \delta_j^i.$$

Hence $s_r^i = (t^{-1})_r^i$. Therefore in V^*, contrariwise,

$$\bar{\varepsilon}^i = (t^{-1})_j^i \varepsilon^j, \qquad \bar{f}_1 = f_j t_i^j.$$

We can also now see that

$$\underline{f} \cdot \underline{x} = \bar{f}_i \bar{x}^i = \left(f_j t_i^j \right) \left(x^k \left(t^{-1} \right)_k^i \right) = f_j t_i^j \left(t^{-1} \right)_k^i x^k$$
$$= f_j x^j$$

is invariant under change of bases. On the other hand, it is now obvious that an inner product such as $\Sigma_i x^i y^i$ is not. There will be special systems under which it will, as we shall see later.

A.6   Bilinear Forms and Tensor Products

We can now generalize and deepen what we have done by

Definition III.A.6.1. A bilinear form H on $V_1 \times V_2$ is such that

1. $H(\underline{x}, \underline{y}) = a, \quad \underline{x} \in V, \quad \underline{y} \in V, \quad a \in F.$
2. $H(a\underline{x}_1 + b\underline{x}_2, \underline{y}) = aH(\underline{x}_1, \underline{y}) + bH(\underline{x}_2, \underline{y})$
   $H(\underline{x}, a\underline{y}_1 + b\underline{y}_2) = aH(\underline{x}, \underline{y}_1) + bH(\underline{x}, \underline{y}_2), \qquad b \in F.$

Now taking $\{\underline{e}_i\}$ as a basis of V_1 with dimension n and $\{\underline{\varepsilon}_\sigma\}$ as a basis of V_2 with dimension m, $\{\varepsilon^i\}$ as the corresponding dual basis on V_1^* and $\{\gamma^\sigma\}$ as that on V_2^*, we wish to show how to construct bilinear form H on $V_1 \times V_2$ in terms of the linear forms $\underline{f}$ and $\underline{g}$ on V_1^* and V_2^*.

Definition III.A.6.2. Let $\underline{f} \in V_1^*, \underline{g} \in V_2^*, \underline{x} \in V_1, \underline{y} \in V_2$. Then

$$(\underline{f} \otimes \underline{g})(\underline{x}, \underline{y}) = \underline{f}(\underline{x})\underline{g}(\underline{y}) = (\underline{f} \cdot \underline{x})(\underline{g} \cdot \underline{y}).$$

It is easily seen that $\underline{f} \otimes \underline{g}$ is bilinear, since

$$(\underline{f} \otimes \underline{g})(\underline{x}_1 + \underline{x}_2, \underline{y}) = f(\underline{x}_1 + \underline{x}_2)g(\underline{y}) = f(\underline{x}_1)g(\underline{y}) + f(\underline{x}_2)g(\underline{y})$$
$$= (\underline{f} \otimes \underline{g})(\underline{x}_1, \underline{y}) + (\underline{f} \otimes \underline{g})(\underline{x}_2, \underline{y})$$

and so on for the rest.

From our definition of H, we have

$$H(\underline{x}, \underline{y}) = H(x^i \underline{e}_i, y^\sigma \underline{c}_\sigma) = x^i y^\sigma H(\underline{e}_i, \underline{c}_\sigma)$$

and

$$H(\underline{e}_i, \underline{c}_\sigma) = H_{i\sigma},$$

so that

$$H(\underline{x}, \underline{y}) = x^i y^\sigma H_{i\sigma}.$$

Suppose we now define $\underline{\varepsilon}^i \otimes \underline{\gamma}^\sigma$ as a bilinear form on $V_1 \times V_2$. By Definition III.A.6.2.,

$$\underline{\varepsilon}^i \otimes \underline{\gamma}^\sigma(\underline{e}_j, \underline{c}_\tau) = (\underline{\varepsilon}^i \cdot \underline{e}_j)(\underline{\gamma}^\sigma \cdot \underline{c}_\tau) = \delta^i_j \delta^\sigma_\tau.$$

Then, if we take

$$H = H_{i\sigma}(\underline{\varepsilon}^i \otimes \underline{\gamma}^\sigma),$$

we have

$$\begin{aligned}
H(\underline{x}, \underline{y}) &= H_{i\sigma}(\underline{\varepsilon}^i \otimes \underline{\gamma}^\sigma)(x^j \underline{e}_j, y^\tau \underline{c}_\tau), \\
&= x^j y^\tau H_{i\sigma}(\underline{\varepsilon}^i \otimes \underline{\gamma}^\sigma)(\underline{e}_j, \underline{c}_\tau), \\
&= x^j y^\tau H_{i\sigma} \delta^i_j \delta^\sigma_\tau, \\
&= x^i y^\sigma H_{i\sigma}.
\end{aligned}$$

So that every H can be expressed as a linear combination of $\underline{\varepsilon}^i \otimes \underline{\gamma}^\sigma$. Further the $\{\underline{\varepsilon}^i \otimes \underline{\gamma}^\sigma\}$ is an independent set. For, if

$$\lambda_{i\sigma}(\underline{\varepsilon}^i \otimes \underline{\gamma}^\sigma) = \underline{0}, \lambda_{i\sigma} \in F,$$

then

$$0 = \lambda_{i\sigma}(\underline{\varepsilon}^i \otimes \underline{\gamma}^\sigma)(\underline{e}_j, \underline{c}_\rho) = \lambda_{i\sigma} \delta^i_j \delta^\sigma_\rho = \lambda_{j\rho},$$

for all j and ρ. Hence the $\{\underline{\varepsilon}^i \otimes \underline{\gamma}^\sigma\}$ are a basis for the vector space of the bilinear forms on $V_1 \times V_2$, of dimension nm. Hence the uniqueness of the representation of H in terms of the $\{\underline{\varepsilon}^i \otimes \underline{\gamma}^\sigma\}$.

Since the $\underline{f} \otimes \underline{g}$ are bilinear forms, they can also be expressed as a linear combination of the basis $\{\underline{\varepsilon}^i \otimes \underline{\gamma}^\sigma\}$.
Let $\underline{f} \in V_1^*$ and $\underline{g} \in V_2^*$. Then

$$\underline{f} = f_i \underline{\varepsilon}^i, \underline{g} = g_\sigma \underline{\gamma}^\sigma,$$

and using the previous definitions and the definition of linear form,

$$\begin{aligned}
(\underline{f} \otimes \underline{g})(\underline{x}, \underline{y}) &= \underline{f}(\underline{x})\underline{g}(\underline{y}) = (\underline{f} \cdot \underline{x})(\underline{g} \cdot \underline{y}) \\
&= (f_i \underline{\varepsilon}^i)(x^j \underline{e}_j)(g_\sigma \underline{\gamma}^\sigma)(y^\rho \underline{c}_\rho) \\
&= (f_i g_\sigma)\underline{\varepsilon}^i(x^j \underline{e}_j)\underline{\gamma}^\sigma(y^\rho \underline{c}_\rho) \\
&= (f_i g_\sigma)(\underline{\varepsilon}^i \otimes \underline{\gamma}^\sigma)(\underline{x}, \underline{y}) \quad\quad (\alpha) \\
&= (f_i g_\sigma)(x^j y^\rho)\underline{\varepsilon}^i(\underline{e}_j)\underline{\gamma}^\sigma(\underline{c}_\rho) \\
&= (f_i g_\sigma)(x^j y^\rho)\delta^i_j \delta^\sigma_\rho = (f_i g_\sigma)(x^i y^\sigma)
\end{aligned}$$

which corresponds to the previous formula for bilinear forms,

$$H(\underline{x}, \underline{y}) = H_{i\sigma} x^i y^\sigma, \quad i = 1, \ldots, n; \quad \sigma = 1, \ldots, m.$$

In the derivation, we developed more formulas than we needed to show what maneuvers we can make. Notably, (α) gives us

$$\underline{f} \otimes \underline{g} = f_i g_\sigma (\underline{\varepsilon}^i \otimes \underline{\gamma}^\sigma).$$

It should be noted that the tensor products of linear forms do not form a vector space, since the sum of any two does not in general give a third. On the other hand, the bilinear forms on $V_1 \times V_2$ do span a vector space, since the sum of any two bilinear forms is a bilinear form, and the product of an element of F and a bilinear form is still a bilinear form. We can sum up in

Theorem A.6.1. *The tensor products of linear forms on V_1 and V_2 are bilinear forms on $V_1 \times V_2$. They form a subset of all bilinear forms on $V_1 \times V_2$ which in turn form a vector space of nm dimensions where V_1 is of n dimensions and V_2 of m, and $\underline{\varepsilon}^i \otimes \underline{\gamma}^\sigma$ is a basis of this vector space.*

We mention in passing a particular bilinear form which will become important later. Let $\underline{x} \in V$ and $\underline{f} \in V^*$. Then, if we write,

$$f(\underline{x}) = \underline{f} \cdot \underline{x} = f_i x^i = \langle \underline{f}, \underline{x} \rangle,$$

we see that

$$\langle \underline{f}, \underline{x} + \underline{y} \rangle = \langle \underline{f}, \underline{x} \rangle + \langle \underline{f}, \underline{y} \rangle;$$
$$\langle \underline{f} + \underline{g}, \underline{x} \rangle = \langle \underline{f}, \underline{x} \rangle + \langle \underline{g}, \underline{x} \rangle;$$
$$\langle a\underline{f}, \underline{x} \rangle = a\langle \underline{f}, \underline{x} \rangle;$$
$$\langle \underline{f}, a\underline{x} \rangle = a\langle \underline{f}, \underline{x} \rangle.$$

Theorem A.6.2. *If $\underline{f} \in V^*$ and $\underline{x} \in V$, then $\underline{f} \cdot \underline{x}$ or $\langle \underline{f}, \underline{x} \rangle$ is a bilinear form on $V^* \times V$.*

This should be remembered when we introduce the inner product to induce a metric on our vector spaces. It will be bilinear.

Suppose in V_1 and V_2 respectively

$$\underline{e}_i = a_i^j \underline{e}_j, \qquad \underline{c}_\sigma = b_\sigma^\rho \underline{c}_\rho$$

so that in V_1^* and V_2^* respectively

$$\overline{\underline{\varepsilon}}^i = (a^{-1})_j^i \underline{\varepsilon}^j, \qquad \overline{\underline{\gamma}}^\sigma = (b^{-1})_\rho^\sigma \underline{\gamma}^\rho.$$

Then

$$\overline{H}_{i\sigma}\left(\underline{\xi}^i \otimes \underline{\gamma}^\sigma\right) = \overline{H}_{i\sigma}\left[\left(a^{-1}\right)^i_j \underline{\varepsilon}^j \otimes \left(b^{-1}\right)^\sigma_\rho \underline{\gamma}^\rho\right]$$
$$= \overline{H}_{i\sigma}\left(a^{-1}\right)^i_j \left(b^{-1}\right)^\sigma_\rho \left(\underline{\varepsilon}^j \otimes \underline{\gamma}^\rho\right)$$
$$= H_{j\rho}\left(\underline{\varepsilon}^j \otimes \underline{\gamma}^\rho\right)$$

where

$$H_{j\rho} = \left(a^{-1}\right)^i_j \left(b^{-1}\right)^\sigma_\rho \overline{H}_{i\sigma},$$

and therefore

$$\overline{H}_{i\sigma} = \left(a^j_i\right)\left(b^\rho_\sigma\right)H_{j\rho},$$

so that

Theorem A.6.3. *The tensor which is a bilinear form on $V_1 \times V_2$ has a double appearance of the matrices of the non-singular linear mappings of V_1 and V_2 into themselves in its transformation. It is a twice covariant tensor and its double subscript was chosen with that in mind.*

We constructed V^* on V as

$$f(\underline{x}) = f_i x^i = a; \qquad f_i, x^i, a \in F.$$

Here we considered $\underline{x}$ as variable and f as function or linear form. But we can also consider f as variable and $\underline{x}$ as function. Then, since, for $\underline{f}^1$ and $\underline{f}^2$ in V^*,

$$\underline{f}^1(\underline{x}) + \underline{f}^2(\underline{x}) = f^1_i \underline{\varepsilon}^i(x^j \underline{e}_j) + f^2_i \underline{\varepsilon}^i(x^j \underline{e}_j)$$
$$= f^1_i x^i + f^2_i x^i = \left(f^1_i + f^2_i\right) x^i$$

Therefore we can write

$$\underline{f}^1(\underline{x}) + \underline{f}^2(\underline{x}) = (\underline{f}^1 + \underline{f}^2)(\underline{x}) = \underline{x}(\underline{f}^1 + \underline{f}^2)$$

and likewise

$$\underline{f}(a\underline{x}) = a\underline{f}(\underline{x}), \qquad (a\underline{x})\underline{f} = (\underline{x})a\underline{f}.$$

Therefore $\underline{x} \in V$ can be considered as a set of linear forms in V^{**} on V^*. Then, as before,

$$\underline{e}_i(\underline{\varepsilon}^j) = \delta^j_i$$

and so on.

Likewise we can consider the tensor product, $\underline{x} \in V_1, \underline{y} \in V_2, \underline{f} \in V_1^*, \underline{g} \in V_2^*$,

$$(\underline{x} \otimes \underline{y})(\underline{f}, \underline{g}) = \underline{x}(\underline{f})\underline{y}(\underline{g}) = \underline{f}(\underline{x})\underline{g}(\underline{y})$$

and

$$(\underline{e}_i \otimes \underline{c}_\rho)(\underline{\varepsilon}^j, \underline{\gamma}^\sigma) = \delta_i^j \delta_\rho^\sigma.$$

Then $\{\underline{e}_i \otimes \underline{c}_\sigma\}$ is a basis for $V_1^{**} \times V_2^{**}$. And we have

$$\begin{aligned}
(\underline{x} \otimes \underline{y})(\underline{f}, \underline{g}) &= x^i y^\rho (\underline{e}_i \otimes \underline{c}_\rho) f_j g_\sigma (\underline{\varepsilon}^j, \underline{\gamma}^\sigma) \\
&= x^i y^\rho f_j g_\sigma (\underline{e}_i \otimes \underline{c}_\rho)(\underline{\varepsilon}^j, \underline{\gamma}^\sigma) \\
&= x^i y^\rho f_j g_\sigma \delta_i^j \delta_\rho^\sigma = x^i y^\sigma f_i g_\sigma.
\end{aligned}$$

These are a subset of the bilinear forms on V_1^* and V_2^*,

$$X(\underline{f}, \underline{g}) = X^{i\sigma} f_i g_\sigma.$$

Proceeding as in theorem A.19, we could show

$$X^{i\sigma} = (a^{-1})_j^i (b^{-1})_\rho^\sigma X^{j\rho}$$

and we now have a twice contravariant tensor. It is also seen that in general

$$(\underline{x} \otimes \underline{y}) \neq (\underline{y} \otimes \underline{x}), \underline{y} \neq \underline{x}.$$

Likewise we can form mixed tensors by considering tensor products

$$(\underline{x} \otimes \underline{g})(\underline{f}, \underline{y}) = \underline{x}(\underline{f})\underline{g}(\underline{y}), \qquad \underline{x} \in V_1, \quad \underline{f} \in V_1^*, \quad \underline{g} \in V_2^*, \quad \underline{y} \in V_2,$$

and

$$(\underline{x} \otimes \underline{g}) = x^i g_\rho (\underline{e}_i \otimes \underline{\gamma}^\rho).$$

It should be noted here and earlier that with subscript and superscript notation the order of appearance of the components of the matrices or tensors is not important. We are not, therefore, committed here to using thin matrices for column vectors and row vectors. So that

$$\begin{aligned}
(\underline{x} \otimes \underline{g})(\underline{f}, \underline{y}) &= x^i g_\rho y^\sigma f_j (\underline{e}_i \otimes \underline{\gamma}^\rho)(\underline{\varepsilon}^j, \underline{c}_\sigma) \\
&= x^i g_\rho y^\sigma f_j \delta_i^j \delta_\sigma^\rho = (x^i g_\rho)(y^\rho, f_i).
\end{aligned}$$

This is in the space $V_1 \times V_2^*$ of once-covariant and once-contravariant tensors; and there is also the space $V_1^* \times V_2$ constructed on $(\underline{\varepsilon}^i \otimes \underline{c}_\sigma)$.

Obviously the laws of transformation will be

$$\overline{H}^i_\rho = (a^{-1})^i_j (b^\sigma_\rho) H^j_\sigma$$
$$\overline{H}^\rho_i = (a^j_i)(b^{-1})^\rho_\sigma H^\sigma_j.$$

Since $\otimes$ is associative from the definition, we can form tensors which are ρ-times covariant and q-times contravariant; they are subsets of vector spaces of multilinear forms.

If we now introduce the operation of contraction on a tensor, we can finally gather in the matrices of linear mappings as mixed tensors. For this purpose, let $\underline{x} \in V_1, \underline{y} \in V_2, \underline{f} \in V_1^*, \underline{g} \in V_2^*$ so that

$$\underline{f} \otimes \underline{g}(\underline{x}, \underline{y}) = \underline{f}(\underline{x})\underline{g}(\underline{y}) \in V_1^* \times V_2^*$$
$$\underline{f} \otimes \underline{g} = f_i g_\sigma \underline{\varepsilon}^i \otimes \underline{\gamma}^\sigma.$$

If we take $\underline{x}$ as fixed and $\underline{y}$ as variable, we have the function

$$\underline{f}(\underline{x})\underline{g}(\cdot) = f_i x^i \underline{g}(\cdot) \in V_2^*.$$

Definition III.A.6.3. The vector function

$$\underline{f}(\underline{x})\underline{g}(\cdot) = f_i x^i \underline{g}(\cdot) = f_i x^i g_\sigma \underline{\gamma}^\sigma \in V_2^*$$

is the contraction of the tensor $\underline{f} \otimes \underline{g}$ with the vector $\underline{x}$. Likewise, the contraction with $\underline{y}$ would be

$$\underline{f}(\cdot)\underline{g}(\underline{y}) = f_i \underline{\varepsilon}^i g_\sigma y^\sigma \in V_1^*.$$

The reader can easily generalize for any tensor.

Consider now the linear mapping A from V_1 to V_2 (or from V to V)

$$\underline{y} = A(\underline{x}).$$

Let the matrix $[A]$ correspond to the tensor $a^\sigma_i \underline{\varepsilon}^i \otimes \underline{c}_\sigma$ on $V_1^* \times V_2$. Then we have the contraction of this tensor with $\underline{x} \in V_1$

$$\underline{y} = a^\sigma_i x^i \underline{c}_\sigma \in V_2$$

or the coordinate formula

$$y^\sigma = a^\sigma_i x^i.$$

Furthermore, if we subject both $\underline{y} \in V_2$ and $\underline{x} \in V_1$ to a linear mapping each in its own space to another basis in that space, we get

$$\overline{y}^\sigma \underline{\overline{c}}_\sigma = y^\sigma \underline{c}_\sigma, \quad \overline{x}^i \underline{\overline{e}}_i = x^i \underline{e}_i, \quad y^\sigma \underline{c}_\sigma = a_i^\sigma x^i \underline{c}_\sigma,$$

$$y^\sigma = b_\rho^\sigma \overline{y}^\rho, \quad x^i = c_j^i \overline{x}^j,$$

so that

$$b_\rho^\sigma \overline{y}^\rho \underline{c}_\sigma = a_j^\sigma c_i^j \overline{x}^i \underline{c}_\sigma,$$

$$b_\rho^\sigma \overline{y}^\rho = a_j^\sigma c_i^j \overline{x}^i,$$

$$(b^{-1})_\sigma^\lambda b_\rho^\sigma \overline{y}^\rho = (b^{-1})_\sigma^\lambda a_j^\sigma c_i^j \overline{x}^i,$$

$$\delta_\rho^\lambda \overline{y}^\rho = (b^{-1})_\sigma^\lambda a_j^\sigma c_i^j \overline{x}^i,$$

$$\overline{y}^\lambda = (b^{-1})_\sigma^\lambda a_j^\sigma c_i^j \overline{x}^i.$$

Then

$$\overline{a}_i^\lambda = (b^{-1})_\sigma^\lambda c_i^j a_j^\sigma,$$

and so transforms in the pattern of the mixed tensor, once-covariant, once-contravariant. The choice of the pattern of indices is vindicated. Letting $V_1 = V_2 = V$, we get A as a linear mapping relating two bases of V if A is non-singular. Again the choice of indices is vindicated for the matrix of coordinate transformations in the same vector space.

Restricting ourselves to tensor products of a space with itself and its dual, we shall have tensors of the form

$$\underline{T} = t_{\rho\sigma}^{ijk} \underline{e}_i \otimes \underline{e}_j \otimes \underline{e}_k \otimes \underline{\varepsilon}^\rho \otimes \underline{\varepsilon}^\sigma \text{ etc.}$$

In general, such a tensor will be on the space $(\otimes^p V^*) \otimes (\otimes^q V)$ and we can contract it with a tensor $\underline{S}$ r-times covariant and s-times contravariant where $r \leq p$ and $s \leq q$.

The Kronecker delta δ_j^i is simply the coordinate tensor corresponding to the identity mapping U of the unit matrix u_j^i. We already anticipated this in defining the dual of any bases $\{\underline{e}_i\}$ of V as $\{\underline{\varepsilon}^i\}$ of V^* where

$$\underline{\varepsilon}^i(\underline{e}_j) = \delta_j^i.$$

Consider the tensor on $V \otimes V^*$ on bases $\{\underline{\overline{e}}_i\}$ and $\{\underline{\overline{\varepsilon}}^i\}$ transformed from $\{\underline{e}_i\}$ and $\{\underline{\varepsilon}^i\}$ by a non-singular linear mapping

$$\overline{\delta}_j^i(\underline{\overline{e}}_i \otimes \underline{\overline{\varepsilon}}^j) = \overline{\delta}_j^i(a_i^k \underline{e}_k) \otimes ((a^{-1})_\ell^j \underline{\varepsilon}^\ell)$$

$$= \overline{\delta}_j^i a_i^k (a^{-1})_\ell^j (\underline{e}_k \otimes \underline{\varepsilon}^\ell),$$

and

$$\bar{\delta}_\ell^k = \bar{\delta}_j^i a_i^k (a^{-1})_\ell^j,$$
$$\bar{\delta}_j^i = \delta_\ell^k (a^{-1})_k^i a_j^\ell = \delta_j^i.$$

Theorem A.6.4. *The trace of the mixed tensor t_j^i, that is, t_i^i is invariant under a transformation of basis in V.*

Proof. By the law of transformation of the mixed tensor

$$\bar{t}_j^i = t_\ell^k (a^{-1})_k^i a_j^\ell, \qquad \bar{t}_i^i = t_\ell^k (a^{-1})_k^i a_i^\ell = t_\ell^k \delta_k^\ell = t_k^k.$$

$\square$

A.7   The Exterior Product and Determinants

In this section, all our vectors are on V and linear forms on V^*.

Definition III.A.7.1. If $\underline{x}$ and $\underline{y}$ are in V, then the exterior product $\wedge$ of $\underline{x}$ and $\underline{y}$ is

$$\underline{x} \wedge \underline{y} = \underline{x} \otimes \underline{y} - \underline{y} \otimes \underline{x}.$$

Since, in general, $\underline{x} \otimes \underline{y} \neq \underline{y} \otimes \underline{x}$, therefore

$$\underline{x} \wedge \underline{y} \neq \underline{y} \wedge \underline{x}, \quad \underline{x} \wedge \underline{y} = -(\underline{y} \wedge \underline{x})$$
$$(\underline{x} \otimes \underline{y})(\underline{f}, \underline{g}) = \underline{x}(\underline{f})\underline{y}(\underline{g})$$
$$(\underline{y} \otimes \underline{x})(\underline{f}, \underline{g}) = \underline{y}(\underline{f})\underline{x}(\underline{g}).$$

To extend this definition to k vectors in n-space, we first generalize δ_j^i to **Definition III.A.7.2.**

$$\delta_{j_1 j_2 \ldots j_k}^{i_1 i_2 \ldots i_k} = \begin{cases} 0 & \text{if } i_1 \ldots i_k \text{ is not a permutation of } j_1 \ldots j_k. \\ 1 & \text{if } i_1 \ldots i_k \text{ is an even permutation of } j_1 \ldots j_k \\ -1 & \text{if } i_i \ldots i_k \text{ is an odd permutation of } j_1, \ldots j_k. \end{cases}$$

Then **Definition III.A.7.3.** The exterior product of k vectors is defined

$$K : (\underline{x}_1 \wedge \underline{x}_2 \wedge \cdots \wedge \underline{x}_k) = \delta_{1 \ldots k}^{i_1 \ldots i_k} \underline{x}_{i_1} \otimes \underline{x}_{i_2} \otimes \cdots \otimes \underline{x}_{i_k}.$$

Since these are multilinear forms (obvious generalization of bilinear), this product is distributive with respect to addition. And these products of k vectors, each from n-space, form themselves, with their finite sums, a vector space which we will denote by $\wedge^k V^n$, calling the elements of this space k-vectors. We can find its dimension. For, if $\{\underline{e}_i\}, i = 1, 2, \ldots, n$ is a basis in V^n, we have seen that $\{\underline{e}_{i_1} \otimes \underline{e}_{i_2}\}$ is a basis for $V^n \otimes V^n = \otimes^2 V^n$, a space of n^2 dimensions, likewise $\{\underline{e}_{i_1} \otimes \underline{e}_{i_2} \otimes \cdots \otimes \underline{e}_{i_k}\}$ is a basis for $\otimes^k V^n, k \leq n$, a space of n^k dimensions. For

$$
\begin{aligned}
0 &= \lambda^{i,\ldots,i_k}(\underline{e}_{i_1} \otimes \cdots \otimes \underline{e}_{i_k})(\varepsilon^{j_1}, \ldots, \varepsilon^{j_k}) \\
&= \lambda^{i,\ldots,i_k} \delta_{i_1}^1 \cdots \delta_{i_k}^{j_k} \\
&= \lambda^{j_i \cdots j_k}.
\end{aligned}
$$

This, in turn, induces a basis on $\wedge^k V^n$: -

$$
\{\underline{e}_{i_1} \wedge \cdots \wedge \underline{e}_{i_k}\} = \{\underline{E}_{i_1 \ldots i_k}\}, i_1 < i_2 < \cdots < i_k.
$$

We have to order the subscripts here, for otherwise we would get elements the same except for sign, by Def. III.A.7.3; they would not form an independent set. That they form an independent set follows from

$$
\underline{0} = \lambda^{i_1 \ldots i_k}(\underline{e}_{i_1} \wedge \cdots \wedge \underline{e}_{i_k}) = \lambda^{i_1 \ldots i_k}\left(\delta_{i_1 \ldots i_k}^{j_1 \ldots j_k} \underline{e}_{j_1} \otimes \cdots \otimes \underline{e}_{j_k}\right),
$$

since these last have already been proved independent. Then the vector space $\wedge^k V^n$ is formed by all linear combinations over F of $\underline{E}_{i_1 \ldots i_k}$. The number of these $\underline{E}_{i_1 \ldots i_k}$ is evidently the number of combinations of n things taken k by k, that is, $\binom{n}{k}$ which equals $\binom{n}{n-k}$. Therefore the vector spaces $\wedge^k V^n$ and $\wedge^{n-k} V^n$ have the same dimension and are isomorphic; we shall study an important isomorphism between them later.

What we have done with vectors in V^n to give us $\wedge^k V^n$, a space of so called k-vectors, can be done with the dual space to give us a space of k-forms, $\wedge^k (V^*)^n$.

We notice that the $\wedge^k V^n$ space contains two different kinds of k-vectors: (1) those which can be expressed simply as a product of k-vectors from V^n, called split vectors; and (2) those k-vectors which are sums of split k-vectors which, however, cannot be reduced to a split vector. We shall prove the existence of such non-splitting vectors later.

The definition of the exterior product is such that it is easily extended to the exterior product of any two k-vectors from $\wedge^k V^n$ or, more generally, from $\wedge^k V^n$ and $\wedge^\ell V^n$ so that, if, for split vectors

$$
\underline{A} = \underline{x}_1 \wedge \cdots \wedge \underline{x}_k, \qquad \underline{B} = \underline{y}_1 \wedge \cdots \wedge \underline{y}_\ell, \qquad k + \ell \leq n,
$$

then

$$
\underline{A} \wedge \underline{B} = \underline{x}_1 \wedge \cdots \wedge \underline{x}_k \wedge \underline{y}_1 \wedge \cdots \wedge \underline{y}_\ell.
$$

And, more generally, if

$$\underline{A} = \lambda^{i_1 \dots i_k} \underline{x}_{i_1} \wedge \cdots \wedge \underline{x}_{i_k}, \qquad \underline{B} = \mu^{j_1 \dots j_\ell} \underline{y}_{j_1} \wedge \cdots \wedge \underline{y}_{j_\ell},$$

then

$$\underline{A} \wedge \underline{B} = \lambda^{i_1 \dots i_k} \mu^{j_1 \dots j_\ell} \underline{x}_{i_1} \wedge \cdots \wedge \underline{x}_{i_k} \wedge \underline{y}_{j_1} \wedge \cdots \wedge \underline{y}_{j_\ell}.$$

This makes the exterior product associative.

As we have already noticed from the definition of the exterior product, a split vector changes sign when two of the vector components of the product are interchanged. Hence, if a split k-vector has two equal component vectors, it is the zero-vector in $\wedge^k V^n$. That is, if $\underline{A}$ splits,

$$\underline{A} \wedge \underline{A} = \underline{0}.$$

In general, if $\underline{A}$ and $\underline{B}$ split, one a k-vector and the other an ℓ-vector,

$$\underline{A} \wedge \underline{B} = (-1)^{k\ell} \underline{B} \wedge \underline{A}.$$

For there are $k\ell$ transpositions from one to the other.

Example. In V^4 we have, as a basis for $\wedge^2 V^4$,

$$\{\underline{e}_1 \wedge \underline{e}_2, \underline{e}_1 \wedge \underline{e}_3, \underline{e}_1 \wedge \underline{e}_4, \underline{e}_2 \wedge \underline{e}_3, \underline{e}_2 \wedge \underline{e}_4, \underline{e}_3 \wedge \underline{e}_4\};$$

and we can consider, as a 2-vector in $\wedge^2 V^4$,

$$\underline{e}_1 \wedge \underline{e}_2 + \underline{e}_3 \wedge \underline{e}_4 = \underline{A}.$$

But

$$\underline{A} \wedge \underline{A} = 2\underline{e}_1 \wedge \underline{e}_2 \wedge \underline{e}_3 \wedge \underline{e}_4 \neq \underline{0}$$

and therefore $\underline{A}$ does not split.

On the other hand, any 1-vector in $\lambda^1 V^n = V^n$ splits since any 1-vector

$$\lambda^i \underline{e}_i = \underline{e}, \qquad \underline{e}_i \in V^n, \quad \underline{e} \in V^n.$$

Taking the other limit, $\wedge^n V^n$, obviously any n-vector

$$\lambda^{i_1 \dots 1n} \underline{e}_{i_1} \wedge \cdots \wedge \underline{e}_{i_n} = \lambda \underline{e}_1 \wedge \cdots \wedge \underline{e}_n$$

and splits. It can be shown that any $(n-1)$-vector also splits.

Theorem A.7.1. *If $\underline{x}_1, \dots, \underline{x}_k$ are k linearly dependent vectors in $V^n, k \leq n$, then*

$$\underline{x}_1 \wedge \cdots \wedge \underline{x}_k = \underline{0}.$$

Proof: Since they are dependent,

$$\underline{x}_1 = \lambda^i \underline{x}_i, \quad i = 2, \ldots, k,$$

and

$$\underline{x}_1 \wedge \cdots \wedge \underline{x}_k = (\lambda^2 \underline{x}_2 + \cdots + \lambda^k \underline{x}_k) \wedge (\underline{x}_2 \wedge \cdots \wedge \underline{x}_k)$$
$$= \underline{0}.$$

The converse can also be proved later on.

Returning to the spaces $\wedge^k V^n$ and $\wedge^{n-k} V^n$, we establish the following isomorphism: -

$$\underline{E}_{i_1 \ldots i_k} = \delta^{1 \ldots n}_{i_1 \ldots i_k j_1 \ldots j_{n-k}}, \quad \underline{E}_{i_1 \ldots i_k} \in \wedge^k V^n,$$
$$\widehat{E}_{j_1 \ldots j_{n-k}} \in \wedge^{n-k} V^n$$

where

$$i_1 < \cdots < i_k, \quad j_1 < \cdots < j_{n-k}.$$

We can consider $\wedge^O V^n = F, \wedge V^n = V^n$. Obviously, $\wedge^k V^n, k > n$, gives zero-basis vectors by the definitions. From now on, the n is understood for V.

Examples on 3-Space and n-Space. Let us take $\wedge^k V^n$ where $n = 3, k = 2$ and develop its isomorphism with $\wedge^{n-k} V^n$. We get

$$\underline{E}_{i_1 i_2} = \delta^{123}_{i_1 i_2 j_1} \widehat{\underline{E}}_{j_1},$$

$$\underline{E}_{12} = \delta^{123}_{123} \widehat{\underline{E}}_3 = \widehat{\underline{E}}_3, \quad \underline{E}_{13} = \delta^{123}_{132} \widehat{\underline{E}}_2 = -\widehat{\underline{E}}_2, \quad \underline{E}_{23} = \delta^{123}_{231} \widehat{\underline{E}}_1 = \widehat{\underline{E}}_1.$$

Since the $\underline{E}_1$'s are $\underline{e}_i$'s of V, we can write

$$\underline{E}_{12} = \underline{e}_1 \times \underline{e}_2 = \underline{e}_3, \quad \underline{E}_{13} = \underline{e}_i \times \underline{e}_3 = -\underline{e}_2, \quad \underline{E}_{23} = \underline{e}_2 \times \underline{e}_3 = \underline{e}_1.$$

This can be simply interpreted as the well-known right-hand screw rule for vector cross-products. Hence for $\underline{x}$ and $\underline{y}$ any two vectors in V,

$$\begin{aligned}
\underline{x} \wedge \underline{y} =&(x^1 \underline{e}_1 + x^2 \underline{e}_2 + x^3 \underline{e}_3) \wedge (y^1 \underline{e}_1 + y^2 \underline{e}_2 + y^3 \underline{e}_3) \\
=&x^1 y^2 \underline{e}_1 \wedge \underline{e}_2 + x^2 y^1 \underline{e}_2 \wedge \underline{e}_1 + x^1 y^3 \underline{e}_1 \wedge \underline{e}_3 + x^3 y^1 \underline{e}_3 \wedge \underline{e}_1 \\
&+ x^3 y^2 \underline{e}_3 \wedge \underline{e}_2 + x^2 y^3 \underline{e}_2 \wedge \underline{e}_3 \\
=&(x^1 y^2 - x^2 y^1)\underline{e}_1 \wedge \underline{e}_2 + (x^1 y^3 - x^3 y^1)\underline{e}_1 \wedge \underline{e}_3 + (x^2 y^3 - x^3 y^2)\underline{e}_2 \wedge \underline{e}_3 \\
=&(x^1 y^2 - x^2 y^1)\underline{e}_3 - (x^1 y^3 - x^3 y^1)\underline{e}_2 + (x^2 y^3 - x^3 y^2)\underline{e}_1.
\end{aligned}$$

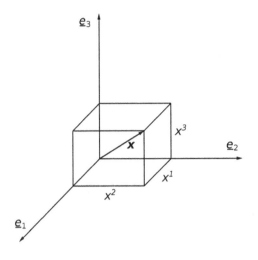

Fig. 56

In view of the notion of determinants to follow, this formula is best remembered as a symbolic determinant

$$\underline{x} \wedge \underline{y} = \begin{vmatrix} \underline{e}_1 & \underline{e}_2 & \underline{e}_3 \\ x^1 & x^2 & x^3 \\ y^1 & y^2 & y^3 \end{vmatrix}$$

But we must also remember

$$\underline{x} \wedge \underline{y} = \left[x^i y^i (\underline{e}_i \otimes \underline{e}_j) - y^k x^\ell (\underline{e}_k \otimes \underline{e}_\ell) \right]$$

where $i, j, k, \ell = 1, 2, 3$. This is a second order antisymmetrical tensor. If we expand, we get the matrix of this tensor for this basis,

$$\begin{pmatrix} 0 & x^1 y^2 - x^2 y^1 & x^1 y^3 - x^3 y^1 \\ x^2 y^1 - x^1 y^2 & 0 & x^2 y^3 - x^3 y^2 \\ x^3 y^1 - x^1 y^3 & x^3 y^2 - x^2 y^3 & 0 \end{pmatrix}$$

By antisymmetrical, we mean $a^{ij} = -a^{ji}$. This is often reported as the equivalence of the axial vector (as opposed to the polar vector) and the antisymmetrical tensor of second order. See, for instance, Sommerfeld, **Mechanics of Deformable Bodies, Lectures on Theoretical Physics**, Vol, II, p.5.

An immediate generalization for $n-1$ vectors in $\wedge^{n-1} V$ where V is of n dimensions, is at hand, to give us a generalization of the cross-product in n dimensions.

$$\underline{E}_{i_1 \ldots i_{n-1}} = \delta^{12 \ldots n}_{i_1 \ldots i_{n-1} j_1} \widehat{\underline{E}}_{j_1},$$

so that

$$\underline{E}_{12\ldots(n-1)} = \widehat{\underline{E}}_n = \underline{e}_n,$$

$$\underline{E}_{12\ldots(n-2)n} = \delta^{12\ldots n}_{12\ldots(n-2)n(n-1)}\widehat{\underline{E}}_{n-1} = -\underline{e}_{n-1},$$

$$\underline{E}_{12\ldots(n-3)(n-1)n} = \delta^{12\ldots n}_{1\ldots(n-3)(n-1)n(n-2)}\widehat{\underline{E}}_{n-2} = \underline{e}_{n-2},$$

$$\underline{E}_{12\ldots(n-4)(n-2)(n-1)n} = \delta^{12\ldots n}_{1\ldots(n-4)(n-2)(n-1)n(n-3)}\widehat{\underline{E}}_{n-3} = -\underline{e}_{n-3}$$

$$\vdots$$

$$\underline{E}_{13\ldots(n-1)n} = \delta^{12\ldots n}_{13\ldots(n-1)(n)2}\widehat{\underline{E}}_2 = (-1)^{n-2}\underline{e}_2$$

$$\underline{E}_{23\ldots(n-1)n} = \delta^{1\ldots n}_{23\ldots(n-1)(n)1}\widehat{\underline{E}}_1 = (-1)^{n-1}\underline{e}_1.$$

Since then

$$(x^i_1\underline{e}_i) \wedge \cdots \wedge (x^i_{n-1}\underline{e}_i) = \sum_{j_1} \delta^{12\ldots n}_{i_1 i_2 \ldots i_{n-1} j_1}\, x^{i_1}_1 x^{i_2}_2 \ldots x^{i_{n-1}}_{n-1} \underline{e}_{j_1},$$

Therefore, in view of the definition to follow, this can evidently be represented by the symbolic determinant

$$\underline{x}_1 \wedge \cdots \wedge \underline{x}_{n-1} = \begin{vmatrix} \underline{e}_1 & \underline{e}_2 & \cdots & \underline{e}_n \\ x^1_1 & x^2_1 & \cdots & x^n_1 \\ \vdots & \vdots & \ddots & \vdots \\ x^1_{n-1} & x^2_{n-1} & \cdots & x^n_{n-1} \end{vmatrix}$$

In passing we should note another form which will arise in the mechanics of a rigid body in 3-space, Gibbs' dyad, $\underline{x} \otimes \underline{y} = x^i y^j (\underline{e}_i \otimes \underline{e}_j)$, $j = 1, 2, 3$.

Determinants. We now define determinants in terms of the exterior product of n vectors in $\wedge^n V$, V being n-dimensional.

Definition III.A.7.4. Given a basis $\{\underline{e}_i\}$ on V and $\underline{x}_i = x^j_i \underline{e}_j$

$$\underline{x}_1 \wedge \cdots \wedge \underline{x}_n = \det(x_1, \ldots, x_n)\underline{E}_{12\ldots n},$$

that is, the determinant of the n vectors in V is the coefficient in F of the exterior product of the n base elements $\underline{e}_i$.

     Since

$$(x^i_1\underline{e}_i) \wedge \cdots \wedge (x^i_n\underline{e}_i) = (x^{i_1}_1 \ldots x^{i_n}_n)\underline{E}_{i_1 \ldots i_n}$$

and

$$\underline{E}_{i_1 \ldots i_n} = \delta^{12\ldots n}_{i_1 i_2 \ldots i_n}\underline{E}_{12\ldots n}$$

so that

$$\underline{x}_1 \wedge \cdots \wedge \underline{x}_n = \left(\delta_{i_1 i_2 \ldots i_n}^{12 \ldots n} x_1^{i_1} x_2^{i_2} \ldots x_n^{i_n} \right) \underline{E}_{12 \ldots n},$$

therefore

Theorem A.7.2. $\det(\underline{x}_1, \ldots, \underline{x}_n) = \left(\delta_{i_1 i_2 \ldots i_n}^{12 \ldots n} x_1^{i_1} x_2^{i_2} \ldots x_n^{i_n} \right).$

This is the familiar rule that the determinant of $(\underline{x}_1, \ldots, \underline{x}_n)$,

$$\begin{vmatrix} x_1^1 & x_1^2 & \cdots & x_1^n \\ x_2^1 & x_2^2 & \cdots & x_2^n \\ \vdots & \vdots & \ddots & \vdots \\ x_n^1 & x_n^2 & \cdots & x_n^n \end{vmatrix} \tag{α}$$

is the sum of all possible terms where each term is the product of elements, one from each different row and column, prefixed by the sign $(-1)^\lambda$ where λ is the number of transpositions of the superscripts from $(1, 2, \ldots, n)$. It is to be noted that we have interchanged rows and columns here with respect to our usual notation in writing the matrix (a_j^i). This will make no difference. For it is not too difficult to show that also

$$\det(\underline{x}_1, \ldots, \underline{x}_n) = \delta_{12 \ldots n}^{i_1 \ldots i_n} x_{i_1}^1 x_{i_2}^2 \ldots x_{i_n}^n.$$

For purposes of calculation, we introduce the notion of the cofactor of a term of the determinant. This essentially depends on the idea of finding a coordinate vector $\underline{y}_i = (y_i^1, \ldots, y_i^n)$ such that

$$\sum_{k=1}^n x_k^j y_k^i = \det(\underline{x}_1, \ldots, \underline{x}_n)$$

for every i and j where $i, j = 1, \ldots, n$. But

$$\det(\underline{x}_1, \ldots, \underline{x}_n) = \delta_{i_1 i_2 \ldots i_n}^{12 \ldots n} x_1^{i_1} x_2^{i_2} \ldots x_n^{i_n}$$

$$= x_1^1 \left[\delta_{i_2 \ldots i_n}^{2 \ldots n} x_2^{i_2} \ldots x_n^{i_n} \right] + (-1)x_1^2 \left[\delta_{i_2 \ldots i_n}^{13 \ldots n} x_2^{i_2} \ldots x_n^{i_n} \right]$$

$$+ \cdots + (-1)^{n-1} x_1^n \left[\delta_{i_2 \ldots i_n}^{12 \ldots (n-1)} x_2^{i_2} \ldots x_n^{i_n} \right].$$

But $\delta_{i_2 \ldots i_n}^{2 \ldots n} x_2^{i_2} \ldots x_n^{i_n} = y_1^1$ is the determinant we get by crossing out the first row and the first column of the matrix (α) of the determinant; $\delta_{i_2 \ldots n}^{13 \ldots n} x_2^{i_2} \ldots x_n^{i_n} = y_1^2$ is the determinant we get by crossing out the first row and the second column since the superscript 2 no longer appears; and so on. The exponents of

(-1) express the fact that from the first term to the second we have interchanged 1 and 2 and so on. This is called the development of the determinant by its first row. The coefficients y_1^i of the x_1^i are called their cofactors.

By interchanging the rows of the determinant, we can construct the development by the jth row. But in interchanging, we interchange successively each row to keep the order $1, 2, \ldots, (j-1), (j+1), \ldots, n$ so that we have to multiply each of the coefficients by $(-1)^{j-1}$. To find a general formula for the exponent of (-1), we consider the coefficient of x_1^1 as multiplied by $(-1)^{1+1}$ corresponding to the elimination of first row and first column, the coefficient of x_1^n as multiplied by $(-1)^{n+1}$, and in general the coefficient or cofactor of x_k^j as

$$ y_k^j = (-1)^{k+j} \delta_{i_2\ldots i_n}^{1\ldots(j-1)(j+1)\ldots n} x_1^{i_2} \ldots x_{k-1}^{i_k} x_{k+1}^{i_{k+1}} \ldots x_n^{i_n}, $$

the determinant gotten by leaving out the jth and kth row.

We can now see how to compute determinants by developing each in turn until we reach 2nd order determinants, which we compute directly

$$ \left(x_1^1 \underline{e}_1 + x_1^2 \underline{e}_2\right) \wedge \left(x_2^1 \underline{e}_1 + x_2^2 \underline{e}_2\right) = x_1^1 x_2^2 \underline{e}_1 \wedge \underline{e}_2 + x_1^2 x_2^1 \underline{e}_2 \wedge \underline{e}_1 $$
$$ = \left(x_1^1 x_2^2 - x_1^2 x_2^1\right) \underline{e}_1 \wedge \underline{e}_2. $$

We can immediately deduce the following from Def. III.A.7.4:

Theorem A.7.3. *If the $\underline{x}_i$ are n vectors in n-dimensional V,*

1. *When, for some i, $\underline{x}_i = \lambda\underline{a} + \mu\underline{b}$, $\lambda, \mu \in F$, then*

   $$ \det(\underline{x}_1, \ldots, \underline{x}_i, \ldots, \underline{x}_n) = \lambda \det(\underline{x}_1, \ldots, \underline{a}, \ldots, \underline{x}_n) + \mu \det(\underline{x}_1, \ldots, \underline{b}, \ldots, \underline{x}_n) $$

   that is, it is a multilinear form of its n vectors since $\wedge^n \underline{x}_i$ is also.

2. *If any two $\underline{x}_i$ are equal, then the determinant is zero since $\underline{E}_{1,\ldots,n}$ is not the zero-vector. The same is true if any two vectors are dependent, and in general if the set $\{\underline{x}_i\}$ is dependent, or any $\underline{x}_i = \underline{0}$.*

3. *$\det(\underline{e}_1, \ldots, \underline{e}_n) = 1$; for given $\{\underline{e}_i\}$.*

4. *If we interchange any two vectors, the sign of the determinant changes.*

5. *One has*

   $$ \det(\underline{x}_1, \ldots, (\underline{x}_i + \underline{x}_j), \ldots, \underline{x}_j, \ldots, \underline{x}_n) = \det(\underline{x}_1, \ldots, \underline{x}_i, \ldots, \underline{x}_j, \ldots, \underline{x}_n) $$
   $$ + \det(\underline{x}_1, \ldots, \underline{x}_j, \ldots, \underline{x}_j, \ldots, \underline{x}_n) $$
   $$ = \det(\underline{x}_1, \ldots, \underline{x}_n). $$

6. If the $\{\underline{x}_i\}$ is an independent set, the determinant is not zero.

Proof of 6: Since $\{\underline{x}_i\}$ are independent, they form a basis for V. Hence applying (1), since $\{\underline{e}_i\}$ is also a basis with determinant equal to 1. and

$$\underline{e}_1 = a^{i_x}_{1-i}, \ldots, \underline{e}_n = a^{i_x}_{n-1}.$$

Therefore

$$1 = \det(\underline{e}_1, \ldots, \underline{e}_n) \ = \det\left(a^i_1 \underline{x}_i, \ldots, a^i_n \underline{x}_i\right)$$
$$= a^{i_1}_1 a^{i_2}_2 \ldots a^{i_n}_n \det(\underline{x}_{i_1}, \ldots, \underline{x}_{i_n}),$$

eliminating any terms where any two i's are equal. But, by (4),

$$\det(\underline{x}_{i_1}, \ldots, \underline{x}_{i_n}) = \pm \det(\underline{x}_1, \ldots, \underline{x}_n).$$

If, therefore, $\det(\underline{x}_1, \ldots, \underline{x}_n) = 0, 1 = 0$ which is false.

The preceding properties are those which one associates with a volume (positive or negative) as we shall now show for a two dimensional V.

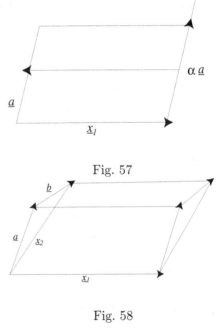

1. If $\underline{x}_2 = \alpha \underline{a}$, the area contained by $\underline{x}_1$ and $\underline{x}_2$ is α times that contained by $\underline{x}_1$ and $\underline{a}$. If $\underline{x}_2 = \underline{a} + \underline{b}$, the area contained by $\underline{x}_1$ and $\underline{x}_2$ is equal to the area contained by $\underline{x}_1$ and $\underline{a}$ plus that contained $\underline{x}_2$ and $\underline{b}$. The truth of this depends on Euclids's Theorem that areas on equal bases between the same parallels are equal.

Fig. 57

2. If $\underline{x}_1$ and $\underline{x}_2$ are equal or parallel, the area contained by them is zero.

3. Arbitrary choice of unit.

4. If we orient the area, letting the area from $\underline{x}_1$ to $\underline{x}_2$ be positive and that from $\underline{x}_2$ to $\underline{x}_1$ be negative.

Fig. 58

5. The area contained by $\underline{x}_1$ and $\underline{x}_2$ is equal to the area contained by $\underline{x}_1$ and $\underline{x}_1 + \underline{x}_2$ by the same theorem of Euclid cited above.

6. If $\underline{x}_1$ and $\underline{x}_2$ are independent (neither being zero), then they are not collinear and the area is not zero.

It can be proved that the function of $(\underline{x}_1, \ldots, \underline{x}_n)$ having these properties is unique. We do not stop to do so. It is therefore taken as the directed volume for n-space.

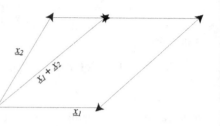

A.8   Applications of Determinants.

Suppose an $n \times n$ matrix a_j^i of a linear mapping from the basis $\{\underline{e}_i\}$ to the basis $\{\underline{\bar{e}}_i\}$ in V, so that

Fig. 59

$$\bar{x}^i \underline{\bar{e}}_i = x^i \underline{e}_i, \quad \underline{\bar{e}}_i = a_i^j \underline{e}_j, \bar{x}^i = (a^{-1})_j^i x^j$$

where $(a^{-1})_j^i$ exists, we know, if the set of $\underline{a}_i = (a_i^1, a_i^2, \ldots, a_i^n)$ is independent (Sec. 3.1.5). How to compute these $(a^{-1})_j^i$ in terms of a_j^i?
We want

$$a_j^i (a^{-1})_k^j = \delta_k^i = u_k^i.$$

That is, we want

$$a_j^i (a^{-1})_k^j = 1, \quad i = k \tag{α}$$

$$a_j^i (a^{-1})_k^j = 0, \quad i \neq k. \tag{β}$$

This would be accomplished in (α) by having as denominators for the $(\underline{a}^{-1})_k^j$ the $\det(\underline{a}^1, \ldots, \underline{a}^n)$ where the $\underline{a}^i$ are the row vectors of the matrix (a_j^i) and for numerators the components of the column vectors $\underline{A}_i$ of a matrix so that

$$\sum_i a_i^1 A_i^1 = \sum_i a_i^2 A_i^2 = \cdots = \sum_i a_i^n A_i^n = \det(\underline{a}^1, \ldots, \underline{a}^n).$$

The problem is solved if the A_i^j are the cofactors of the a_i in the matrix of the a_i^j. We shall always use this notation for the cofactors.
But this also solves the problem of (β). For, $i \neq k$,

$$\sum_j a_j^i A_j^k = a_1^i (-1)^{k+1} \delta_{2 \ldots n}^{i_2 \ldots i_n} a_{i_2}^1 \ldots a_{i_k}^{k-1} a_{i_{k+1}}^{k+1} \ldots a_{i_n}^n + \ldots$$

$$+ a_n^i (-1)^{k+n} \delta_{1 \ldots (n-1)}^{i_2 \ldots i_n} a_{i_2}^1 \ldots a_{i_k}^{k-1} a_{i_{k+1}}^{k+1} \ldots a_{i_n}^n.$$

But this is the determinant with ith row replacing the kth row. Hence it will have two rows the same and is zero.

Therefore

$$\sum_j a^i_j \frac{[A^k_j]}{\det(\underline{a}^1,\ldots,\underline{a}^n)} = \delta^i_k, \qquad (a^{-1})^j_k = \frac{1}{\det(\underline{a}^1,\ldots,\underline{a}_n)} A^k_j.$$

And so, if

$$(a^i_j) = \begin{pmatrix} a^1_1 & \cdots & a^1_n \\ \vdots & \ddots & \vdots \\ a^n_1 & \cdots & a^n_n \end{pmatrix}, \qquad ((a^{-1})^i_j) = \frac{1}{\det(\underline{a}^1,\ldots,\underline{a}^n)} \begin{pmatrix} A^1_1 & \cdots & A^n_1 \\ \vdots & \ddots & \vdots \\ A^1_n & \cdots & A^n_n \end{pmatrix}.$$

We can now see that the transformation is non-singular if and only if the determinant of the matrix of the transformation is not zero; and also the necessary and sufficient condition that the set of row vectors (or column vectors) of an $n \times n$ matrix be independent is that the determinant of the matrix be not equal to zero.

Application to Systems of Linear Equations. We can consider the system

$$x^k a^i_k = 0, \quad k = 1,\ldots,n; \quad i = 1,\ldots,m \tag{α}$$

Let S_0 be the set of vectors $\underline{x}$ which are solutions. It is easy to see it is a vector space. For, if $\underline{x}_1$ and $\underline{x}_2$ are solutions so is $\underline{x}_1 + \underline{x}_2$. And if $\underline{x}$ is a solution, so is $a\underline{x}, a \in F$. It is a subspace of all n-dimensional vectors $\underline{x}$ of V. Now if the number of independent row vectors $\underline{a}^i$, which are linear forms in V^* with respect to the $\underline{x}$ in V is s, then the dimension of the space of the solutions $\underline{x}$ is $n - s$, by Theorem A.3.4, where $s \leq n$.

Next consider the column vectors $\underline{a}_j$ of the matrix (a^i_j) and suppose the number of independent ones is r (the first r for convenience). Therefore

$$\underline{a}_k = -\lambda^i_k \underline{a}_i, \quad i = 1, 2, \ldots, r; \quad k = r+1, \ldots, n.$$

or

$$\lambda^i_k \underline{a}_i + \underline{a}_k = \underline{0},$$

so that each of the n-dimensional $n - r$ coordinate vectors

$$\underline{y}_1 = \left(\lambda^1_{r+1}, \lambda^2_{r+1}, \ldots, \lambda^r_{r+1}, 1, 0, \ldots, 0\right),$$

$$\vdots$$

$$\underline{y}_{n-r} = \left(\lambda^1_n, \lambda^2_n, \ldots, \lambda^r_n, 0, 0, \ldots, 1\right),$$

is a vector-solution $\underline{x}$ of (α), since we can write them in the form

$$x^1 \underline{a}_1 + x^2 \underline{a}_2 + \cdots + x^n \underline{a}_n = \underline{0}.$$

These vectors $\underline{y}_i$ are linearly independent. For, if there exist μ^i such that

$$\mu^i \underline{y}_i = \underline{0}, \ i = 1, \ldots n - r,$$

then we have, considering the components of the $\underline{y}_i$ from $r + 1$ to n,

$$\mu^i = 0.$$

Furthermore any solution $\underline{x}$ can be expressed as a linear combination of these vectors. For any $\underline{x}$, we have

$$x^i \underline{a}_i = \underline{0}, \ i = 1, \ldots, n.$$

Take

$$x^{r+i} \underline{y}_i \text{ where } i = 1, \ldots, n - r.$$

Since the $\underline{y}_i$ are solutions, therefore $x^{r+i} \underline{y}_i$ is a solution and

$$\underline{x} - x^{r+i} \underline{y}_i = (z^1, \ldots, z^r, 0, \ldots, 0)$$

is also a solution, so that

$$z^j \underline{a}_j = \underline{0}, \ j = 1, \ldots, r.$$

But $\{\underline{a}_j\}$ is independent and therefore

$$z^j = 0 \text{ and } \underline{x} = x^{r+i} \underline{y}_i,$$

so that the vector-space of solutions $\underline{x}$ has dimension $n - r$.

But it is also $n - s$ where s is a maximum number of independent row vectors. Hence

$$s = r,$$

and we define the rank of (a^i_j) as the number of independent column or row vectors.

We can immediately deduce the following:

1. If the matrix $A = (a^i_j)$ is $n \times n$ and there are n independent vectors, then $\det A \neq 0$, and

$$x^i a^j_i (a^{-1})^k_j = 0 = x^i,$$

and $\underline{x} = \underline{0}$ is the only solution.

2. If the matrix A is $n \times n$ and there are less than n independent vectors, then $\det[A] = 0$ and the vector spaces of solutions $\underline{x}$ is of dimension greater than zero and there are other solutions than $\underline{x} = \underline{0}$.

3. If the matrix A is $n \times m$ (m is number of rows), where $m < n$, then the number of independent row vectors $s \leq m < n$ so that $n - s \neq 0$ and there are always solutions $\underline{x} \neq \underline{0}$.

4. If the matrix A is $n \times m$ and $m > n$, then there may be no solution besides $\underline{x} = \underline{0}$.

The Linear Mapping of a Determinant. We remember that, if $\{\underline{e}_i\}$ and $\{\underline{x}_i\}$ are two different bases in V and A is a linear mapping of V into V, then there exists a non-singular mapping X of V into V such that

$$\underline{x}_i = x_i^j \underline{e}_j$$

where x_j^i is the representation of X for the basis $\{\underline{e}_i\}$, and

$$A(\underline{x}_i) = \bar{a}_i^j \underline{x}_j$$

where $\bar{a}_i^j$ is the representation of A for the basis $\{\underline{x}_i\}$ in V. Hence

$$
\begin{aligned}
A(\underline{x}_1) \wedge \cdots \wedge A(\underline{x}_n) &= \left(\bar{a}_1^{j_1} \underline{x}_{j_1} \right) \wedge \left(\bar{a}_2^{j_2} \underline{x}_{j_2} \right) \wedge \cdots \wedge \left(\bar{a}_n^{j_n} \underline{x}_{j_n} \right) \\
&= \left[\delta_{i_1 i_2 \ldots i_n}^{12 \ldots n} \, \bar{a}_1^{i_1} \bar{a}_2^{i_2} \ldots \bar{a}_n^{i_n} \right] \underline{E}_{\underline{x}_1 \underline{x}_2 \ldots \underline{x}_n} \\
&= \det[A] \det[X] \underline{E}_{12 \ldots n} \\
&= \left(\bar{a}_1^{j_1} x_{j_1}^{k_1} \underline{e}_{k_1} \right) \wedge \cdots \wedge \left(\bar{a}_n^{j_n} x_{j_n}^{k_n} \underline{e}_{k_n} \right) \\
&= \delta_{i_1 i_2 \ldots i_n}^{12 \ldots n} \left[\left(x_{j_1}^{i_1} \bar{a}_1^{j_1} \right) \ldots \left(x_{j_n}^{i_n} \bar{a}_n^{j_n} \right) \right] \underline{E}_{12 \ldots n} \\
&= \det([X])[\bar{A}]) \underline{E}_{12 \ldots n}
\end{aligned}
$$

where $[\bar{A}]$ is the representation of linear mapping A and $[X]$ of non-singular linear mapping X. Hence

$$\det[\bar{A}] \det[X] = \det([X][\bar{A}]), \quad \det[X] \neq 0.$$

This can be extended to the case where $\det[X] = 0$, if we remember that the singular linear mapping X followed by the linear mapping A is singular. Therefore $A \cdot X$ is singular and

$$\det([\bar{A}][X]) = \det[\bar{A}] \det[X] = 0.$$

A.9   Inner Products and Orthogonality.

To furnish a metric on a vector space V, that is, to furnish a method of assigning lengths to vectors (hence distances between points) and of measuring angles between them, we impose an inner product. An inner product (or scalar product) $\underline{x} \cdot \underline{y}$ must satisfy the following conditions:

Definition III.A.9.1.

1. $\underline{x} \cdot \underline{y} \in F, \quad \underline{x}, \underline{y} \in V$;

2. $\underline{x} \cdot (\lambda \underline{y} + \mu \underline{z}) = \lambda(\underline{x} \cdot \underline{y}) + \mu(\underline{x} \cdot \underline{z})$;

   $(\lambda \underline{x} + \mu \underline{y}) \cdot \underline{z} = \lambda(\underline{x} \cdot \underline{z}) + \mu(\underline{y} \cdot \underline{z}); \quad \lambda, \mu \in F; \quad \underline{x}, \underline{y}, \underline{z} \in V$;

3. $\underline{x} \cdot \underline{y} = \underline{y} \cdot \underline{x}$.

Essentially, inner products are symmetric bilinear forms. It is usual to impose the further condition that $\underline{x} \cdot \underline{x} \geq 0$ and equals zero if and only if $\underline{x} = \underline{0}$. In this case, we say the corresponding quadratic function $\underline{x} \cdot \underline{x}$ of the bilinear form $\underline{x} \cdot \underline{y}$ is positive definite. But we shall only realize this last condition for the inner product we impose on V when F is the field of real numbers. This shall not be the case in the four-dimensional Minkowskian space-time of Special Relativity where F is field of complex numbers and where $\underline{x} \cdot \underline{x}$ can also be negative and be zero for $\underline{x} \neq \underline{0}$. For this reason, we do not use the condition

$$\underline{x} \cdot \underline{y} = \overline{\underline{y} \cdot \underline{x}}$$

where $\overline{\underline{y} \cdot \underline{x}}$ is the conjugate of $\underline{y} \cdot \underline{x}$, but use the simpler condition (3).

Inner products (or symmetric bilinear forms) are also classified as degenerate, if $\underline{x} \cdot \underline{y} = 0$ for fixed $\underline{x} \neq \underline{0}$ and all $\underline{y}$ or for fixed $\underline{y} \neq \underline{0}$ and all $\underline{x}$. Otherwise they are non-degenerate. We shall always be dealing with the latter, although there will be sometimes subspaces of our spaces where the inner product is degenerate.

Theorem A.9.1. *If the quadratic function $\underline{x} \cdot \underline{x} \geq 0$ and $\underline{x} \cdot \underline{x} = 0$ only for $\underline{x} = \underline{0}$, then for F the field of real numbers, $(\underline{x} \cdot \underline{y})$ satisfies the Schwarz Inequality*

$$(\underline{x} \cdot \underline{y})^2 \leq (\underline{x} \cdot \underline{x})(\underline{y} \cdot \underline{y}), \quad \underline{x}, \underline{y} \in V.$$

Proof: Consider $\underline{x} + \underline{y}$ not $\underline{0}$, $(\underline{x} \cdot \underline{y})$ not degenerate,

$$(\underline{x} + \lambda \underline{y}) \cdot (\underline{x} + \lambda \underline{y}) \equiv (\underline{x} \cdot \underline{x}) + 2\lambda(\underline{x} \cdot \underline{y}) + \lambda^2(\underline{y} \cdot \underline{y})$$

from the symmetry and bilinearity. But we have assumed

$$(\underline{x} + \lambda \underline{y}) \cdot (\underline{x} + \lambda \underline{y}) \geq 0.$$

Then, since $\underline{y} \cdot \underline{y} > 0$,

$$\lambda^2 + 2\lambda \frac{(x \cdot y)}{\underline{y} \cdot \underline{y}} + \frac{x \cdot x}{\underline{y} \cdot \underline{y}} \geq 0,$$

$$\left(\lambda + \frac{x \cdot y}{\underline{y} \cdot \underline{y}}\right)^2 + \left[\frac{x \cdot x}{\underline{y} \cdot \underline{y}} - \left(\frac{x \cdot y}{\underline{y} \cdot \underline{y}}\right)^2\right] \geq 0$$

for all $\lambda \in F$, the field of real numbers, and in particular for

$$\lambda = \frac{-\underline{x} \cdot \underline{y}}{\underline{y} \cdot \underline{y}}$$

so that

$$\frac{x \cdot x}{\underline{y} \cdot \underline{y}} - \left(\frac{x \cdot y}{\underline{y} \cdot \underline{y}}\right)^2 \geq 0,$$

$$(\underline{x} \cdot \underline{x})(\underline{y} \cdot \underline{y}) \geq (\underline{x} \cdot \underline{y})^2 \qquad (\alpha)$$

If $\underline{x} = \underline{0}$ or $\underline{y} = \underline{0}$, the equality obviously holds.

If $\underline{x} \cdot \underline{x} = 0$ for certain $\underline{x}_i \neq \underline{0}$ (the semi-definite case), then the Schwartz' Inequality still holds and by (α),

$$(\underline{x}_1 \cdot \underline{y})^2 \leq 0$$

and obviously, if F is field of real numbers,

$$(\underline{x}_1 \cdot \underline{y}) = 0$$

for all $\underline{y}$. This is the degenerate case.

For the indefinite case, the Schwarz Inequality does not hold.

Theorem A.9.2. *If the quadratic function is positive definite as in the previous theorem, then*

$$\left[(\underline{x} + \underline{y}) \cdot (\underline{x} + \underline{y})\right]^{\frac{1}{2}} \leq (\underline{x} \cdot \underline{x})^{\frac{1}{2}} + (\underline{y} \cdot \underline{y})^{\frac{1}{2}},$$

the triangle inequality.

Proof: $(\underline{x} + \underline{y}) \cdot (\underline{x} + \underline{y}) = (\underline{x} \cdot \underline{x}) + (\underline{y} \cdot \underline{y}) + 2(\underline{x} \cdot \underline{y}).$

By Schwarz' Inequality

$$(\underline{x} \cdot \underline{y}) \leq (\underline{x} \cdot \underline{x})^{\frac{1}{2}} (\underline{y} \cdot \underline{y})^{\frac{1}{2}}.$$

Hence

$$(\underline{x} + \underline{y}) \cdot (\underline{x} + \underline{y}) \leq (\underline{x} \cdot \underline{x}) + (\underline{y} \cdot \underline{y}) + 2(\underline{x} \cdot \underline{x})^{\frac{1}{2}} (\underline{y} \cdot \underline{y})^{\frac{1}{2}}$$
$$\leq \left[(\underline{x} \cdot \underline{x})^{\frac{1}{2}} + (\underline{y} \cdot \underline{y})^{\frac{1}{2}} \right]^2.$$

Definition III.A.9.2. $(\underline{x} \cdot \underline{x})^{\frac{1}{2}}$ is the length of the vector $\underline{x}$, written $|\underline{x}|$.

Definition III.A.9.3. If $(\underline{x} \cdot \underline{y}) = 0$, $\underline{x} \neq \underline{0}$, $\underline{y} \neq \underline{0}$, then $\underline{x}$ and $\underline{y}$ are said to be orthogonal.

It is to be remembered, we have already met a more general form of an inner product on $V \times V^*$,

$$\underline{f} \cdot \underline{x} = f_i x^i, \qquad \varepsilon^i(\underline{e}_j) = \delta^i_j,$$

where a basis $\{\underline{e}_i\}$ has been chosen in V and a dual basis $\{\varepsilon^i\}$ in V^*. We have seen that, if we transform the basis $\{\underline{e}_i\}$ to $\{\overline{\underline{e}}_i\}$ by non-singular A which induces a corresponding transformation $\{\varepsilon^i\}$ to $\{\overline{\varepsilon}^i\}$ by A^{-1}, then

$$\overline{f}_i \overline{x}^i = f_i x^i.$$

For, then,

$$\overline{x}^i = (a^{-1})^i_j x^j, \qquad \overline{f}_i = a^j_i f_j.$$

Now any inner product $\underline{x} \cdot \underline{y}$, not degenerate, where F is field of real numbers and $\{\underline{e}_i\}$ is a basis, gives by Def. III.A.9.1,

$$\underline{x} \cdot \underline{y} = (x^i \underline{e}_i)(y^j \underline{e}_j) = x^i y^j (\underline{e}_i \cdot \underline{e}_j) = x^i y^j g_{ij}$$
$$\underline{x} \cdot \underline{x} = x^i x^j g_{ij}, \qquad x^i, y^i, g_{ij} \in F.$$

so that

Theorem A.9.3. *Given a basis $\{\underline{e}_i\}$ on V, an inner product $\underline{x} \cdot \underline{y}$ is determined on V when the n^2 numbers*

$$g_{ij} = \underline{e}_i \cdot \underline{e}_j \in F$$

are assigned. Then

$$\underline{x} \cdot \underline{y} = x^i y^j g_{ij}.$$

Now we can find a way of considering V as self-dual so that from a certain point of view $V = V^*$. To do this, we suppose an inner product determined on V. Then we consider the equation, for all $\underline{x} \in V$,

$$f_a(\underline{x}) = \underline{f}_a \cdot \underline{x} = \underline{a} \cdot \underline{x} = \lambda(\underline{a}, \underline{x}), \quad \lambda \in F, \tag{α}$$

where $\underline{a}$ is a fixed element of V and $\underline{f}_a$ a fixed element of V^*. The mapping from $\underline{a}$ to $\underline{f}_a$ we shall show to be single-valued, linear, and non-singular.

1. For every $\underline{a}$ there is only one $\underline{f}_a$. For suppose two, $\underline{f}_a^1$ and $\underline{f}_a^2$. Then

$$\lambda(\underline{a}, \underline{x}) = \underline{f}_a^1 \cdot \underline{x} = \underline{f}_a^2 \cdot \underline{x}, \quad \left(\underline{f}_a^1 - \underline{f}_a^2\right) \cdot \underline{x} = 0,$$

for all $\underline{x} \in V$. Therefore

$$\underline{f}_a^1 = \underline{f}_a^2.$$

2. The mapping $\underline{a} \to \underline{f}_a$ is a linear mapping. For

$$\underline{f}_a = C(\underline{a})$$

can be shown to fulfill the conditions of linearity, using the equation (α) and linear properties of $\underline{f}_a$ and $\underline{a}$. The details are left to the reader.

3. The linear mapping C is non-singular. For, if $\underline{a} \cdot \underline{x} = 0$ for all $\underline{x}$, $\underline{a} = \underline{0}$, since $\underline{a} \cdot \underline{x}$ is assumed non-degenerate. Therefore, if $C(\underline{a}) = \underline{f}_0$ where $\underline{f}_0$ is the identically zero function in V^*, then $\underline{a} = \underline{0}$. But, if C is singular, by Theorem A.4.2, $C(\underline{a}) = \underline{f}_0$ and $\underline{a} \neq \underline{0}$ for some $\underline{a}$.

We therefore have an isomorphism between every $\underline{a} \in V$ and $\underline{f} \in V^*$. But, if we wish a complete analog of

$$\underline{f} \cdot \underline{x} = f_i x^i,$$

invariant under the transformation of coordinates in V and the corresponding transformation in V^*, then we shall have to define bases in V and their transformations so that

$$\underline{x} \cdot \underline{y} = \sum_i x^i y^i; \quad \underline{x}, \underline{y} \in V, \ x^i, y^i \in F,$$

is invariant. We remember

$$\overline{x}^i = (a^{-1})^i_j x^j, \quad \overline{f}_i = a^j_i f_j, \quad \overline{y}^i = (a^{-1})^i_j y^j.$$

If, therefore, y^i and $\bar{y}^i$ are to take the place of f_i and $\bar{f}_i$, and $\underline{x} \cdot \underline{y}$ is to remain invariant, then we must have

$$(a^{-1})^i_j = a^j_i, \quad [A^{-1}] = [A]',$$

so that

$$\sum_i \bar{x}^i \bar{y}^i = \sum_k (a^{-1})^i_j x^j a^k_i y^k = \sum_k x^j y^k \delta^i_j = \sum_j x^j y^j,$$

where $[A]'$ is transpose of matrix $[A]$. Hence the inverse of the matrix of A must be the transpose of the matrix of A. But

$$(a^{-1})^i_j = \frac{1}{\det[A]} A^j_i = a^j_i,$$

and, if $\det[A] = 1$, then $A^j_i = a^j_i$. We can take our cue from

$$\varepsilon^i(\underline{e}_j) = \delta^i_j$$

and define

Definition III.A.9.4. An orthonormal basis in V is a set of n vectors $\{\underline{e}_i\}$ such that

$$\underline{e}_i \cdot \underline{e}_j = \gamma_{ij} = \delta^i_j; \ i, j, = 1, \ldots, n.$$

If we wish to emphasize that V is now being considered self-normal, then we could write

$$\underline{e}^i \cdot \underline{e}_j = \gamma^i_j = \delta^i_j.$$

This would allow us to use our summation convention, but it is not usual.

   That such a set $\{\underline{e}_i\}$ is in fact a basis can be shown by proving such a set independent. For, if

$$\lambda^i \underline{e}_i = \underline{0},$$

then

$$\lambda^i \underline{e}_i \cdot \underline{e}_j = \lambda^i \delta^j_i = \lambda^j = 0, \text{ for every } j.$$

Theorem A.9.4. *If $\{\underline{e}_i\}$ is an orthonormal basis on V, then an inner product $\underline{x} \cdot \underline{y}$ defined on V is such that*

$$\underline{x} \cdot \underline{y} = \sum_i x^i y^i,$$

where $\underline{x} = x^i \underline{e}_i$ and $\underline{y} = y^i \underline{e}_i$.

Proof: By Def. III.A.9.1,

$$\underline{x} \cdot \underline{y} = (x^i \underline{e}_i) \cdot (y^j \underline{e}_j) = x^i y^j (\underline{e}_i \cdot \underline{e}_j) = \sum_i x^i y^i.$$

Suppose $\{\underline{e}_i\}$ is an orthonormal basis in n-dimensional V so that

$$\underline{x} \cdot \underline{e}_j = (x^i \underline{e}_i) \cdot \underline{e}_j = x^j,$$

and any $\underline{x} \in V$ can be expressed as

$$\underline{x} = \sum_i (\underline{x} \cdot \underline{e}_i) \underline{e}_i. \tag{α}$$

Suppose $\{\underline{\bar{e}}_i\}$ is another orthonormal basis of V; then, by (α),

$$\underline{\bar{e}}_i = \sum_{j=1}^{n} (\underline{\bar{e}}_i \cdot \underline{e}_j) \underline{e}_j \tag{β}$$

and also, by (α),

$$\underline{e}_i = \sum_j (\underline{\bar{e}}_j \cdot \underline{e}_i) \underline{\bar{e}}_j. \tag{γ}$$

Considering (β),

$$\underline{\bar{e}}_i \cdot \underline{\bar{e}}_i = 1 = \sum_j (\underline{\bar{e}}_i \cdot \underline{e}_j)^2$$

$$\underline{\bar{e}}_i \cdot \underline{\bar{e}}_k = 0 = \sum_j (\underline{e}_i \cdot \underline{e}_j)(\underline{e}_j \cdot \underline{e}_k), \quad k \neq i,$$

or

$$\sum_j (\underline{e}_j \cdot \underline{\bar{e}}_i)(\underline{e}_j \cdot \underline{\bar{e}}_k) = \delta^i_k. \tag{δ}$$

If we write $\underline{e}_j \cdot \underline{\bar{e}}_i = t^j_i$, denoted also on $[T]$, then we can write (δ) as

$$t^j_i (t')^k_j = \delta^k_i \text{ or } [T]' = [T] = [U]$$

where $(t')^k_j$ and $[T]'$ are the transpose of t^k_j and $[T]$ respectively. And $[T]' = [T^{-1}]$ as we set out to prove. We can sum this up in

Theorem A.9.5. *If $\{\underline{e}_i\}$ and $\{\underline{\bar{e}}_i\}$ are orthonormal bases in V, then*

$$\underline{\bar{e}}_i = t^j_i \underline{e}_j, \quad \underline{e}_i = (t')^j_i \underline{\bar{e}}_j$$

where

$$t_i^j (t')_j^k = \delta_i^k = \sum_j (t_i^j)(t_k^j)$$

and

$$\det(t_i^j) \det((t')_i^j) = \left[\det(t_i^j)\right]^2 = 1.$$

It follows immediately from this theorem that the determinant of such matrices (t_j^i), often called orthogonal, is either $+1$ or -1, and these matrices can in this way be divided into two classes.

Although the existence of an orthonormal basis in V is guaranteed by the isomorphism which gives V as its own dual, there is an obvious method of constructing such a basis given any basis $\{\underline{v}_i\}$ of V. Thus, first we take

$$\underline{u}_1 = \underline{v}_1 \neq \underline{0};$$

then determine

$$\underline{u}_2 = \underline{v}_2 + \lambda \underline{v}_1, \ \lambda \in F,$$

so that $\underline{u}_1$ and $\underline{u}_2$ are orthogonal, that is,

$$\underline{u}_1 \cdot \underline{u}_2 = \underline{u}_1 \cdot \underline{v}_2 + \lambda \underline{u}_1 \cdot \underline{v}_1 = 0$$
$$\lambda = -\frac{\underline{v}_2 \cdot \underline{u}_1}{\underline{u}_1 \cdot \underline{v}_1}$$

where we assume the values $\underline{v}_i \cdot \underline{v}_j = g_{ij}$ as determining the inner product along with the basis $\{\underline{v}_i\}$. It is clear that $\underline{u}_2 \neq \underline{0}$; for then $\underline{v}_1$ and $\underline{v}_2$ would be dependent. Again take

$$\underline{u}_3 = \underline{v}_3 + \mu \underline{u}_1 + \nu \underline{u}_2$$

so that $\underline{u}_3$ is orthogonal to $\underline{u}_1$ and $\underline{u}_2$, that is,

$$\underline{u}_3 \cdot \underline{u}_1 = 0 = \underline{v}_3 \cdot \underline{u}_1 + \mu(\underline{u}_1 \cdot \underline{u}_1)$$
$$\underline{u}_3 \cdot \underline{u}_2 = 0 = \underline{v}_3 \cdot \underline{u}_2 + \nu(\underline{u}_2 \cdot \underline{u}_2)$$

and μ and ν are determined. This method can be continued until a set of n neutrally orthogonal non-zero vectors $\{\underline{u}_i\}$ is constructed. We can then reduce them to vectors of length 1 by taking

$$\underline{e}_i = \frac{1}{|\underline{u}_i|} \underline{u}_i.$$

A.10   Geometrical Interpretation of the Inner Product in Euclidean n-Space.

$$\underline{a} \cdot \underline{a} = (a^i \underline{e}_i) \cdot (a^j \underline{e}_j) = (a^1)^2 + \cdots + (a^n)^2.$$

In 3-dimensions, this is the usual Pythagorean Theorem, giving the square of the length of the vector $\underline{a}$. We extend this interpretation to n-space.

Again, in the triangle

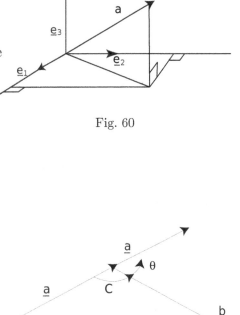

Fig. 60

$$\underline{a} + \underline{b} = \underline{c}.$$

By the laws of the inner product,

$$\underline{a} \cdot \underline{a} + \underline{b} \cdot \underline{b} + 2\underline{a} \cdot \underline{b} = \underline{c} \cdot \underline{c}.$$

But the usual cosine law of trigonometry states

$$|\underline{a}|^2 + |\underline{b}|^2 - 2|\underline{a}||\underline{b}| \cos C = |\underline{c}|^2.$$

Hence

$$2\underline{a} \cdot \underline{b} = -2|\underline{a}||\underline{b}| \cos C,$$
$$\underline{a} \cdot \underline{b} = |\underline{a}||\underline{b}| \cos \theta,$$

or

$$\cos \theta = \frac{\underline{a} \cdot \underline{b}}{(\underline{a} \cdot \underline{a})^{\frac{1}{2}} (\underline{b} \cdot \underline{b})^{\frac{1}{2}}}.$$

Fig. 61

The Schwarz Inequality shows that in n-space also

$$-1 \le \frac{\underline{a} \cdot \underline{b}}{(\underline{a} \cdot \underline{a})^{\frac{1}{2}} (\underline{b} \cdot \underline{b})^{\frac{1}{2}}} \le 1,$$

and suggests that we can define the cosine of the angle between two vectors $\underline{a}$ and $\underline{b}$ in n-space in the same way. This holds, of course, for any positive definite inner product determined on the vector space.

The Inner Product and Exterior Product.
Again, if we recall

$$\underline{x}_2 \wedge \underline{x}_3 \wedge \cdots \wedge \underline{x}_n = \begin{vmatrix} \underline{e}_1 & \underline{e}_2 & \cdots & \underline{e}_n \\ x_1^2 & x_2^2 & \cdots & x_n^2 \\ \vdots & \vdots & \ddots & \vdots \\ x_1^n & x_2^n & \cdots & x_n^n \end{vmatrix}$$

$$= X_1^1 \underline{e}_1 + X_2 \underline{e}_2 + \cdots + X_n^1 \underline{e}_n = \underline{X}$$

where X_i^1 is the cofactor of x_i^1 as if x_i^1 were replacing $\underline{e}_i$.
Then

$$\underline{X} \cdot \underline{x}_i = \det(\underline{x}_2, \underline{x}_3, \ldots, \underline{x}_n, \underline{x}_i) = 0, \ \ i = 2, \ldots, n,$$

and so $\underline{X}$ is orthogonal to any $\underline{x}_i$ of the set. Hence, in particular, in 3-space

$$\underline{a} \times \underline{b} \cdot \underline{a} = \underline{a} \times \underline{b} \cdot \underline{b} = 0,$$

and $\underline{a} \times \underline{b}$ is orthogonal to $\underline{a}$ and $\underline{b}$.

In 3-space, we can easily show (Lagrange's Identity) that

$$(\underline{a} \cdot \underline{b})^2 + (\underline{a} \times \underline{b}) \cdot (\underline{a} \times \underline{b}) = (\underline{a} \cdot \underline{a})(\underline{b} \cdot \underline{b}),$$

by expressing the vectors in terms of an orthonormal basis and multiplying out. Hence

$$\begin{aligned}
|\underline{a} \times \underline{b}|^2 &= (\underline{a} \cdot \underline{a})(\underline{b} \cdot \underline{b}) - (\underline{a} \cdot \underline{b})^2 \\
&= |\underline{a}|^2|\underline{b}|^2(1 - \cos^2 \theta) = |\underline{a}|^2|\underline{b}|^2 \sin^2 \theta,
\end{aligned}$$

and therefore we have the direction and magnitude of $\underline{a} \times \underline{b}$.

The sense of $\underline{a} \times \underline{b}$ will be determined by convention on the cross products between the elements of the orthonormal basis. Suppose $\underline{a}$ and $\underline{b}$ independent; they determine a 2-dimensional subspace of the 3-space. Take $\underline{e}_1$ so that

$$\underline{a} = |\underline{a}|\underline{e}_1,$$

and $\underline{e}_2$ so that

$$\underline{e}_2 \cdot \underline{e}_1 = 0, \quad \underline{e}_2 \cdot \underline{e}_2 = 1,$$

$$\underline{e}_2 = \lambda \underline{e}_1 + \mu \underline{b};$$

and therefore

$$0 = \lambda \underline{e}_1 \cdot \underline{e}_1 + \mu \underline{b} \cdot \underline{e}_1$$

$$\mu = -\frac{\lambda}{\underline{b} \cdot \underline{e}_1}.$$

But also

$$\underline{e}_2 \cdot \underline{e}_2 = 1 = \lambda \underline{e}_1 \cdot \underline{e}_2 + \mu \underline{b} \cdot \underline{e}_2$$

$$\mu = \frac{1}{\underline{b} \cdot \underline{e}_2} = -\frac{\lambda}{\underline{b} \cdot \underline{e}_1}, \qquad \lambda = \frac{-\underline{b} \cdot \underline{e}_1}{\underline{b} \cdot \underline{e}_2}, \ \ \underline{b} \cdot \underline{e}_1 \text{ and } \underline{b} \cdot \underline{e}_2 \neq 0;$$

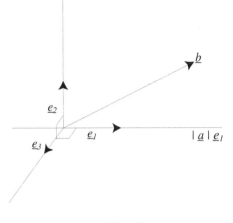

Fig. 62

and

$$\underline{e}_2 = \frac{-\underline{b} \cdot \underline{e}_1}{\underline{b} \cdot \underline{e}_2}\underline{e}_1 + \frac{1}{\underline{b} \cdot \underline{e}_2}\underline{b},$$
$$\underline{b} = (\underline{b} \cdot \underline{e}_1)\underline{e}_1 + (\underline{b} \cdot \underline{e}_2)\underline{e}_2,$$
$$\underline{a} \times \underline{b} = |\underline{a}|(\underline{b} \cdot \underline{e}_1)\underline{e}_1 \times \underline{e}_1 + |\underline{a}|(\underline{b} \cdot \underline{e}_2)\underline{e}_1 \times \underline{e}_2$$
$$= |\underline{a}|(\underline{b} \cdot \underline{e}_2)\underline{e}_1 \times \underline{e}_2.$$

Hence the sense of $\underline{a} \times \underline{b}$ is determined by the sense we give $\underline{e}_1 \times \underline{e}_2$ which is arbitrary and the sign of $\underline{b} \cdot \underline{e}_2$ which is not. We shall assign the sense of $\underline{e}_1 \times \underline{e}_2$ according to the right-hand screw rules, that is,

$$\underline{e}_1 \times \underline{e}_2 = \underline{e}_3$$

where $\underline{e}_3$ is orthogonal to $\underline{e}_1$ and $\underline{e}_2$ as we have just proved; and, of two possible senses, we assign that of an observer with his feet at the common initial point of $\underline{e}_1$ and $\underline{e}_2$ and his head looking down on the plane of $\underline{e}_1$ and $\underline{e}_2$ so that the angle from $\underline{e}_1$ to $\underline{e}_2$ is counterclockwise. The sign of $\underline{b} \cdot \underline{e}_2$ is positive if and only if $\underline{b}$ makes a positive angle θ (counterclockwise) from $\underline{a}$ less than π radians from the point of view of $\underline{e}_3$, otherwise negative. Hence $\underline{a} \times \underline{b}$ has the sense of $\underline{e}_3$ if θ is positive from $\underline{a}$ to $\underline{b}$ from point of view of $\underline{e}_3$ and $0 < \theta < \pi$; otherwise the sense of $-\underline{e}_3$.

We can summarize this

$$\underline{a} \times \underline{b} = |\underline{a}||\underline{b}|\sin\theta\,\underline{e}_3.$$

where θ is angle with initial side $\underline{a}$ and terminal side $\underline{b}, 0 \leq \theta \leq \pi$, where $\underline{e}_3$ is unit vector orthogonal to $\underline{a}$ and $\underline{b}$ in sense of right-hand screw from $\underline{a}$ to $\underline{b}$.

The $\det(\underline{e}_1, \underline{e}_2, \underline{e}_3) = \underline{e}_1 \times \underline{e}_2 \cdot \underline{e}_3 = 1$. If now we take another orthonormal set $\{\underline{\bar{e}}_i\}, i = 1, 2, 3$, then

$$\underline{\bar{e}}_1 = t_1^i \underline{e}_i, \quad \underline{\bar{e}}_2 = t_2^i \underline{e}_i, \quad \underline{\bar{e}}_3 = t_3^i \underline{e}_i$$

$$\underline{\bar{e}}_1 \times \underline{\bar{e}}_2 = (t_1^i \underline{e}_i) \times (t_2^j \underline{e}_j) = t_1^i t_2^j (\underline{e}_i \times \underline{e}_j)$$
$$= (t_1^2 t_2^3 - t_2^2 t_1^3)\underline{e}_1 + (t_2^1 t_1^3 - t_1^1 t_2^3)\underline{e}_2$$
$$+ (t_1^1 t_2^2 - t_2^1 t_1^2)\underline{e}_3,$$
$$= T_3^1 \underline{e}_1 + T_3^2 \underline{e}_2 + T_3^3 \underline{e}_3.$$

But $[T]$ is an orthogonal matrix so that $[T]' = [T^{-1}]$. Hence the cofactors T_j^i of $[T]$ are such that

$$\frac{1}{\det[T]}\, T_j^i = t_j^i$$

and

$$\underline{\bar{e}}_1 \times \underline{\bar{e}}_2 = \left(t_3^1 \underline{e}_1 + t_3^2 \underline{e}_2 + t_3^3 \underline{e}_3\right) \det[T]$$
$$= \pm\underline{\bar{e}}_3$$

according as $\det[T] = +1$ or -1. If $\det[T] = +1$, we call it orientation preserving. Further

$$\underline{\bar{e}}_1 \times \underline{\bar{e}}_2 \cdot \underline{\bar{e}}_3 = \det[\underline{\bar{e}}_1 \underline{\bar{e}}_2 \underline{\bar{e}}_3] = \left(T_3^i \underline{e}_i\right) \cdot \left(t_3^j \underline{e}_j\right) = \sum_i t_3^i T_3^i$$
$$= \det[T]$$

And, assuming $\det[\underline{e}_1 \underline{e}_2 \underline{e}_3] = 1$, we get $\det[\underline{\bar{e}}_1 \underline{\bar{e}}_2 \underline{\bar{e}}_3] = 1$ also if $\det[T]$ is $+1$ and orientation-preserving. Hence an orthonormal set which is right-hand screw remains such when transformed by an orthogonal matrix $[T]$ which is orientation-preserving.

We can also give the following elementary geometric representations:

1. $|\underline{a} \times \underline{b}|$ = area of parallelogram contained by $\underline{a}$ and $\underline{b}$

   $|\underline{a}||\underline{b}| \sin \theta$ = (base) $\cdot$ (height).

2. $\underline{a} \cdot \underline{b} = |\underline{a}||\underline{b}| \cos \theta$ = (orthogonal projection of $\underline{b}$ on $\underline{a}$) (length of $\underline{a}$)

We have seen that the equation of a plane through $\underline{x}_0$ in n-space is

$$\underline{x} = \underline{x}_0 + \lambda^i \underline{a}_i, \ i = 1, 2, \ldots, (n-1),$$

where $\{\underline{a}_i\}$ is independent. Or

$$\underline{x} - \underline{x}_0 = \lambda^i \underline{a}_i.$$

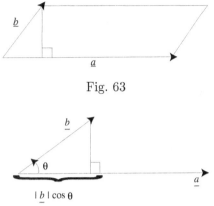

Fig. 63

Now if $\underline{n}$ is orthogonal to each $\underline{a}_i$, then

$$(\underline{x} - \underline{x}_0) \cdot \underline{n} = 0$$

Fig. 64

and $\underline{n}$ is orthogonal to every vector $\underline{x} - \underline{x}_0$ in that plane. In 3-space we can take $\underline{n}$ so that

$$\underline{n} = \underline{a}_1 \times \underline{a}_2,$$

and in n-space

$$\underline{n} = \begin{vmatrix} \underline{e}_1 & \cdots & \cdots & \underline{e}_n \\ a_1^1 & a_2^1 & \cdots & a_n^1 \\ \vdots & \vdots & \ddots & \vdots \\ a_1^{n-1} & a_2^{n-1} & \cdots & a_n^{n-1} \end{vmatrix}$$

Triple Products. There are two forms of triple products for 3-dimensional Euclidean space.

The product $\underline{a} \times \underline{b} \cdot \underline{c} = \det(\underline{a}\ \underline{b}\ \underline{c})$ is, as we have already seen, the volume (positive or negative) of the parallelotope contained by $\underline{a}, \underline{b}$, and $\underline{c}$. This can be seen directly:

Fig. 65

$$\underline{a} \times \underline{b} \cdot \underline{c} = (\text{area of base})(\text{projection of } \underline{c} \text{ on orthogonal to base})$$
$$= (\text{area of base})(\text{height})(\pm 1).$$

Since $\underline{a} \times \underline{b} \cdot \underline{c} = \det(\underline{a}, \underline{b}, \underline{c})$, it can be written $[\underline{a}, \underline{b}, \underline{c}]$ since position of $\times$ and $\cdot$ can be interchanged.

The product $\underline{a} \times (\underline{b} \times \underline{c})$ is most useful in mechanics. In fact, E.A. Milne almost dedicates his **Vectorial Mechanics** to the triple cross product as it appears in the following identity. Since $\underline{a} \times (\underline{b} \times \underline{c})$ is orthogonal to $\underline{b} \times \underline{c}$ which is orthogonal to $\underline{b}$ and $\underline{c}$, therefore it lies in the plane of $\underline{b}$ and $\underline{c}$ and

$$\underline{a} \times (\underline{b} \times \underline{c}) = \lambda \underline{b} + \mu \underline{c}.$$

Then

$$\underline{a} \cdot [\underline{a} \times (\underline{b} \times \underline{c})] = \lambda \underline{a} \cdot \underline{b} + \mu \underline{a} \cdot \underline{c}.$$

Assuming $\underline{a} \cdot \underline{b} \neq 0$,

$$\lambda = -\mu \frac{\underline{a} \cdot \underline{c}}{\underline{a} \cdot \underline{b}}$$

$$\underline{a} \times (\underline{b} \times \underline{c}) = -\mu \frac{\underline{a} \cdot \underline{c}}{\underline{a} \cdot \underline{b}} \underline{b} + \mu \underline{c},$$

$$= \frac{-\mu}{\underline{a} \cdot \underline{b}} [(\underline{a} \cdot \underline{c}) \underline{b} - (\underline{a} \cdot \underline{b}) \underline{c}].$$

We have yet to find μ. If we express in terms of an orthonormal basis and expand to find the coefficient of say $\underline{e}_i$, we find

$$\frac{-\mu}{\underline{a} \cdot \underline{b}} = 1.$$

Theorem A.10.1.

$$\underline{a} \times (\underline{b} \times \underline{c}) = (\underline{a} \cdot \underline{c}) \underline{b} - (\underline{a} \cdot \underline{b}) \underline{c}.$$

Since the "dot" and "cross" can be interchanged, there follows the useful formula

$$(\underline{a} \times \underline{b}) \cdot (\underline{c} \times \underline{d}) = \underline{a} \cdot [\underline{b} \times (\underline{c} \times \underline{d})]$$
$$= \underline{a}[(\underline{b} \cdot \underline{d}) \underline{c} - (\underline{b} \cdot \underline{c}) \underline{d}]$$
$$= (\underline{ac})(\underline{bd}) - (\underline{bc})(\underline{ad}).$$

B Elementary Differential Geometry of Curves in Euclidean 3-Space.

In this part we shall be in Euclidean 3-dimensional space as a realization of a vector-space.

B.1 Curves Given by Vector Functions of a Real Variable and Derivatives of Vectors.

We shall define a curve in 3-space.

Definition III.B.1.1. $\underline{x}$ is a curve in 3-space if, for every t in a real interval $a \leq t \leq b$, there corresponds one $\underline{x}$ (position-vector representing a point X) so that

$$\underline{x}(t) = \varphi^i(t)\underline{e}_i, \quad i = 1, 2, 3,$$

where $\varphi_i(t)$ are continuous functions of t and coordinates of an orthonormal bases $\{\underline{e}_i\}$, often written $\{\underline{i}, \underline{j}, \underline{k}\}$.

We assume an inner product defined on reals as given in Part A, so that also

$$\underline{x} = \sum_i (\underline{x} \cdot \underline{e}_i)\underline{e}_i.$$

We distinguish between a curve and its trace. The trace is the set of all points of the curve. Different sets $\{\varphi_i\}$ may obviously have the same trace. Thus

$$\{t, \sqrt{1 - t^2}, 0\}, \quad -1 \leq t \leq 1,$$

and

$$\{\cos t, \sin t, 0\}, \quad 0 \leq t \leq \pi,$$

give the same semi-circle, but in a different way. When t is time, we can say that the trace is the geometrical configuration and the different sets furnish the different types of motion of a point on that configuration.

On the other hand, it is possible to have the same motions on the same trace with different parameterizations. We would call these the same curves. Thus we might simply change the time scale. This occurs when the parameter t is replaced by a monotonically increased function of u, $\chi(u)$, so that

$$t = \chi(u), \quad a = \chi(\alpha), \quad b = \chi(\beta), \quad \alpha \leq u \leq \beta,$$

$$\underline{x} = \varphi^i(t)\underline{e}_i = \varphi^i[\chi(u)]\underline{e}_i = \psi^i(u)\underline{e}_i.$$

We define the limit of the vector-function of a real variable analogously to real-valued functions of a real variable.

Definition III.B.1.2. Suppose $\underline{x}$ and $\underline{b}$ in V, with $F = R$ (real numbers) and an inner product defined as in Part A. If, when for every ε in $R > 0$, there always exists a δ in R such that for t in an interval of R and

$$0 < |t - t_0| < \delta,$$
$$|\underline{x}(t) - \underline{b}| < \varepsilon,$$

then $\lim\limits_{t \to t_0} \underline{x} = \underline{b}$. The function $\underline{x}(t)$ is continuous at $t = t_0$, if $\lim\limits_{t \to t_0} \underline{x}(t) = \underline{x}(t_0)$.

There follow the usual rules

1. $\lim\limits_{t \to t_0} [\underline{x}(t) + \underline{y}(t)] = \lim\limits_{t \to t_0} \underline{x} + \lim\limits_{t \to t_0} \underline{y} = \underline{a} + \underline{b}$

2. $\lim\limits_{t \to t_0} (\lambda(t)\underline{x}(t)) = \lim\limits_{t \to t_0} \lambda \lim\limits_{t \to t_0} \underline{x} = m\underline{a}$

3. $\lim\limits_{t \to t_0} (\underline{x} \cdot \underline{y}) = (\lim\limits_{t \to t_0} \underline{x}) \cdot (\lim\limits_{t \to t_0} \underline{y}) = \underline{a} \cdot \underline{b}$

4. $\lim\limits_{t \to t_0} (\underline{x} \times \underline{y}) = (\lim\limits_{t \to t_0} \underline{x}) \times (\lim\limits_{t \to t_0} \underline{y}) = \underline{a} \times \underline{b}$

We outline the proof of 4. Consider

$$
\begin{aligned}
|(\underline{x} \times \underline{y}) - (\underline{a} \times \underline{b})| &= |(\underline{x} \times \underline{y}) - (\underline{x} \times \underline{b}) + (\underline{x} \times \underline{b}) - (\underline{a} \times \underline{b})| \\
&= |\underline{x} \times (\underline{y} - \underline{b}) + (\underline{x} - \underline{a}) \times \underline{b}| \\
&\leq |\underline{x} \times (\underline{y} - \underline{b})| + |(\underline{x} - \underline{a}) \times \underline{b}| \\
&\leq |\underline{x}||\underline{y} - \underline{b}| + |\underline{x} - \underline{a}||\underline{b}|.
\end{aligned}
$$

But $|\underline{y} - \underline{b}|$ and $|\underline{x} - \underline{a}|$ are ε's and $|\underline{x}|$ is bounded. Therefore the theorem follows.

It is easily proved that $\underline{x} = \varphi^i(t)\underline{e}_i$ is continuous if and only if the $\varphi^i(t)$ are.

Definition III.B.1.3. The derivative of the vector $\underline{x}(t)$ is

$$
\lim_{h \to 0} \frac{\underline{x}(t + h) - \underline{x}(t)}{h} = \frac{d\underline{x}}{dt} = \underline{x}'.
$$

Evidently $\underline{x}(t)$ as position-vector must represent a curve in the sense defined above.

The geometrical picture is as follows.

$$
\frac{1}{h}[\underline{x}(t + h) - \underline{x}(t)]
$$

is a vector parallel to the vector from the point on the curve represented by $\underline{x}(t)$ to the point on the curve represented by $\underline{x}(t + h)$. For curves where the limit (not zero) exists as $h \to 0$, $\frac{d\underline{x}}{dt}$ is taken as a vector tangent to the curve at $\underline{x}(t)$.

There follow the usual theorems on derivatives of vector functions which can be easily proved in the manner of their real-valued functions analogs.

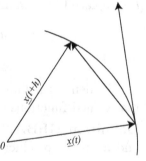

Fig. 66

1. $(\underline{x} + \underline{y})' = \underline{x}' + \underline{y}'$

2. $(\lambda \underline{x})' = \lambda' \underline{x} + \lambda \underline{x}'$,

3. $(\underline{x} \cdot \underline{y})' = \underline{x}' \cdot \underline{y} + \underline{x} \cdot \underline{y}'$,

4. $(\underline{x} \times \underline{y})' = \underline{x}' \times \underline{y} + \underline{x} \times \underline{y}'$;
and therefore, immediately

5. for fixed $\{\underline{e}_i\}$,

$$\frac{d\underline{x}}{dt} = \frac{d}{dt}(\varphi^i \underline{e}_i) = (\varphi^i)' \underline{e}_i,$$

and, using chain rule,

6.

$$\frac{d|\underline{x}|}{dt} = \frac{d(\underline{x} \cdot \underline{x})^{\frac{1}{2}}}{dt} = \frac{\frac{d\underline{x}}{dt} \cdot \underline{x}}{|\underline{x}|}.$$

B.2 Arc-length and Curvature

Since we define the length of the arc of a curve as the limit of the sum of the lengths of the sides of inscribed polygons as the limit of the length of the longest side goes to zero, and since the distance from the point on the arc given by $\underline{x}(t)$ to the point on the arc given by $\underline{x}(t + h)$ is

$$|\underline{x}(t + h) - \underline{x}(t)| = |\Delta \underline{x}|,$$

therefore the arc-length on the curve from $t = a$ to $t = b$ is s where

$$s = \int_a^b \left(\frac{d\underline{x}}{dt} \cdot \frac{d\underline{x}}{dt} \right)^{\frac{1}{2}} dt$$

when such a limit exists. It obviously does for piecewise continuous $\frac{d\underline{x}}{dt}$ or piecewise smooth $\underline{x}(t)$. Therefore

$$\frac{ds}{dt} = \left(\frac{d\underline{x}}{dt} \cdot \frac{d\underline{x}}{dt} \right)^{\frac{1}{2}} = \left| \frac{d\underline{x}}{dt} \right|,$$

which means we have chosen the sense of arc-length on the curve so that s is a monotonic increasing function of t. The opposite could be true, and, therefore, if no stipulation is made

$$\frac{ds}{dt} = \pm \left| \frac{d\underline{x}}{dt} \right|.$$

It should be noted that under change of parameter, s is invariant. For, if $t = t(u), a = t(\alpha), b = t(\beta)$, and $t(u)$ is one-to-one,

$$s = \int_a^b \left(\frac{d\underline{x}}{dt} \cdot \frac{d\underline{x}}{dt} \right)^{\frac{1}{2}} dt = \int_\alpha^\beta \left(\frac{d\underline{x}}{dt} \cdot \frac{d\underline{x}}{dt} \right)^{\frac{1}{2}} \frac{dt}{du} du$$

$$= \int_\alpha^\beta \left(\frac{d\underline{x}}{dt} \frac{dt}{du} \cdot \frac{d\underline{x}}{dt} \frac{dt}{du} \right)^{\frac{1}{2}} du$$

$$= \int_\alpha^\beta \left(\frac{d\underline{\hat{x}}}{du} \cdot \frac{d\underline{\hat{x}}}{du} \right)^{\frac{1}{2}} du$$

where $\underline{x}(t) = \underline{x}(t(u)) = \underline{\hat{x}}(u)$. Often, by abuse of notation, since $\underline{x}$ and $\underline{\hat{x}}$ are same vector, one writes $\underline{x}(t(u)) = \underline{x}(u)$. s is also invariant under a transformation represented by an orthogonal matrix. Thus $\underline{\overline{x}} = A(\underline{x})$, $\frac{d\underline{\overline{x}}}{dt} = A\left(\frac{d\underline{x}}{dt} \right)$, since A is independent of t.[2] But such an inner product has already been proved invariant.

If t is time, then $\frac{ds}{dt}$ is instantaneous rate of change of arc-length per unit of time and is called speed. Now, from the definition, when $\underline{x}$ is position-vector of the curve, $\frac{d\underline{x}}{dt}$ is the tangent vector at the point $\underline{x}$ for any parameter. It is, therefore, proper to call $\frac{d\underline{x}}{dt}$, when t is time, the velocity vector, since it represent the direction and sense of the motion as a vector and the speed or magnitude of the motion in length.

If we take s as the parameter for our curve so that $\underline{x}(s)$, then, by chain rule, considering

$$\underline{\hat{x}}[t(s)] = \underline{x}(s),$$

we have

$$\frac{d\underline{x}}{ds} = \frac{d\underline{\hat{x}}}{dt} \frac{dt}{ds} = \frac{d\underline{\hat{x}}}{dt} \Big/ \frac{ds}{dt} \text{ and } \left| \frac{d\underline{x}}{ds} \right| = 1, \frac{ds}{dt} \neq 0,$$

so that $\frac{d\underline{x}}{ds}$ is the **unit tangent vector**.

Therefore, considering $\underline{x}(s)$,

$$\frac{d\underline{x}}{ds} \cdot \frac{d\underline{x}}{ds} = 1,$$

and, taking derivative of this again with respect to s,

$$\frac{d^2\underline{x}}{ds^2} \cdot \frac{d\underline{x}}{ds} = 0,$$

---

[2] $\underline{\overline{x}} = x^j(a_j^i \underline{e}_i)$, $\frac{d\underline{\overline{x}}}{dt} = \frac{dx^j}{dt}(a_j^i \underline{e}_i) = A(\frac{d\underline{x}}{dt})$.

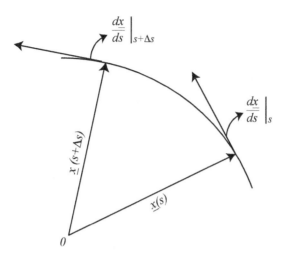

Fig. 67

so that $\frac{d^2x}{ds^2}$ is orthogonal to $\frac{dx}{ds}$ when neither is 0 It should be noted in passing that this is a general result for any variable vector whose inner product with itself is a constant, not zero. $\frac{d^2x}{ds^2}$ is said to be in the direction of the principal normal to the curve for reasons that will appear. For the length, sense, and direction of this vector have intrinsic geometric meanings.

Intuitively, $\frac{d^2x}{ds^2}$ is the rate of change of the unit tangent vector per unit arc-length, that is, the rate of change of the **direction** of the unit tangent vector per unit arc-length (since its length remains constant). This means it measures the rate of turning of the curve per unit arc-length. It is said to measure the curvature of the curve. Thus

$$\frac{\left(\frac{dx}{ds}\Big|_{s+\Delta s} \right) - \left(\frac{dx}{ds}\Big|_{s} \right)}{\Delta s}$$

is the average rate of change of direction of tangent to the curve over distance of arc Δs. We then pass to the limit. Intuitively it is seen that $\frac{d^2x}{ds^2}$ always points to the convex side of the curve. We therefore define,

Definition III.B.2.1. The curvature of the curve at $\underline{x}(s)$ is

$$\left| \frac{d^2x}{ds^2} \right| = \kappa(s), \qquad \frac{d^2x}{ds^2} = \kappa(s)\underline{n}.$$

$\frac{d^2x}{ds^2}$ is called the curvature vector and $\underline{n}$ the principal normal.

In the case of a circle,

$$\underline{x} \cdot \underline{x} = r^2, \quad r^2 \text{ constant,}$$

$$\frac{d\underline{x}}{ds} \cdot \underline{x} = 0$$

so that tangent is orthogonal to radius-vector. Further

$$\frac{d^2\underline{x}}{ds^2} \cdot \underline{x} + \frac{d\underline{x}}{ds} \cdot \frac{d\underline{x}}{ds} = 0$$

$$\frac{d^2\underline{x}}{ds^2} \cdot \underline{x} = -1, \kappa(\underline{n} \cdot \underline{x}) = -1, \underline{n} \cdot \underline{x} = -\frac{1}{\kappa}.$$

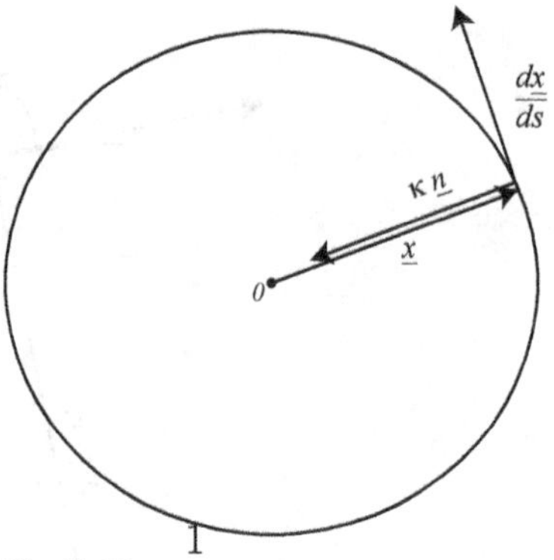

Fig. 68

But

$$\frac{d^2\underline{x}}{ds^2} \cdot \frac{d\underline{x}}{ds} = 0$$

so that $\dfrac{d^2\underline{x}}{ds^2}$ is in direction of $\underline{x}$ from 0, center of circle. Since their inner product is -1, cosine of angle between them is negative, and $\frac{d^2\underline{x}}{ds^2}$ is in opposite sense of $\underline{x}$; it points to the center of the circle. And therefore

$$\underline{n} \cdot \underline{x} = -r, \quad \kappa = \frac{1}{r}, \quad \frac{1}{\kappa} = r.$$

For this reason $\frac{1}{\kappa} = r$ is called the radius of curvature of any curve at the given point. This means, $0 < \frac{1}{r} < \alpha$, the curvature of a curve at a point can always be compared to a circle with the same curvature as curve at that point.

Since curves are rarely expressed explicitly in terms of the parameter s, and we shall be interested in the parameter t as time, we should like an expression for $\frac{d^2\underline{x}}{ds^2}$ in terms of any parameter t. Thus

$$\underline{x}(t) = \underline{\hat{x}}[s(t)],$$

$$\frac{d\underline{\hat{x}}}{ds}\frac{ds}{dt} = \frac{d\underline{x}}{dt}, \qquad \frac{d\underline{\hat{x}}}{ds} = \frac{d\underline{x}}{dt}/(\pm)\left|\frac{d\underline{x}}{dt}\right|$$

$$\frac{d^2\underline{\hat{x}}}{ds^2}\frac{ds}{dt} = (\pm)\frac{\dfrac{d^2\underline{x}}{dt^2}\left|\dfrac{d\underline{x}}{dt}\right| - \dfrac{d\underline{x}}{dt}\left(\dfrac{d^2\underline{x}}{dt^2}\dfrac{d\underline{x}}{dt}\right)/\left|\dfrac{d\underline{x}}{dt}\right|}{\dfrac{\left|d\underline{x}\right|^2}{dt}}$$

$$= (\pm)\frac{\left(\dfrac{d^2\underline{x}}{dt^2}\left(\dfrac{d\underline{x}}{dt} \cdot \dfrac{d\underline{x}}{dt}\right) - \dfrac{d\underline{x}}{dt}\left(\dfrac{d^2\underline{x}}{dt^2} \cdot \dfrac{d\underline{x}}{dt}\right)\right)}{\left|\dfrac{d\underline{x}}{dt}\right|^3}$$

$$\frac{d^2\hat{x}}{ds^2} = \frac{\frac{dx}{dt} \times \left(\frac{dx}{dt^2} \times \frac{dx}{dt} \right)}{\left| \frac{dx}{dt} \right|^4}$$

by Theorem A.10.1 and the fact that $\frac{ds}{dt} = \pm |\frac{dx}{dt}|$. While we are about it, since $\frac{d^2x}{dt^2}$ with, t as time, is the acceleration vector, we express it in terms of $\frac{d^2x}{ds^2}$ and $\frac{dx}{ds}$. Thus

$$\frac{dx}{dt} = \frac{d\hat{x}}{ds}\frac{ds}{dt}, \qquad \frac{d\hat{x}}{ds} = \underline{t},$$

$$\frac{d^2x}{dt^2} = \frac{d^2\hat{x}}{ds^2}\left(\frac{ds}{dt}\right)^2 + \frac{d\hat{x}}{ds}\frac{d^2s}{dt^2} = \kappa\underline{n}\left(\frac{ds}{dt}\right)^2 + \underline{t}\frac{d^2s}{dt^2}$$

And so we have broken the acceleration vector up into two orthogonal components, one in the tangential direction and one in the direction of the principal normal. The coefficients $\left(\frac{ds}{dt}\right)^2$ and $\frac{d^2s}{dt^2}$ are also physically meaningful. The former is proportional to the kinetic energy and the latter is the rate of change of speed per unit time. It is to be remembered this formula holds, of course, even when t is not time. We sum this up in

Theorem B.2.1. *If* $\underline{x}(t) = \hat{x}[s(t)]$, *then*

$$\frac{d^2\hat{x}}{ds^2} = \frac{\frac{dx}{dt} \times \left(\frac{d^2x}{dt^2} \times \frac{dx}{dt} \right)}{\left| \frac{dx}{dt} \right|^4},$$

$$\frac{d^2x}{dt^2} = \frac{d^2\hat{x}}{ds^2}\left(\frac{ds}{dt}\right)^2 + \frac{d\hat{x}}{ds}\frac{d^2s}{dt^2} = \kappa\underline{n}\left(\frac{ds}{dt}\right)^2 + \underline{t}\frac{d^2s}{dt^2}.$$

But the direction and sense also have a particular geometric meaning which already appeared in the case of the circle where $\frac{d^2\hat{x}}{ds^2}$ pointed inwards to the center of the circle. To this end we introduce

Definition III.B.2.2. The osculating plane at a point of a smooth curve is the plane which is the limit of the set of planes each of which is determined by three points of the curve as two of the distinct points approach the third as a limit. The osculating circle at a point of a smooth curve is the circle which is the limit of the set of circles each determined by three points of the curve as two of the distinct points approach the third as a limit.

Theorem B.2.2.

A. *The osculating plane of a curve* $\underline{x}(t)$ *is determined by the vectors* $\frac{dx}{dt}$ *and* $\frac{d^2x}{dt^2}$ *that is, lies in the plane of the vectors* $\underline{n}$ *and* $\underline{t}$.

B. *The osculating circle has its center on the line from the point represented by $\underline{x}(t)$ on the curve along the vector $\underline{n}$ with radius $r = \frac{1}{\kappa}$.*

Proof: A. Take three distinct points $\underline{x}(t)$, $\underline{x}(t+h)$, $\underline{x}(t+\alpha h)$, $\alpha \neq 0$ or 1, but fixed, on a smooth curve. Consider the normal to the plane determined by these three points

$$[\underline{x}(t+\alpha h) - \underline{x}(t)] \times [\underline{x}(t+h) - \underline{x}(t)] = \underline{A}(h),$$

where t is fixed. Let us expand by Taylor's Theorem. We can suppose a fixed orthonormal basis with variable coefficients each expanded in the usual way to give us the vector expansion. Hence

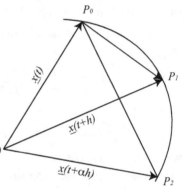

Fig. 69

$$\underline{A}(h) = \left[(\alpha h)\frac{d\underline{x}}{dt} + \frac{(\alpha h)^2}{2!}\frac{d^2\underline{x}(t+\theta\alpha h)}{dt^2} \right] \times \left[h\frac{d\underline{x}}{dt} + \frac{h^2}{2!}\frac{d^2\underline{x}(t+\theta_1 h)}{dt^2} \right]$$

where $0 < \theta < 1$, $0 < \theta_1 < 1$, and θ and θ_1 each represent a different θ and θ_1 for each component of the vector. Then

$$\frac{1}{h^3}\underline{A}(h) = \left[\frac{\alpha}{2!}\left(\frac{d\underline{x}}{dt} \times \frac{d^2\underline{x}(t+\theta_1 h)}{dt^2} \right) + \frac{\alpha^2}{2!}\left(\frac{d^2\underline{x}(t+\theta_\alpha h)}{dt^2} \times \frac{d\underline{x}}{dt} \right) \right.$$
$$\left. + \frac{h\alpha^2}{(2!)^2}\left(\frac{d^2\underline{x}(t+\theta_\alpha h)}{dt^2} \right) \times \frac{d^2\underline{x}(t+\theta_1 h)}{dt^2} \right]$$

and

$$\lim_{h \to 0} \frac{1}{h^3}\underline{A}(h) = \left[(\alpha^2 - \alpha)\frac{d\underline{x}}{dt} \times \frac{d^2\underline{x}}{dt^2} \right]$$

if $\frac{d^2\underline{x}}{dt^2}$ is continuous at the point. Hence the limiting plane contains both $\frac{d\underline{x}}{dt}$ and $\frac{d^2\underline{x}}{dt^2}$, and therefore $\frac{d^2\underline{x}}{ds^2}$.

B. Let $\underline{x}(s)$ be positive-vector of points on curve from 0. Let $\underline{x}(s_0)$, $\underline{x}(s_1)$, $\underline{x}(s_2)$ be three distinct points on curve, and let C be the center of the circle through these three points with position-vector $\underline{x}_c$ from 0. Then

Fig. 70

$$|\underline{x}(s_0) - \underline{x}_c| = |\underline{x}(s_1) - \underline{x}_c| = |\underline{x}(s_2) - \underline{x}_c| = r(s_1, s_2)$$

where s_0 will remain fixed and $\underline{x}_c$, temporarily fixed, is a function of s_1 and s_2. This suggests we consider the function

$$f(s) \equiv (\underline{x} - \underline{x}_c) \cdot (\underline{x} - \underline{x}_c) - r^2$$

which is zero at $\underline{x}(s_0)$, $\underline{x}(s_1)$, $\underline{x}(s_2)$. Then, for the moment keeping $\underline{x}_c$ and r fixed and considering $\underline{x}$ as a function of s, we take derivatives with respect to s,

$$f'(s) \equiv 2(\underline{x} - \underline{x}_c) \cdot \frac{d\underline{x}}{ds}, \qquad f''(s) \equiv 2\left(\frac{d\underline{x}}{ds}\frac{d\underline{x}}{ds}\right) + 2(\underline{x} - \underline{x}_c) \cdot \frac{d^2\underline{x}}{ds^2}.$$

By Rolle's Theorem, since

$$f(s_0) = f(s_1) - f(s_2) = 0,$$
$$f'[s_0 + \theta_1(s_1 - s_0)] = f'[s_1 + \theta_2(s_2 - s_1)] = 0, \quad 0 < \theta_1, \theta_2 < 1;$$

and again

$$f''(\xi) = 0 \text{ where } \xi = s_0 + \theta_1(s_1 - s_0) + \theta_3[s_1 + \theta_2(s_2 - s_1) - s_0 - \theta_1(s_1 - s_0)].$$

Hence, at ξ,

$$1 + (\underline{x} - \underline{x}_c) \cdot \frac{d^2\underline{x}}{ds^2} = 0, \quad (\underline{x} - \underline{x}_c) \cdot \frac{d^2\underline{x}}{ds^2} = -1.$$

As s_2 and s_1 approach s_0, ξ approaches s_0, $\underline{x} \to \underline{x}(s_0)$, $\frac{d^2\underline{x}}{ds^2} \to \frac{d^2\underline{x}(s_0)}{ds^2}$, and $\underline{x} - \underline{x}_c$ being always in the plane through 3 points, will lie at the limit in the osculating plane so that

$$\underline{x} - \underline{x}_c = \lambda\frac{d^2\underline{x}(s_0)}{ds^2} + \mu\frac{d\underline{x}(s_0)}{ds}$$
$$(\underline{x} - \underline{x}_c) \cdot \frac{d^2\underline{x}(s_0)}{ds^2} = \lambda\kappa^2 = -1,$$
$$(\underline{x} - \underline{x}_c) \cdot \frac{d\underline{x}(s_0)}{ds} = \mu = 0.$$

Hence, at limit s_0, limit of $\underline{x} - \underline{x}_c$ is

$$(\underline{x} - \underline{x}_c) = -\frac{d^2\underline{x}}{ds^2}\frac{1}{\kappa^2} = \frac{-1}{\kappa}\underline{n}.$$

And so the principal normal and curvature vector are directed to the center of the osculating circle where $\frac{d^2\underline{x}}{ds^2}$ is continuous.

B.3   The Moving Trihedral and Frenet Formulas

We take the binormal $\underline{b} = \underline{t} \times \underline{n}$ to complete the orthonormal basis $\{\underline{t}, \underline{n}, \underline{b}\}$ of the vector space as realized in Euclidean 3-space, a basis which is called the moving trihedral of the curve $\underline{x}(s)$, to be thought as having its origin at the point on the curve given by $\underline{x}(s)$. If we introduce the methods of Part A of this chapter, we can find the relations between $\underline{t}, \frac{dt}{ds}, \frac{dn}{ds}$, and $\frac{db}{ds}$.

Let us first introduce the derivative of a square matrix $[A]$, $\frac{d[A]}{dt} = \frac{da^i_j}{dt}$, the matrix of the derivative of each element of the matrix. To avoid too many symbols, we shall denote, for the time being, the matrix A without brackets.

Definition III.B.3.1. The Cartan matrix of a square non-singular differentiable matrix A is a matrix function of A,

$$K(A) = \left(\frac{dA}{dt} \right) (A^{-1}).$$

Theorem B.3.1. $K(AB) = K(A) + A[K(B)]A^{-1}$.

Proof: By def.

$$K(AB) = \frac{d(AB)}{dt}(AB)^{-1}.$$

But the derivative of the product of two matrices follows the general rule for products, since the matrix is a bilinear form like the inner product. The reader can check in detail for himself. Hence

$$K(AB) = \left(\frac{dA}{dt}B + A\frac{dB}{dt} \right)(B^{-1}A^{-1}),$$

since

$$(AB)^{-1}(AB) = U = (B^{-1}A^{-1})(AB) = B^{-1}(AA^{-1})B.$$

Therefore

$$K(AB) = \frac{dA}{dt}A^{-1} + A\left(\frac{dB}{dt}B^{-1} \right)A^{-1}$$
$$= K(A) + A[K(B)]A^{-1}.$$

Theorem B.3.2. *The Cartan matrix of an orthogonal matrix A is skew symmetric, that is, its transpose is its negative.*

Proof: Since A is orthogonal, its transpose is its inverse,

$$AA' = AA^{-1} = U,$$

and therefore,

$$\frac{d(AA')}{dt} = \frac{dA}{dt}A' + A\frac{dA'}{dt} = \underline{0}$$

$$\frac{dA}{dt}A' = \frac{dA}{dt}A^{-1} = -A\frac{dA'}{dt}$$

$$K(A) = -A\frac{dA'}{dt} \tag{α}$$

But we can easily show

$$(AB)' = B'A',$$

using the elements:

$$\left(a_j^i b_k^j\right)' = (c_k^i)' = c_i^k = \left(b_j^k a_i^j\right) = B'A'.$$

Therefore

$$[K(A)]' = \left(\frac{dA}{dt}A^{-1}\right)' = \left(\frac{dA}{dt}A'\right)'$$

$$= (A')'\left(\frac{dA}{dt}\right)' = (A^{-1})^{-1}\left(\frac{dA}{dt}\right)'$$

$$= A\left(\frac{dA}{dt}\right)' = A\frac{dA'}{dt}, \tag{β}$$

since obviously

$$\left(\frac{dA}{dt}\right)' = \frac{dA'}{dt}.$$

Putting (α) and (β) together, we get

$$K(A) = -(K(A))'.$$

Now we can consider the relation of a fixed basis $\{\underline{e}_i\}$ and the basis $\{\underline{t}, \underline{n}, \underline{b}\}$ in two ways. (1) In a Euclidean 3-space so that

$$\{\underline{t}, \underline{n}, \underline{b}\} = A(s)\{\underline{e}_1, \underline{e}_2, \underline{e}_3\} + \underline{d}$$

where $\underline{d}$ is the position-vector of the origin of $\{\underline{t}, \underline{n}, \underline{b}\}$ with respect to the origin of $\{\underline{e}_i\}$. Or (2) in 3-dimensional vector space where all vectors of same direction, sense, and length are identical so that $\underline{d}$ has no meaning. Since we are here interested only in the relation of rotation of the two frames, it is the second interpretation we want here. That is

$$\{\underline{t}, \underline{n}, \underline{b}\} = A(s)\{\underline{e}_1, \underline{e}_2, \underline{e}_3\}$$

where $\{\underline{e}_i\}$ is fixed with respect to s, and all others are functions of s. For compactness, let us write this as

$$\underline{t}_i = a_i^j \underline{e}_j, \quad \underline{e}_j = (a_j^i)^{-1}\underline{t}_i, \quad \underline{t}_1 = \underline{t}, \quad \underline{t}_2 = \underline{n}, \quad \underline{t}_3 = \underline{b}.$$

Then

$$\frac{d\underline{t}_i}{ds} = \frac{d(a_i^j)}{ds}\underline{e}_j.$$

Since $\{\underline{e}_i\}$ is constant with respect to s. And

$$\frac{d\underline{t}_i}{ds} = \frac{d(a_i^j)}{ds}(a_j^k)^{-1}\underline{t}_k$$

or

$$\frac{d\{\underline{t},\underline{n},\underline{b}\}}{ds} = K(A)\{\underline{t},\underline{n},\underline{b}\},$$

or

$$\frac{d\underline{t}_i}{ds} = k_i^j \underline{t}_j.$$

But $K(A)$ is skew symmetric by Theorem B.3.2 so that, writing it another way,

$$\frac{d}{ds}\{\underline{t},\underline{n},\underline{b}\} = \begin{pmatrix} 0 & k_2^1 & k_3^1 \\ -k_2^1 & 0 & k_3^2 \\ -k_3^1 & -k_3^2 & 0 \end{pmatrix}\{\underline{t},\underline{n},\underline{b}\};$$

and we get, since $\frac{d^2\underline{x}}{ds^2} = \kappa(s)\underline{n}$,

Theorem B.3.3.

1. $\dfrac{d\underline{t}}{ds} = \dfrac{d^2\underline{x}}{ds^2} = k_2^1\underline{n} + k_3^1\underline{b} - \kappa(s)\underline{n}$, and $k_3^1 = 0$;

2. $\dfrac{d\underline{n}}{ds} = -\kappa(s)\underline{t} + k_3^2\underline{b} = -\kappa(s)\underline{t} + \tau\underline{b}$;

3. $\dfrac{d\underline{b}}{ds} = -k_3^2\underline{n} = -\tau\underline{n}$;

4. $\dfrac{d^3\underline{x}}{ds^3} = \dfrac{d\kappa}{ds}\underline{n} + \kappa\dfrac{d\underline{n}}{ds} = \dfrac{d\kappa}{ds}\underline{n} + \kappa[-\kappa\underline{t} + \tau\underline{b}]$

$\qquad = -\kappa^2\underline{t} + \dfrac{d\kappa}{ds}\underline{n} + \kappa\tau\underline{b}.$

These are the Frenet formulas.

Definition III.B.3.2. The coefficient τ in

$$\frac{d\underline{b}}{ds} = -\tau\underline{n}$$

is called the torsion of the curve and measures the rate of change of the direction of the binormal $\underline{b}$ per unit arc-length.

From (1) of Theorem B.3.3, we get

$$\frac{d^2\underline{x}}{ds^2} = \underline{n} = \left(\frac{d\underline{t}}{ds}\right)^2 \frac{1}{|\frac{d\underline{t}}{ds}|} = |\frac{d\underline{t}}{ds}| = |\frac{d^2\underline{x}}{ds^2}|,$$

and κ, the curvature, is essentially positive. But from (4), dotting with $\frac{d\underline{x}}{ds} \times \frac{d^2\underline{x}}{ds^2} = \underline{t} \times \kappa\underline{n}$, we get

$$\frac{d\underline{x}}{ds} \times \frac{d^2\underline{x}}{ds^2} \cdot \frac{d^3\underline{x}}{ds^3} = (\underline{t} \times \kappa\underline{n}) \cdot \kappa\tau\underline{b} = \kappa^2\tau(\underline{t} \times \underline{n} \cdot \underline{b}) = k^2\tau$$

since $\underline{t} \times \underline{n} \cdot \underline{b} = \det(\underline{t}, \underline{n}, \underline{b}) = 1$; and

Theorem B.3.4.

$$\tau = \frac{[\frac{d\underline{x}}{ds}, \frac{d^2\underline{x}}{ds^2}, \frac{d^3\underline{x}}{ds^3}]}{\kappa^2};$$

and, obviously, τ can be either positive or negative.

Chapter 4

The Dynamics of a Particle

We now proceed to deduce, from the set of axioms formulated at the end of Chapter II, the equations of motion of a particle subject to a field of accelerations produced by a fixed center of force. In other words, we assume in this chapter (with a few exceptions) a center of force which is unaffected by the attracted particle and is fixed so that the vectors of acceleration of the attracted particle are directed all to one point and vary usually as some function of the distance from that one fixed point.

A The First Integrations.

We can integrate the equation given by Axiom 2 in two different ways.

A.1 The Energy-Work Equation.

If we integrate, from Axiom. 2, dotting with $\frac{d\underline{x}}{dt}$ and $\underline{x}$ as the position-vector from the center of force to a particle with mass m,

$$m\frac{d^2\underline{x}}{dt^2} \cdot \frac{d\underline{x}}{dt} = \underline{F} \cdot \frac{d\underline{x}}{dt},$$

we get,

$$\frac{1}{2}m\left(\frac{d\underline{x}}{dt}\right)_t^2 - \frac{1}{2}m\left(\frac{d\underline{x}}{dt}\right)_{t_0}^2 = \int_{t_0}^{t} \underline{F} \cdot \frac{d\underline{x}}{dt}\, dt = \int_{\underline{x}_0}^{\underline{x}} \underline{F} \cdot d\underline{x} \qquad (\textbf{A.1.1})$$

The left-hand side is the difference in kinetic energy at times t and t_0, where $\frac{1}{2}m\left(\frac{d\underline{x}}{dt}\right)^2$ is called the kinetic energy for reasons which will become more apparent later. The right-hand side is a line-integral and is called the work done over the path taken by the particle from $\underline{x}_0$ to $\underline{x}$ or from t_0 to t.

The question arises: when does the integral $\int_{\underline{x}_0}^{\underline{x}} \underline{F} \cdot d\underline{x}$ depend on the path between $\underline{x}_0$ and $\underline{x}$ and when is it the same for all paths between the same points. The answer is given in analysis. In any connected open region where the path is sectionally smooth, if $\int_{\underline{x}_0}^{\underline{x}} \underline{F} \cdot d\underline{x}$ depends only on $\underline{x}_0$ and $\underline{x}$, then

$$
\nabla \times \underline{F} = \begin{vmatrix} \underline{e}_1 & \underline{e}_2 & \underline{e}_3 \\ \frac{\partial}{\partial x} & \frac{\partial}{\partial y} & \frac{\partial}{\partial z} \\ F_1 & F_2 & F_3 \end{vmatrix} = \underline{0}.
$$

Or, for any such **closed** curve

$$
\oint_C \underline{F} \cdot d\underline{x} = 0 \Rightarrow \nabla \times \underline{F} = 0.
$$

For the converse, if in any open simply connected region $\nabla \times \underline{F} = \underline{0}$, then $\oint_C \underline{F} \cdot d\underline{x} = 0$. [1] When the integral is independent of the curve, we write

$$
\int_{\underline{x}_0}^{\underline{x}} \underline{F} \cdot d\underline{x} = U(\underline{x}) - U(\underline{x}_0)
$$

where $U(\underline{x})$ is called the force-function. If now we write

$$
\frac{1}{2}mv^2 - \frac{1}{2}mv_0^2 = \int_{\underline{x}_0}^{\underline{x}} \underline{F} \cdot d\underline{x} = U(\underline{x}) - U(\underline{x}_0) \tag{A.1.2}
$$

and let $U(\underline{x}) = -V(\underline{x})$ where $V(\underline{x})$ is called the potential energy,
Then we write

$$
\frac{1}{2}mv^2 + V(\underline{x}) = \frac{1}{2}mv_0^2 + V(\underline{x}_0) = \lambda \tag{A.1.3}
$$

where λ is a constant; and we say in such fields of acceleration, the kinetic energy plus the potential energy is constant. Such fields of acceleration are called naturally conservative fields. We shall see later why Leibniz, who first seized on the kinetic energy as the important entity in mechanics, named it and used it in a way which was prophetic for the history of physics.

Suppose we are in a field of constant acceleration g such as a small area on the surface of the Earth. For we have seen that for the solar system, in view of Kepler's law on the distances and periodic times of the planets, the

---

[1]We can, of course, use Stoke's Theorem $\oint_C \underline{F} \cdot d\underline{x} = \int_S \nabla \times \underline{F} \cdot d\underline{S}$.

field of acceleration is the inverse-square field. Hence, assuming that we have proved a spherical body, having the same density at the same distance from its center, attracts as if its total mass were in a point at the center, for a small area of the surface of the Earth and for a small height the acceleration vectors can be treated as everywhere parallel, of the same sense and magnitude. Then the energy equation becomes

$$\frac{1}{2}mv^2 - \frac{1}{2}mv_0^2 = \int_{\underline{x}_0}^{\underline{x}} m\underline{g} \cdot d\underline{x} = m\underline{g} \cdot (\underline{x} - \underline{x}_0) = m\underline{g} \cdot \underline{x} - m\underline{g} \cdot \underline{x}_0 \quad (A.1.4)$$

if the particle moves freely with only the force of gravity acting. But $m\underline{g} \cdot (\underline{x} - \underline{x}_0)$ can also be written

$$m\underline{g} \cdot (\underline{x} - \underline{x}_0) =$$
$$m\underline{g} \cdot (\text{projection of } \underline{x} - \underline{x}_0 \text{ on } \underline{g})$$
$$= mgh$$

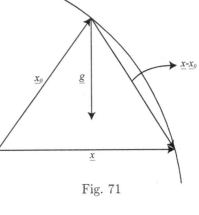

Fig. 71

where h is the height of fall in the direction of $\underline{g}$. So we have

$$mv^2 - mv_0^2 = 2mgh$$
$$v^2 - v_0^2 = 2gh \qquad (A.1.5)$$

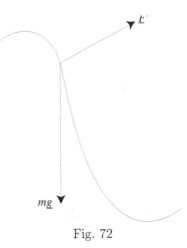

Fig. 72

Suppose, however, the particle is constrained to move on a prescribed smooth curve without friction. We must now make certain assumptions in order to formulate this mathematically. We assume that friction is a force acting on the particle in the direction of its velocity vector but in opposite sense. The constraint to move on a certain curve is assumed to be a force normal to the curve. Thus, if a bead slides on a wire bent to a given curve, the wire exerts on the bead a force-vector normal to the tangent vector to the curve at that point.

Hence, if a particle is constrained to move on a certain curve under gravity, there act on it not only the force $m\underline{g}$ but also a force $\underline{F}'$ normal to the curve which represents the force of constraint of the curve. Then

$$\int_{\underline{x}_0}^{\underline{x}} \underline{F} \cdot d\underline{x} = \int_{\underline{x}_0}^{\underline{x}} (m\underline{g} + \underline{F}') \cdot d\underline{x} = \int_{\underline{x}_0}^{\underline{x}} m\underline{g} \cdot d\underline{x} \qquad (A.1.6)$$

since $\underline{F}' \cdot d\underline{x} = 0$. Therefore, we say that the forces of geometric constraint on a particle arising from the prescription of a fixed curve do no work.

Incidentally, we can verify that

$$\nabla \times m\underline{g} = \nabla \times m(0, g, 0) = \underline{0}.$$

If we look at the work-integral $\int_{\underline{x}_0}^{\underline{x}} \underline{F} \cdot d\underline{x}$ geometrically, we shall see that it follows the usual notion of work done by a force, namely, the distance the particle is displaced by a constant force in a straight line multiplied by the component of the force in the direction of the displacement. For a more general curve, we approximate the curve by vector-chords $\Delta \underline{x}_i$ and take the force-vector $\underline{F}(\xi_i, \eta_i, \zeta_i)$ at some point (ξ_i, η_i, ζ_i) of curve subtended by the $\Delta \underline{x}_i$. Then the approximate work done is

$$\sum_{i=1}^{n} \underline{F}(\xi_i, \eta_i, \zeta_i) \cdot \Delta \underline{x}_i$$

and the limit as $n \to \infty$ is

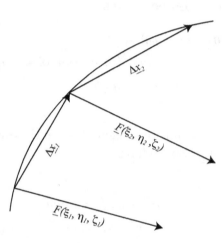

Fig. 73

$$\lim_{n \to \infty} \sum_{i=1}^{n} \underline{F}(\xi_i, \eta_i, \zeta_i) \cdot \Delta \underline{x}_i = \int_{\underline{x}_0}^{\underline{x}} \underline{F} \cdot d\underline{x}.$$

A.2   The Impulse-Momentum Equation.

We can integrate the equation of Axiom 2 in another way. The first way gave us a scalar equation or an equation in the field elements of the vector space. The second way gives us a vector equation. Thus

$$m\frac{d\underline{x}}{dt} - m\frac{d\underline{x}}{dt}\Big|_{t_0} = \int_{t_0}^{t} \underline{F} dt$$

or

$$mv - mv_0 = \int_{t_0}^{t} F\,dt \qquad (A.2.1)$$

The left-hand side is the difference in the momentum of the particle at t and t_0, and the right-hand side is called the impulse. It is instructive, at this point, to consider a system of a finite number of particles, and to distinguish between the forces acting on the particles within the system and from without the system.

Then, by Axiom 3, for every force exerted by one particle on another there is an equal and opposite one, so that, if F'_i is the internal force acting on the ith particle,

$$\sum_i F'_i = 0,$$

and

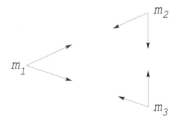

Fig. 74

$$\sum_i m_i v_i - \sum_i m_i v_{0_i} = \int_{t_0}^{t} \left(\sum_i F_i + \sum_i F'_i \right) dt = \int_{t_0}^{t} \sum_i F_i\,dt \qquad (A.2.2)$$

where F_i is the external force acting on the ith particle.

A.3   The Ballistic Pendulum.

Let us apply the two equations we have derived by a first integration to the ballistic pendulum. We have already indicated its historical importance; we shall now develop it more thoroughly and show in more detail why Leibniz considered the results from it so important. If we suppose the two pendulums to fall from the same height, at the bottom they will meet with equal and opposite velocities, since by energy equation,

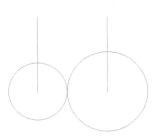

Fig. 75

$$\frac{1}{2}v^2 = gh, \qquad v_0 = 0,$$

and the force of constraint due to the thread is normal to the path and does no work. At the point of impact at the bottom, the velocities of the two pendula are in the same horizontal straight line. What happens at the point of impact? we do not look microscopically but only macroscopically, that is, in the large. But we notice that, at that point, the external forces acting on the system of the two pendulums are normal to the velocities at that point and have no effect on them, or, in other words, the accelerations at that point are normal to the velocities in the horizontal line and do not affect them. The forces of friction are neglected. From this point of view, we can assume that we have conservation of momentum for these velocities at the instant of impact.

$$m_1 v_1 + m_2 v_2 = m_1 v_1' + m_2 v_2' \qquad (A.3.1)$$

where v_1 and v_2 are velocities on impact and v_1' and v_2' velocities from impact. In the special case of this pendulum experiment,

$$m_1 v_1 - m_2 v_1 = m_1 v_1' + m_2 v_2' \qquad (A.3.2)$$

since $v_1 = -v_2$. Hence

$$(m_1 - m_2)v_1 = m_1 v_1' + m_2 v_2'.$$

From the energy equation, we get

$$v_1^2 = 2gh$$

where h is the height of fall along the circle. Further we can measure v_1' and v_2' from the same equation by the heights to which each pendulum rises after impact. From the point of view of Huygens and Newton, assuming the masses are known from the balance, the conservation of momentum could be verified and this would serve to justify Axiom 3. But we can also consider this as a method of measuring relative mass from the known velocities. For

$$\left(\frac{m_1}{m_2} - 1\right) v_1 = \frac{m_1}{m_2} v_1' + v_2'$$

$$\frac{m_1}{m_2} = \frac{v_2' + v_1}{v_1 - v_1'} \qquad (A.3.3)$$

More instructive for further developments is to consider that the particles obey not only the more general case of (A.3.1) at impact but, also conserve their kinetic energy before and after. This case is approximately that of very hard substances such as steel, ivory, etc. Thus

Perfectly Elastic Collision. We now have

$$\frac{1}{2}m_1v_1^2 + \frac{1}{2}m_2v_2^2 = \frac{1}{2}m_1(v_1')^2 + \frac{1}{2}m_2(v_2')^2 \tag{A.3.4}$$

$$m_1\left[v_1^2 - (v_1')^2\right] = m_2\left[(v_2')^2 - v_2^2\right] \tag{A.3.5}$$

From (A.3.1) we have

$$m_1(v_1 - v_1') = m_2(v_2' - v_2).$$

Dividing, we get

$$v_1 + v_1' = v_2' + v_2, v_1 - v_2 = v_2' - v_1' \tag{A.3.6}$$

From this equation, which only holds exceptionally, we can see that (A.3.4) does not in general hold. In fact, if we take the well known case where the two particles move off together after collision, we get

Perfectly Inelastic Collision:

$$m_1v_1 + m_2v_2 = (m_1 + m_2)v_1' \tag{A.3.7}$$

Let us now compare the kinetic energy before collision with the kinetic energy after collision.

$$\frac{1}{2}[m_1v_1^2 + m_2v_2^2] = \text{kinetic energy before,}$$

$$\frac{1}{2}(m_1 + m_2)(v_1')^2 = \text{kinetic energy after.}$$

Subtracting the second from the first, after substitution from (A.3.7),

$$\frac{1}{2}\left[m_1v_1^2 + m_2v_2^2 - \frac{(m_1v_1 + m_2v_2)^2}{m_1 + m_2}\right] = \frac{m_1m_2(v_2 - v_1)^2}{2(m_1 + m_2)} \tag{A.3.8}$$

which is an essentially positive quantity, showing there is a loss of kinetic energy.

Leibniz' metaphysical thesis of the existence of a plenum of centers of force, the monads, corresponding to a spatial plenum of atoms, led him to assume that the loss of kinetic energy is only apparent and that the "lost" kinetic energy is simply the vibrations of the atoms in the pendulums. For kinetic energy, these small vibrations are raised to the second power and do not cancel out; for the conservation of momentum, they are vectors. Since heat, for Leibniz, was a phenomenon resulting from the rapid vibration of small particles, there is little doubt that Leibniz thought of this situation as presenting a case of the equivalence of heat and kinetic energy. It was not taken up and treated thoroughly until more than a century later after the strange vagaries of the phlogiston theory. Equation (A.3.8) will reappear in the Special Theory of Relativity as a consequence of the conservation of momentum as formulated there.

B Different Fields of Acceleration for Central Forces.

We shall now examine the different fields of acceleration in some depth.

B.1   The Field of Constant Acceleration

We now derive the general equation of a particle moving without constraints in a field of constant acceleration g. Let the initial conditions be as follows. At $t = 0$, $\frac{d\underline{x}}{dt} = \underline{v}_0, \underline{x} = \underline{x}_0$. Then

$$m\frac{d^2\underline{x}}{dt^2} = m\underline{g}, \qquad\qquad \frac{d\underline{x}}{dt} = \underline{g}t + \underline{c}, \qquad\qquad \underline{c} = \underline{v}_0,$$

$$\frac{d\underline{x}}{dt} = \underline{g}t + \underline{v}_0, \qquad\qquad \underline{x} = \frac{1}{2}\underline{g}t^2 + \underline{v}_0 t + \underline{c}_1, \qquad\qquad \underline{c}_1 = \underline{x}_0.$$

$$\underline{x} = \frac{1}{2}\underline{g}t^2 + \underline{v}_0 t + \underline{x}_0 \qquad\qquad\qquad\qquad \text{(B.1.1)}$$

The position-vector $\underline{x}$ traces a parabola with a vertical axis; this can be seen as follows, if we choose the vertical plane containing $\underline{v}_0$ and $\underline{g}$ as through $\underline{e}_1$ and $\underline{e}_2$.

$$\underline{x} = x\underline{e}_1 + y\underline{e}_2 + z\underline{e}_3$$

$$= \frac{1}{2}\underline{g}\underline{e}_3 t^2 + [(v_0)_x\underline{e}_1 + (v_0)_y\underline{e}_3]\, t + x_0\underline{e}_1 + y_0\underline{e}_2 + z_0\underline{e}_3$$

$$x = (v_0)_x t + x_0,$$

$$t = \frac{x - x_0}{(v_0)_x}$$

$$z = -\frac{1}{2}gt^2 + (v_0)_z t + z_0$$

$$= -\frac{1}{2}g\frac{(x - x_0)^2}{(v_0)_x^2} + (v_0)_z\frac{(x - x_0)}{(v_0)_x} + z_0.$$

The vertex of this parabola is determined as

$$\frac{dz}{dx} = \frac{-g}{(v_0)_x^2}(x - x_0) + \frac{(v_0)_z}{(v_0)_x} = 0, \quad x - x_0 = \frac{(v_0)_x(v_0)_z}{g}.$$

We can make a more interesting equation, if we add to the field of constant acceleration, a force of friction proportional in magnitude to the velocity of the particle. Then, by Axiom 2,

$$m\frac{d^2\underline{x}}{dt^2} = m\underline{g} - (\lambda')^2\frac{d\underline{x}}{dt}$$

$$\frac{d^2\underline{x}}{dt^2} = \underline{g} - \lambda^2\frac{d\underline{x}}{dt}, \quad (\lambda')^2/m = \lambda^2 \tag{B.1.2}$$

$$\frac{d^2\underline{x}}{dt^2} + \lambda^2\frac{d\underline{x}}{dt} = \underline{g} \tag{B.1.3}$$

This differential equation can be solved in the usual way by first solving the homogeneous part

$$\frac{d^2\underline{x}}{dt^2} + \lambda^2\frac{d\underline{x}}{dt} = \underline{0} \tag{B.1.4}$$

by letting

$$\underline{x} = \underline{c}e^{\mu}$$

and determining μ after substitution in (B.1.4).

$$\mu^2 + \lambda^2\mu = 0, \quad \mu_1 = 0, \quad \mu_2 = -\lambda^2$$

so that the general solution of (B.1.4) is

$$\underline{x} = \underline{c}_1 + \underline{c}_2 e^{-\lambda^2 t}.$$

We find next a particular solution of (B.1.4) which is obviously at hand as

$$\underline{x} = \frac{1}{\lambda^2}\underline{g}t.$$

The general solution of (B.1.2) is

$$\underline{x} = \underline{c}_1 + \underline{c}_2 e^{-\lambda^2 t} + \frac{1}{\lambda^2}\underline{g}t, \tag{B.1.5}$$

$$\frac{d\underline{x}}{dt} = -\lambda^2\underline{c}_2 e^{-\lambda^2 t} + \frac{1}{\lambda^2}\underline{g}, \tag{B.1.6}$$

As $t \to \infty$, $\frac{d\underline{x}}{dt} \to \frac{1}{\lambda^2}\underline{g}$, a constant velocity. This is obviously not a conservative field of force.

Milne, in his **Vectorial Mechanics**, gives another interesting vectorial development of (B.1.3). We integrate directly to get

$$\frac{d\underline{x}}{dt} + \lambda^2\underline{x} = \underline{g}t + \underline{c}.$$

If $\underline{x_0} = 0$, $t_0 = 0$, $\underline{v} = \underline{v_0}$, we have

$$\frac{d\underline{x}}{dt} + \lambda^2\underline{x} = \underline{g}t + \underline{v_0} \tag{B.1.7}$$

Then, if $\underline{e_1}$ is a horizontal unit-vector and $\underline{e_3}$ vertical, since $\underline{g} = -g\underline{e_3}$, $\underline{g} \cdot \underline{e_1} = 0$,

$$(\underline{v} - \underline{v_0}) \cdot \underline{e_1} = -\lambda^2\underline{x} \cdot \underline{e_1}$$

$$\frac{(-\underline{v} + \underline{v_0})\underline{e_1}}{\lambda^2} = \underline{x} \cdot \underline{e_1} \tag{B.1.8}$$

is the horizontal range. And

$$\lim_{t \to \infty} \underline{x} \cdot \underline{e_1} = \frac{\left(-\frac{1}{\lambda^2}\underline{g} + \underline{v_0}\right)}{\lambda^2} \cdot \underline{e_1} = \frac{-\underline{g} + \lambda^2\underline{v_0}}{\lambda^4}$$

and is limited.

For the maximum height, we take from (B.1.7),

$$\frac{d\underline{x}}{dt} \cdot \underline{e_3} = -\lambda^2\underline{x} \cdot \underline{e_3} + \underline{g} \cdot \underline{e_3}t + \underline{v_0} \cdot \underline{e_3} = 0$$

$$\lambda^2 z = -gt + \underline{v_0} \cdot \underline{e_3}$$

$$z = \frac{\underline{v_0} \cdot \underline{e_3} - gt}{\lambda^2} \tag{B.1.9}$$

where t is still undetermined. But, multiplying by $e^{\lambda^2 t}$ in (B.1.7) , we get

$$\frac{d\underline{x}}{dt}e^{\lambda^2 t} + \lambda^2 e^{\lambda^2 t}\underline{x} = \underline{g}e^{\lambda^2 t}t + \underline{v_0}e^{\lambda^2 t}$$

$$\frac{d}{dt}(e^{\lambda^2 t}\underline{x}) = \underline{g}e^{\lambda^2 t}t + \underline{v_0}e^{\lambda^2 t}.$$

Integrating and using the initial conditions, $\underline{x_0} = 0$, $\underline{v} = \underline{v_0}$ at $t = 0$, we have

$$e^{\lambda^2 t}\underline{x} = \frac{1}{\lambda^4}\underline{g}\left[\lambda^2 e^{\lambda^2 t}t - e^{\lambda^2 t} + 1\right] + \frac{1}{\lambda^2}\underline{v}_0\left[e^{\lambda^2 t} - 1\right]$$

$$\underline{x} = \frac{1}{\lambda^4}\underline{g}\left[\lambda^2 t - 1 + e^{-\lambda^2 t}\right] + \frac{1}{\lambda^2}\underline{v}_0\left[1 - e^{-\lambda^2 t}\right]$$

$$\frac{d\underline{x}}{dt} = \frac{1}{\lambda^4}\underline{g}\left[\lambda^2 - \lambda^2 e^{-\lambda^2 t}\right] + \underline{v}_0 e^{-\lambda^2 t}.$$

Therefore the time of flight to the maximum height is gotten as the root of

$$\frac{d\underline{x}}{dt}\cdot\underline{e}_3 = \frac{1}{\lambda^4}\underline{g}\cdot\underline{e}_3\left[\lambda^2 - \lambda^2 e^{-\lambda^2 t}\right] + \underline{v}_0\cdot\underline{e}_3 e^{-\lambda^2 t} = 0.$$

$$= \frac{-g}{\lambda^2}\left[1 - e^{-\lambda^2 t}\right] + \underline{v}_0\cdot\underline{e}_3 e^{-\lambda^2 t} = 0.$$

or

$$\lambda^2\underline{v}_0\cdot\underline{e}_3 + g = g e^{\lambda^2 t}$$

$$t = \frac{1}{\lambda^2}\log\left[\lambda\underline{v}_0\cdot\underline{e}_3 + g\right] \tag{B.1.10}$$

So that substituting (B.1.10) in (B.1.9), we get

$$z = \frac{\underline{v}_0\cdot\underline{e}_3 - \frac{g}{\lambda^2}\log[\lambda^2\underline{v}_0\cdot\underline{e}_3 + g]}{\lambda^2} \tag{B.1.11}$$

The Simple Pendulum. We now consider in detail a case of constrained motion in the field of constant acceleration without friction, the case of a particle constrained to move on a vertical circle.

Let θ be the angle that OQ, the line from center of circle to particle, makes with the vertical line OR from center to lowest point of circle. Then $\underline{n}$ is toward the center from the particle and $\underline{t}$ is in the sense of increasing θ. There are two forces acting: $m\underline{g}$, vertically down, and $\underline{P}$, the force of constraint along OQ, normal to the tangent to the curve. Hence

$$\underline{P} = P\underline{n}$$

where P can be positive or negative. Then, by Axiom 2,

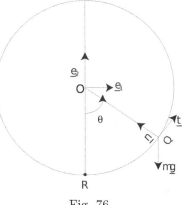

Fig. 76

$$m\frac{d^2\underline{x}}{dt^2} = m\frac{d^2\underline{x}}{ds^2}\left(\frac{ds}{dt}\right)^2 + m\frac{d\underline{x}}{ds}\frac{d^2 s}{dt^2} = m\underline{g} + \underline{P}.$$

But for circle $s = a\theta$, $\frac{ds}{dt} = a\frac{d\theta}{dt}$, $\frac{d^2s}{dt^2} = a\frac{d^2\theta}{dt^2}$ where a is radius of circle. Further

$$\underline{n} = \cos\theta\underline{e}_2 - \sin\theta\underline{e}_1, \quad \underline{t} = \sin\theta\underline{e}_2 + \cos\theta\underline{e}_1.$$

Hence

$$m\frac{1}{a}\underline{n}\left(a\frac{d\theta}{dt}\right)^2 + m\underline{t}a\frac{d^2\theta}{dt^2} = -mg\underline{e}_2 + P\underline{n}.$$

If we now equate the $\underline{n}$-components and $\underline{t}$ components of each side,

1. $ma\left(\dfrac{d\theta}{dt}\right)^2 = -mg(\underline{e}_2 \cdot \underline{n}) + P = -mg\cos\theta + P$

$$P = mg\cos\theta + ma\left(\frac{d\theta}{dt}\right)^2 \tag{B.1.12}$$

2. $$ma\frac{d^2\theta}{dt^2} = -mg\sin\theta.$$

$$\frac{d^2\theta}{dt^2}\frac{d\theta}{dt} = -\frac{g}{a}\sin\theta\frac{d\theta}{dt}$$

$$\frac{1}{2}\left(\frac{d\theta}{dt}\right)^2 - \frac{1}{2}\left(\frac{d\theta}{dt}\right)^2_{t_0} = \frac{g}{a}(\cos\theta - \cos\theta_0) \tag{B.1.13}$$

If now we substitute $\left(\frac{d\theta}{dt}\right)^2$ from (B.1.13) into (B.1.12), we have

$$P = mg\cos\theta + ma\left[\frac{2g}{a}(\cos\theta - \cos\theta_0) + \left(\frac{d\theta}{dt}\right)^2_{t_0}\right]$$

$$= m\left[3g\cos\theta - 2g\cos\theta_0 + a\left(\frac{d\theta}{dt}\right)^2_{t_0}\right] \tag{B.1.14}$$

a formula which gives P in terms of the initial angle, the initial angular speed, and the angle of its present position.

Let us now turn to (B.1.13) in order to determine the periodic time of the pendulum when it swings back and forth or when it resolves about the circle. There are obviously three cases.

Case 1. Here the initial angle θ_0 is $(\theta_0) < \pi$, with $(\frac{d\theta}{dt})_{t_0} = 0$, that is, the pendulum starts at rest at θ_0. Then

$$\frac{d\theta}{dt} = \sqrt{\frac{2g}{a}}(\cos\theta - \cos\theta_0)^{\frac{1}{2}}, \ 0 \le |\theta| \le |\theta_0| < \pi,$$

choosing the positive sign. Then

$$dt = \sqrt{\frac{a}{2g}} \frac{d\theta}{(\cos\theta - \cos\theta_0)^{\frac{1}{2}}}.$$

We convert to the half-angle,

$$\cos\theta = \cos^2\frac{\theta}{2} - \sin^2\frac{\theta}{2} = 1 - 2\sin^2\frac{\theta}{2},$$

$$dt = \sqrt{\frac{a}{2g}} \frac{d\theta}{\left(2\sin^2\frac{\theta_0}{2} - 2\sin^2\frac{\theta}{2}\right)^{\frac{1}{2}}} = \sqrt{\frac{a}{g}} \frac{d\frac{\theta}{2}}{(\sin^2\frac{\theta_0}{2} - \sin^2\frac{\theta}{2})^{\frac{1}{2}}}$$

Let

$$\sin\frac{\theta}{2} = \sin\frac{\theta_0}{2}\sin\varphi, \quad \cos\frac{\theta}{2}d\frac{\theta}{2} = \sin\frac{\theta_0}{2}\cos\varphi d\varphi$$

$$dt = \sqrt{\frac{a}{g}} \frac{\sin\frac{\theta_0}{2}\cos\varphi d\varphi}{\cos\frac{\theta}{2}\sin\frac{\theta_0}{2}(1 - \sin^2\varphi)^{\frac{1}{2}}}$$

$$= \sqrt{\frac{a}{g}} \frac{d\varphi}{(1 - \sin^2\frac{\theta_0}{2}\sin^2\frac{\theta}{2})^{\frac{1}{2}}}$$

And, when $\theta = 0, \varphi = 0$, when $\theta = \theta_0, \varphi = \frac{\pi}{2}$, so that

$$\frac{1}{4}\text{P.T.} = \sqrt{\frac{a}{g}} \int_0^{\frac{\pi}{2}} \frac{d\varphi}{(1 - k^2\sin^2\varphi)^{\frac{1}{2}}} \tag{B.1.15}$$

But for $0 \leq |\theta_0| < \pi, 0 \leq \sin^2\frac{\theta_0}{2} < 1$, and for $0 < |\varphi| \leq \frac{\pi}{2}, \sin^2\varphi \leq 1$. Hence we expand the integrand by the binomial series, and since this series is uniformly convergent for any subinterval of the interval of convergence, we can integrate them by term. Thus

$$\text{P.T.} = 4\sqrt{\frac{a}{g}}\left[\frac{\pi}{2} + \frac{k^2}{4}\frac{\pi}{4} + \frac{3}{8}k^4\frac{1\cdot 3}{2\cdot 4}\frac{\pi}{2} + \ldots\right]$$

$$= 2\pi\sqrt{\frac{a}{g}}\left[1 + \frac{k^2}{4} + \frac{9}{64}k^4 + \ldots\right] \tag{B.1.16}$$

Case 2. Returning again to (B.1.13),

$$\left(\frac{d\theta}{dt}\right)^2 - \left(\frac{d\theta}{dt}\right)^2_{t_0} = \frac{2g}{a}(\cos\theta - \cos\theta_0)$$

Here we suppose θ_0, and let $(\frac{d\theta}{dt})_{t_0}$ be the angular velocity acquired as if falling from a height greater than 2a. This will insure its swinging over the top. Then

$$\left(\frac{d\theta}{dt}\right)^2 = \frac{2g}{a}(\cos\theta - 1) + \left(\frac{d\theta}{dt}\right)^2_{t_0},$$

$$dt = \sqrt{\frac{a}{2g}} \frac{d\theta}{\left[\frac{a}{2g}\left(\frac{d\theta}{dt}\right)^2_{t_0} - 2\sin^2\frac{\theta}{2}\right]^{\frac{1}{2}}}$$

$$= \frac{1}{(\frac{d\theta}{dt})_{t_0}} \frac{d\frac{\theta}{2}}{\left[1 - \frac{4g}{a}\frac{1}{(\frac{d\theta}{dt})^2_{t_0}}\sin^2\frac{\theta}{2}\right]^{\frac{1}{2}}} \qquad\text{(B.1.17)}$$

But, since at $\theta = 0$, the velocity is as if the particle has fallen from a height grater than 2a,
Therefore

$$v_0^2 = a^2\left(\frac{d\theta}{dt}\right)^2_{t_0} = 2gh > 4ga.$$

Hence

$$\frac{4g}{a}\frac{1}{\left(\frac{d\theta}{dt}\right)^2_{t_0}} = k^2 < 1,$$

and so we can integrate (B.1.17) term by term as in the previous case.

Case 3. Here again we suppose $\theta_0 = 0$, and we let $(\frac{d\theta}{dt})_{t_0}$ be the angular velocity as if the particle has fallen from the height 2a.
Then

$$\frac{4g}{a}\frac{1}{(\frac{d\theta}{dt})^2_{t_0}} = 1$$

and (B.1.17) becomes

$$dt = \frac{1}{(\frac{d\theta}{dt})_{t_0}} \frac{d\frac{\theta}{2}}{[1 - \sin^2\frac{\theta}{2}]^{\frac{1}{2}}}.$$

Letting $x = \sin\frac{\theta}{2}$, $dx = \cos\frac{\theta}{2}\frac{d\theta}{2}$, we have, for $0 \le \theta \le \pi$,

$$dt = \frac{1}{(\frac{d\theta}{dt})_{t_0}} \frac{dx}{1 - x^2} = \frac{1}{2(\frac{d\theta}{dt})_{t_0}}\left[\frac{dx}{1 - x} + \frac{dx}{1 + x}\right].$$

The integral of this last is

$$\frac{1}{2(\frac{d\theta}{dt})_{t_0}} \log \frac{1+x}{1-x} = \frac{1}{2(\frac{d\theta}{dt})_{t_0}} \log \left[\frac{1+\sin\frac{\theta}{2}}{1-\sin\frac{\theta}{2}} \right].$$

Therefore integrating for θ from 0 to θ, we have

$$t = \frac{1}{2(\frac{d\theta}{dt})_{t_0}} \log \left[\frac{1+\sin\frac{\theta}{2}}{1-\sin\frac{\theta}{2}} \right].$$

B.2 The Inverse-square Field

We shall first consider the attractive field. This is given by Axiom 2 as

$$\frac{d^2 x}{dt^2} = -M_\gamma \frac{x}{|x|^3} \tag{B.2.1}$$

where x is the position-vector of the particle m from the center of force whose attracting mass is M_γ. We have also the areal law

$$x \times \frac{dx}{dt} = 2k \tag{B.2.2}$$

Then crossing these last two equations together, we get

$$2k \times \frac{d^2 x}{dt^2} = M_\gamma \frac{x \times \left(x \times \frac{dx}{dt} \right)}{|x|^3} = M_\gamma \frac{x \cdot \frac{dx}{dt} x - \frac{(x \cdot x)}{|x|} \frac{dx}{dt}}{|x|^2}$$

$$= M_\gamma \frac{\frac{d|x|}{dt} x - |x| \frac{dx}{dt}}{|x|^2} = M_\gamma \frac{d}{dt} \left[\frac{x}{|x|} \right].$$

Integrating, we get

$$2k \times \frac{dx}{dt} = -M_\gamma \frac{x}{|x|} - C \tag{B.2.3}$$

This is called Hamilton's Integral. From this central equation we can develop nearly the whole theory of this important field.

A. From Hamilton's Integral, we proceed first to find the orbit in polar coordinates by dotting with x,

$$2k \times \frac{dx}{dt} \cdot x = 2k \cdot \left(\frac{dx}{dt} \times x \right) = -4k^2 = -M_\gamma |x| - C \cdot x$$

$$4k^2 = |x|[M_\gamma + |C| \cos\theta] \tag{B.2.4}$$

$$|x| = \frac{4k^2/M_\gamma}{1 + \frac{|C|}{M_\gamma} \cos\theta}$$

where θ is the angle between $\underline{x}$ and $\underline{C}$. It should be noted that, by dotting Hamilton's Integral with $\underline{k}$, we get

$$0 = 0 - \underline{C} \cdot \underline{k},$$

so that $\underline{C}$ is normal to $\underline{k}$ and lies in the plane of $\underline{x}$ and $\frac{d\underline{x}}{dt}$ which are also normal to $\underline{k}$.

Now (B.2.4) is the equation of a conic section in polar coordinates as we can see from what follows. By the definition of conic sections

$$\frac{|\underline{x}|}{d - |\underline{x}| \cos \theta} = e$$

where d is the positive distance between F, the focus, and A, the intersection of the directrix and C, where the directrix is taken normal to $\underline{C}$. Hence

$$|\underline{x}| = \frac{de}{1 + e \cos \theta} \quad \text{(B.2.5)}$$

Comparing (B.2.4) and (B.2.5),

$$de = \frac{4k^2}{M_\gamma}, \qquad e = \frac{|C|}{M_\gamma}$$
$$\text{(B.2.6)}$$

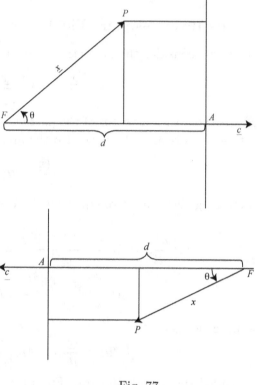

Fig. 77

For the ellipse, $e < 1$, and there is no restriction on θ in (B.2.5). But for the hyperbola, $e > 1$, and, since all the other quantities in (B.2.5) are essentially positive, it is impossible that

$$e \cos \theta < 0 \text{ and } |e \cos \theta| > 1.$$

This restricts the angle θ so that

$$0 \le |\theta| < \delta < \pi,$$

where δ depends on e. This means we have an "inner" focus hyperbola, that is, we have that branch of the hyperbola which contains the focus in its concave side. Thus, in the figures $|\theta|$ cannot be greater than $|\alpha|$ where α is the angle that the asymptote makes with $\underline{C}$, where $\underline{C}$ is in the direction of the major axis of the conic.

It is to be noticed that, if we had written the Hamilton Integral as

$$2\underline{k} \times \frac{d\underline{x}}{dt} = -M_\gamma \frac{\underline{x}}{|\underline{x}|} + \underline{C},$$

reversing the sense of $\underline{C}$ (Thus changing the figures) we would get the other form of the polar coordinate equation for conics,

$$|\underline{x}| = \frac{de}{1 - e\cos\theta}.$$

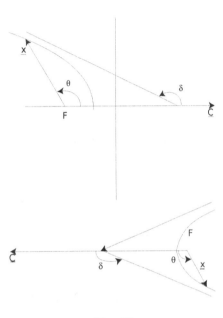

Fig. 78

Then restrictions on θ would be stated differently. Since we will want to compare our equation with the equation of the conics in canonical form, we remember that for the ellipse and hyperbola respectively,

$$ed = \left(\frac{a}{e} - ae\right) e = a(1 - e^2) = \frac{b^2}{a},$$
$$ed = \left(ae - \frac{a}{e}\right) e = a(e^2 - 1) = \frac{b^2}{a},$$

so that

$$\frac{4k^2}{M_\gamma} = \frac{b^2}{a} \qquad (B.2.7)$$

We have thus derived Kepler's law of the ellipse and more. From this we easily derive the periodic time P.T.$= \frac{\pi ab}{|\underline{k}|} \cdot \frac{2\pi a^{3/2}}{\sqrt{M_\gamma}}.$

B. From Hamilton's Integral, we can derive the velocity vector by crossing with $\underline{k}$, so that

$$\underline{k} \times \left(2\underline{k} \times \frac{d\underline{x}}{dt}\right) = -2k^2\frac{d\underline{x}}{dt} = -M_\gamma\frac{\underline{k} \times \underline{x}}{|\underline{x}|} - \underline{k} \times \underline{C}$$

$$\frac{d\underline{x}}{dt} = \frac{\underline{k}}{2k^2} \times \left[M_\gamma\frac{\underline{x}}{|\underline{x}|} + \underline{C}\right] \tag{B.2.8}$$

$$= \frac{\underline{k}}{2k^2} \times [M_\gamma\underline{e} + \underline{C}]$$

where $\underline{e}$ is a unit-vector, variable, in the direction of $\underline{x}$. If one wishes the equation in the form

$$\frac{d\underline{x}}{dt} = \frac{M_\gamma}{2k^2}\underline{k} \times \underline{e} + \frac{\underline{k} \times \underline{C}}{2k^2},$$

then the hodograph of $\frac{d\underline{x}}{dt}$, that is, its graph considered as a position-vector, is a circle with center $\frac{\underline{k} \times \underline{C}}{2k^2}$ and radius

$$\left|\frac{M_\gamma}{2k^2}\underline{k} \times \underline{e}\right| = \frac{M_\gamma}{2|\underline{k}|}.$$

If we remember that, at $\theta = 0$, $\underline{x} = \lambda^2\underline{C}$, and at $\theta = \pi$ (for the ellipse), $\underline{x} = -\lambda^2\underline{C}$, we get

$$\frac{d\underline{x}}{dt}\bigg|_{\theta=0} = \underline{k} \times \left[\frac{M_\gamma}{2k^2}\frac{\underline{C}}{|\underline{C}|} + \frac{1}{2k^2}\underline{C}\right] = \frac{\underline{k} \times \underline{C}}{2k^2}\left[\frac{M_\gamma}{|\underline{C}|} + 1\right],$$

$$\frac{d\underline{x}}{dt}\bigg|_{\theta=\pi} = \frac{\underline{k} \times \underline{C}}{2k^2}\left[-\frac{M_\gamma}{|\underline{C}|} + 1\right];$$

and so

$$\frac{\frac{d\underline{x}}{dt}\big|_{\theta=0}}{\frac{d\underline{x}}{dt}\big|_{\theta=\pi}} = \frac{|[\frac{1}{e} + 1]|}{|[-\frac{1}{e} + 1]|} = \frac{a[1 + e]}{a[1 - e]},$$

that is, the speeds of the planet at the perihelion and aphelion are inversely as the distances from the sun as Kepler saw at the beginning of his investigations.

Returning to (B.2.8), we can get v^2 in an unusual form.

$$\frac{d\underline{x}}{dt} = \underline{k} \times \frac{1}{2k^2}[M_\gamma\underline{e} + \underline{C}] = \underline{k} \times \frac{1}{2k^2}\underline{\alpha}$$

$$\frac{d\underline{x}}{dt} \cdot \frac{d\underline{x}}{dt} = \frac{1}{4k^4}(\underline{k} \times \underline{\alpha}) \cdot (\underline{k} \times \underline{\alpha})$$

$$= \frac{1}{4k^4}\{\underline{k} \cdot [\underline{\alpha} \times (\underline{k} \times \underline{\alpha})]\} = \frac{1}{4k^4}\underline{k} \cdot [\alpha^2\underline{k} - (\underline{\alpha} \cdot \underline{k})\underline{\alpha}]$$

$$= \frac{1}{4k^2}\alpha^2, \qquad \underline{\alpha} = M_\gamma\underline{e} + \underline{C},$$

since $\underline{k}$ is normal to $\underline{x}$ and $\underline{C}$. Hence

$$v^2 = \frac{1}{4k^2}[M^2\gamma^2 + 2M_\gamma\underline{e}\cdot\underline{C} + C^2]$$

$$= \frac{1}{4k^2}[M^2\gamma^2 + 2M_\gamma|\underline{C}|\cos\theta + |\underline{C}|^2]$$

$$= \frac{M^2\gamma^2}{4k^2}\left[1 + 2e\cos\theta + e^2\right], \qquad e = \frac{|\underline{C}|}{M_\gamma}.$$

But, from B.2.5 and B.2.6,

$$e\,\cos\theta = \frac{\frac{4k^2}{m_\gamma} - |\underline{x}|}{|\underline{x}|},$$

so that

$$v^2 = \frac{M^2\gamma^2}{4k^2}\left[1 + 2\frac{4k^2}{M_\gamma|\underline{x}|} - 2 + e^2\right] = M_\gamma\left[\frac{2}{|\underline{x}|} + \frac{a}{b^2}(e^2 - 1)\right].$$

Hence for ellipse and hyperbola respectively,

$$v^2 = M_\gamma\left[\frac{2}{|\underline{x}|} \mp \frac{1}{a}\right].$$

C. We can start from Hamilton's integral again and use the energy-equation to find a useful formula for the eccentricity $e = \frac{|\underline{C}|}{M_\gamma}$. Thus

$$\underline{C} = -M_\gamma\frac{\underline{x}}{|\underline{x}|} - 2\underline{k} \times \frac{d\underline{x}}{dt},$$

$$= -M_\gamma\frac{\underline{x}}{|\underline{x}|} + \frac{d\underline{x}}{dt} \times \left(\underline{x} \times \frac{d\underline{x}}{dt}\right)$$

$$= -M_\gamma\frac{\underline{x}}{|\underline{x}|} + \left[v^2\underline{x} - \left(\underline{x}\cdot\frac{d\underline{x}}{dt}\right)\frac{d\underline{x}}{dt}\right].$$

We know there is an $\underline{x}$, where $\underline{x}_1\cdot\frac{d\underline{x}}{dt}\big|_{\underline{x}_1} = 0$, notably when $\theta = 0$ and $\underline{x} = \lambda^2\underline{C}$. This can be checked in (B.2.8). Hence

$$\underline{C} = -M_\gamma\frac{\underline{x}_1}{|\underline{x}_1|} + [v_1^2\underline{x}_1] \tag{B.2.9}$$

In this field, we compute the energy equation

$$v^2 - v_0^2 = -2 \int_{\underline{x}_0}^{\underline{x}} \frac{M_\gamma \underline{x} \cdot d\underline{x}}{|\underline{x}|^3} = -2M_\gamma \int_{\underline{x}_0}^{\underline{x}} \frac{\frac{\underline{x} \cdot d\underline{x}}{|\underline{x}|}}{|\underline{x}|^2} = -2M_\gamma \int_{\underline{x}_0}^{\underline{x}} \frac{d|\underline{x}|}{|\underline{x}|^2}$$

$$= 2M_\gamma \left[\frac{1}{|\underline{x}|} - \frac{1}{|\underline{x}_0|} \right]. \tag{B.2.10}$$

Hence

$$v_1^2 = 2M_\gamma \left[\frac{1}{|\underline{x}_1|} - \frac{1}{|\underline{x}_0|} \right] + v_0^2.$$

Substituting in (B.2.9), we have

$$\underline{C} = -M_\gamma \frac{\underline{x}_1}{|\underline{x}_1|} + \underline{x}_1 \left[\frac{2M_\gamma}{|\underline{x}_1|} - \frac{2M_\gamma}{|\underline{x}_0|} + v_0^2 \right]$$

$$= \underline{x}_1 \left[\frac{M_\gamma}{|\underline{x}_1|} + h \right], \text{ where } h = v_0^2 - \frac{2M_\gamma}{|\underline{x}_0|}.$$

Since $\underline{C} = \frac{1}{\lambda^2} \underline{x}_1$, we can be sure that, whatever the value of h,

$$\frac{M_\gamma}{|\underline{x}_1|} + h \geq 0.$$

And, when $\theta = 0$,

$$|\underline{x}_1| = \frac{\frac{4k^2}{M_\gamma}}{1 + \frac{|\underline{C}|}{M_\gamma}} = \frac{4k^2}{M_\gamma + |\underline{C}|}.$$

Since

$$|\underline{C}|^2 = |\underline{x}_1|^2 \left[\frac{M_\gamma}{|\underline{x}_1|} + h \right]^2; \qquad |\underline{C}| = M_\gamma + h|\underline{x}_1| \geq 0$$

therefore

$$|\underline{C}| = M_\gamma + h \left[\frac{4k^2}{M_\gamma + |\underline{C}|} \right]$$

$$|\underline{C}|M_\gamma + |\underline{C}|^2 = (M_\gamma)^2 + M_\gamma|\underline{C}| + h(4k^2)$$

$$\frac{|\underline{C}|^2}{(M_\gamma)^2} = e^2 = \left(1 + h\frac{4k^2}{(M_\gamma)^2} \right)$$

$$e = \sqrt{1 + h\frac{4k^2}{(M_\gamma)^2}}, \qquad h = v_0^2 - \frac{2M_\gamma}{|\underline{x}_0|} \tag{B.2.11}$$

We know already the quantity under the radical sign cannot be negative. Therefore, when $h < 0$, $0 \le e < 1$, when $h = 0, e = 1$, when $h > 0, e > 1$. Evidently h depends on the strength of the center of force M_γ and the initial speed squared, v_o^2, and the initial distance $|\underline{x}_o|$ from the center of force. It is notable that the direction of initial velocity has nothing to do with the eccentricity of the resulting conic.

It is to be remarked that, for the inverse-square field, any plane region containing the origin (the center of force) is not simply connected. Any region not containing the origin is simply connected and $\nabla \times F = \underline{0}$ as the reader may easily check for himself. But it is also true that the work-integral is zero about any closed plane curve containing the origin. For the integral is zero over a circle about the origin, since here

$$\underline{x} = a \cos \theta \underline{e}_1 + a \sin \theta \underline{e}_2,$$
$$d\underline{x} = (-a \sin \theta \underline{e}_1 + a \cos \theta \underline{e}_2)d\theta,$$
$$\int_C \frac{\underline{x} \cdot d\underline{x}}{|\underline{x}|^3} = \int_C \frac{0 d\theta}{|\underline{x}|^3} = 0,$$

and it is easily shown that the integral over any other closed curve about the origin is the same.

In II.B., we saw that two of Kepler's laws, which we have deduced from the Newtonian axioms, lead to the equation of time

$$\lambda t = \varphi + e \sin \varphi, \qquad \lambda = \frac{2k}{ab}, \qquad 0 < e < 1,$$

where φ is the eccentric anomaly. We now add a method of approximating a solution φ, given λ and e. Consider

$$\varphi_1 = \lambda t - e \sin \varphi_o$$

for given t. This suggests the operator

$$T(\varphi) \equiv \lambda t - e \sin(\varphi).$$

Let the complete metric space be the set of real numbers φ, with metric

$$|\varphi, \overline{\varphi}| \equiv |\varphi - \overline{\varphi}|.$$

If φ is in the space, so is $T(\varphi)$. Further

$$|T(\varphi), \ T(\overline{\varphi})| \le T|\varphi - \overline{\varphi}| = |e(\sin \varphi - \sin \overline{\varphi})| \le |e| \, |\cos(\varphi + \theta \overline{\varphi})||\varphi - \overline{\varphi}|.$$

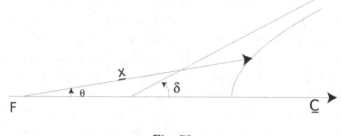

Fig. 79

Hence $T < 1$ and is a contraction on the space. Therefore, by the Fixed Point Theorem, there exists a unique φ such that, for any given φ_0 in the space,

$$\lim_{n \to \infty} T^n(\varphi_0) = \varphi, \ \varphi = T(\varphi)$$

Thus

$$\varphi_1 = \lambda t - e \sin \varphi_0$$
$$\varphi_2 = \lambda t - e \sin \varphi_1$$
$$\vdots$$
$$\varphi_n = \lambda t - e \sin \varphi_{n-1}$$

where

$$\lim_{n \to \infty} \varphi_n = \varphi, \qquad \varphi = \lambda t - e \sin \varphi.$$

The Repulsive Field: To get all the corresponding equations of the repulsive inverse-square field, we have only to replace M by $-M$. Thus

$$|\underline{x}| = \frac{\frac{-4k^2}{M_\gamma}}{1 - \frac{|\underline{C}|}{M_\gamma} \cos \theta} = \frac{-de}{1 - e \cos \theta} \qquad (B.2.11^*)$$

Here $e \cos \theta$ must be greater than 1, Therefore $e > 1$. The orbit of a particle in the repulsive field is always a hyperbola; but is an "outer" focus hyperbola, since $\cos \theta$ cannot be too small, that is,

$$0 \le |\theta| \le \delta < \frac{\pi}{2}.$$

Then Hamilton's Integral is

$$2\underline{k} \times \frac{d\underline{x}}{dt} = M_\gamma \frac{\underline{x}}{|\underline{x}|} - \underline{C}.$$

When $|\underline{x}| \to \infty$ along upper half of hyperbola, then $\underline{v} \to v_\infty \underline{u}$ where $\underline{u}$ is a unit-vector in the direction of asymptote, and

$$2\underline{k} \times v_\infty \underline{u} = M_\gamma \underline{u} - \underline{C},$$
$$\underline{C} \cdot \underline{u} = M_\gamma.$$
$$\underline{C} \times \underline{u} = \underline{u} \times (2\underline{k} \times v_\infty \underline{u}) = v_\infty 2\underline{k},$$

since $\underline{k} \cdot \underline{u} = 0$. Therefore

$$\frac{|\underline{C} \times \underline{u}|}{|\underline{C} \cdot \underline{u}|} = \frac{|2\underline{k}v_\infty|}{M_\gamma} = |\tan \delta|$$

where δ is acute angle between $\underline{C}$ and asymptote. But

$$2\underline{k} = \underline{x}_\infty \times \underline{v}_\infty$$

which is not as indeterminate as it seems.

For, if we take $\bar{\underline{x}}$ as a vector from F, the focus, to any point on the asymptope, and $\underline{p} = \bar{\underline{x}}$ when $\bar{\underline{x}}$ is normal to asymptote, we have by the well-known theorem about parallelograms on equal bases between the same parallel lines,

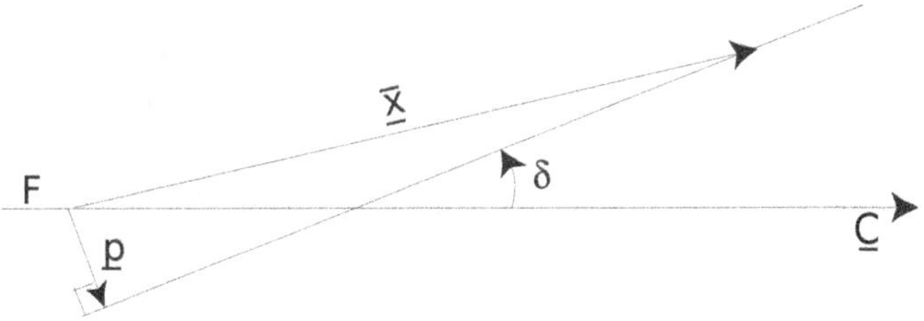

Fig. 80

$$\underline{p} \times \underline{v}_\infty = \bar{\underline{x}} \times \underline{v}_\infty = \underline{x}_\infty \times \underline{v}_\infty,$$

considering $\bar{\underline{x}}_\infty = \underline{x}_\infty$. Hence

$$|2\underline{k}| = |\underline{p} \times \underline{v}_\infty| = |p v_\infty|.$$

Since $\underline{p}$ is normal to $\underline{v}_\infty = v_\infty \underline{u}$. And

$$|\tan \delta| = \frac{v_\infty^2 p}{M_\gamma} \tag{B.2.12}$$

This formula is used in Rutherford's theory. For this theory the pos-
itively charged particles of x-rays (α
- particles) are repulsed by the pos-
itively charged nucleus of the atom
since the negative electrons, assumed
to give a negative spherical region
surrounding the nucleus are supposed
to have a negligible effect in Rutherford's model; the field is therefore
taken to be repulsive inverse-square.
The α-particles therefore move on hy-
perbolas with "outer" focus and approach very rapidly the direction of
the asymptotes, so that (B.2.12) is

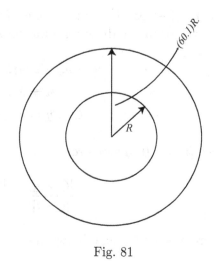

Fig. 81

applicable. But the Rutherford theory cannot be confirmed directly but only
statistically on the percentage of particles experiencing a wide deviation from
a collision.

Before leaving the inverse-square field, we should report an early investigation of Newton (circa 1669): According to the theory of Kepler, the celestial
and terrestrial mechanics should be one and the same. Neither Copernicus
nor Galileo accepted this, we should remember. It was natural for Newton to
consider the motion of the moon about the Earth (ignoring the action of the
sun) with the constant g of terrestrial acceleration as it appears, say, in the
oscillation of a pendulum. For this purpose Newton considered the orbit of
the moon as a circle about the Earth with the Earth's center at the center of
the circle, thus ignoring the several small irregularities of its motion which he
was later to explain in the Principia as a part of the three-body problem of
the sun, Earth, and moon.

We give an exposition of this, using modern numbers, which are not too far
off from those used by Newton, (and agree fairly well with those of Ptolemy).
We use two formulas from the inverse-square field:

$$|\underline{F}| = \frac{mM_\gamma}{r^2}, \quad \text{P.T.} = \frac{2\pi a^{\frac{3}{2}}}{\sqrt{M_\gamma}}, \quad M_\gamma = \text{accelerating mass of Earth.}$$

On surface of Earth, at R distance from the center,

$$mg = \frac{mM_\gamma}{R^2}, \quad g = \frac{M_\gamma}{R^2}.$$

But periodic time of Moon is

$$\text{P.T.} = \frac{2\pi a^{\frac{3}{2}}}{\sqrt{M_\gamma}}, \quad a = 60.1R, \quad R = 637 \cdot (10^6)\text{cms.}$$

and

$$\text{P.T.} = 236 \cdot (10^4)\text{secs}$$

Hence, from the two equations, we can find g and M_γ, and then check for g here on the surface of the Earth. We have

$$236 \cdot (10^4) = 2\pi[(60.1)R]^{\frac{3}{2}}/M_\gamma,$$

$$\sqrt{M_\gamma} = 2\pi[(60.1)R]^{\frac{3}{2}}/(236) \cdot (10^4) = \frac{2\pi[(60.1)^{\frac{3}{2}}(637)^{\frac{3}{2}}(10^9)]}{(236) \cdot (10^4)}.$$

This gives us the accelerating mass of the Earth. Substituting to find g, we have

$$g = \frac{4\pi^2(60.1)^3(637)^3(10)^{18}}{(637)^2(10)^{12}(236)^2(10)^8} = \frac{4\pi^2(60.1)^3(637)}{(236)^2 10^2} \approx 981\text{cm/secs}$$

which agrees with measurements of g on the Earth.

B.3 The Elastic Field

We suppose now that the acceleration induced on a particle varies as the distance from the fixed center of force. We first consider it attractive. Then, from Axiom 2,

$$m\frac{d^2\underline{x}}{dt^2} = -mM_\gamma \underline{x}$$

where M_γ is taken as the measure of the gravitational mass of the center; let it be ω^2. Then

$$\frac{d^2\underline{x}}{dt^2} = -\omega^2 \underline{x},$$

and the general solution is

$$\begin{aligned}
\underline{x} &= \underline{C}_1 e^{+i\omega t} + \underline{C}_2 e^{-i\omega t} \\
&= \underline{C}_1[\cos\omega t + i\sin\omega t] + \underline{C}_2[\cos\omega t - i\sin\omega t] \qquad \text{(B.3.1)} \\
&= (\underline{C}_1 + \underline{C}_2)\cos\omega t + i(\underline{C}_1 - \underline{C}_2)\sin\omega t
\end{aligned}$$

We are obviously taking our field element from the field of complex numbers. But we only wish here to consider the real part. So we have

$$x = \underline{A} \cos \omega t + \underline{B} \sin \omega t \tag{B.3.2}$$

$$\frac{d\underline{x}}{dt} = -\underline{A}\omega \sin \omega t + \omega \underline{B} \cos \omega t \tag{B.3.3}$$

If we take as initial conditions $t = 0$, $\underline{x} = \underline{x}_0$, $\frac{d\underline{x}}{dt} = \underline{v}_0$, we get

$$\underline{x} = \underline{x}_0 \cos \omega t + \frac{1}{\omega} \underline{v}_0 \sin \omega t$$
$$\frac{d\underline{x}}{dt} = -\omega \underline{x}_0 \sin \omega t + \underline{v}_0 \cos \omega t \tag{B.3.4}$$

This is an ellipse with center at origin, and $\underline{A}$ and $\underline{B}$ conjugate semi-diameters. For, when $t = 0$, $\underline{x} = \underline{A}$ and $\frac{d\underline{x}}{dt} = \omega \underline{B}$, that is, the tangent at the point given by $\underline{A}$ is parallel to $\underline{B}$. And $t = \frac{\pi}{2\omega}$, $\underline{x} = \underline{B}$, $\frac{d\underline{x}}{dt} = -\omega \underline{A}$, that is, the tangent at the point given by $\underline{B}$ is parallel to $\underline{A}$. To show it is actually an ellipse, we shall show we can reduce this vector equation, by the proper introduction of an orthonormal basis $\{\underline{e}_1, \underline{e}_2\}$ to the canonical form

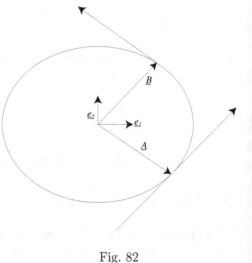

$$\frac{x^2}{a^2} + \frac{y^2}{b^2} = 1.$$

Fig. 82

For any $\{\underline{e}_1, \underline{e}_2\}$ orthonormal,

$$\underline{x} = (\underline{x} \cdot \underline{e}_1)\underline{e}_1 + (\underline{x} \cdot \underline{e}_2)\, \underline{e}_2 = \underline{A} \cos \omega t + \underline{B} \sin \omega t$$
$$\underline{x} \cdot \underline{e}_1 = (\underline{A} \cdot \underline{e}_1) \cos \omega t + (\underline{B} \cdot \underline{e}_1) \sin \omega t = x$$
$$\underline{x} \cdot \underline{e}_2 = (\underline{A} \cdot \underline{e}_2) \cos \omega t + (\underline{B} \cdot \underline{e}_2) \sin \omega t = y.$$

To find what we want, we remember that, when $x = 0, \frac{dy}{dt} = 0, \frac{dx}{dt} \neq 0$. But $\frac{dy}{dt} = -\omega(\underline{A} \cdot \underline{e}_2) \sin \omega t + \omega(\underline{B} \cdot \underline{e}_2) \cos \omega t$.

Therefore, we see that the conditions for being in canonical form are

$$(\underline{A} \cdot \underline{e}_1) \cos \omega t + (\underline{B} \cdot \underline{e}_1) \sin \omega t = 0$$
$$-\omega(\underline{A} \cdot \underline{e}_2) \sin \omega t + \omega(\underline{B} \cdot \underline{e}_2) \cos \omega t = 0$$

$$\tan \omega t = -\frac{\underline{A} \cdot \underline{e}_1}{\underline{B} \cdot \underline{e}_1} = \frac{\underline{B} \cdot \underline{e}_2}{\underline{A} \cdot \underline{e}_2}, \qquad \underline{A} \cdot \underline{e}_1 = \frac{(\underline{B} \cdot \underline{e}_1)(\underline{B} \cdot \underline{e}_2)}{\underline{A} \cdot \underline{e}_2} \tag{B.3.5}$$

Then

$$x^2 = [(\underline{A} \cdot \underline{e}_1)^2 \cos^2 \omega t + (\underline{B} \cdot \underline{e}_1)^2 \sin^2 \omega t + 2(\underline{A} \cdot \underline{e}_1)(\underline{B} \cdot \underline{e}_1) \cos \omega t \sin \omega t]$$
$$y^2 = [(\underline{A} \cdot \underline{e}_2)^2 \cos^2 \omega t + (\underline{B} \cdot \underline{e}_2)^2 \sin^2 \omega t + 2(\underline{A} \cdot \underline{e}_2)(\underline{B} \cdot \underline{e}_2) \cos \omega t \sin \omega t]$$
$$\frac{(\underline{A} \cdot \underline{e}_1)^2}{(\underline{B} \cdot \underline{e}_2)} y^2 = [(\underline{B} \cdot \underline{e}_1)^2 \cos^2 \omega t + (\underline{A} \cdot \underline{e})^2 \sin^2 \omega t - 2(\underline{A} \cdot \underline{e}_1)(\underline{B} \cdot \underline{e}_1) \cos \omega t \sin \omega t]$$

$$x^2 + \left[\frac{\underline{A} \cdot \underline{e}_1}{\underline{B} \cdot \underline{e}_2}\right]^2 y^2 = (\underline{A} \cdot \underline{e}_1)^2 + (\underline{B} \cdot \underline{e}_1)^2.$$

That the condition (B.3.5) can always be satisfied can be shown as follows. Let α be the given angle between $\underline{A}$ and $\underline{B}$ and θ the unknown angle between $\underline{e}_1$ and $\underline{A}$. Then

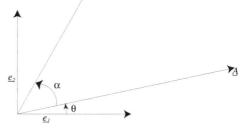

$$\frac{-\underline{B} \cdot \underline{e}_1}{\underline{A} \cdot \underline{e}_1} = \frac{\underline{A} \cdot \underline{e}_2}{\underline{B} \cdot \underline{e}_2}$$

is equivalent to

$$\frac{-|\underline{B}| \cos(\theta + \alpha)}{|\underline{A}| \cos(\theta)} = \frac{|\underline{A}| \cos\left(\frac{\pi}{2} - \theta\right)}{|\underline{B}| \cos\left(\frac{\pi}{2} - (\theta + \alpha)\right)}$$

Fig. 83

which is equivalent to

$$\frac{-|\underline{B}|^2}{|\underline{A}|^2} = \sin(2\alpha) \cot(2\theta) + \cos(2\alpha)$$

$$\cot \theta = \frac{\frac{-|\underline{B}|^2}{|\underline{A}|^2} - \cos(2\alpha)}{\sin(2\alpha)}$$

But this always determines θ for $-\frac{\pi}{2} \le \theta \le \frac{\pi}{2}$.

The periodic time for equation (B.3.4) is obviously $\frac{2\pi}{\omega}$. When $\underline{v}_0$ and $\underline{x}_0$ are in the same straight line, the motion reduces to motion in that straight line in what is called simple harmonic motion.

Example:

Suppose a mass hanging on an elastic spring at rest under gravity (friction is ignored). The stretched spring is of length $d + \ell$, but before being stretched by the mass m it was of length d. Then, by Axiom 2,

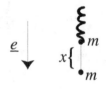

Fig. 84

$$mg\underline{e} - m\omega^2\ell\underline{e} = \underline{0}$$

where ω^2 is the constant of proportionality representing the elastic strength of the spring. Then

$$\omega^2 = \frac{g}{\ell}.$$

If then we stretch the mass down further, but not beyond the limit of elasticity of the spring, we have, by Axiom 2,

$$m\frac{d^2x}{dt^2}\underline{e} = mg\underline{e} - m\omega^2(\ell + x)\underline{e} = mg\underline{e} - mg\underline{e} - m\frac{g}{\ell}x\underline{e}$$

$$\frac{d^2x}{dt^2} = -\frac{g}{\ell}x$$

$$x = x_0 \cos\sqrt{\frac{g}{\ell}}t + \sqrt{\frac{\ell}{g}}v_0 \sin\sqrt{\frac{g}{\ell}}t.$$

Repulsive Elastic Field. In this case, we have

$$\frac{d^2x}{dt^2} = \omega^2\underline{x}$$

$$\underline{x} = \underline{C_1}e^{\omega t} + \underline{C_2}e^{-\omega t}$$

$$= \underline{A}\cos h(\omega t) + \underline{B}\sin h(\omega t),$$

and, as one might guess, this is an hyperbola with the center at the origin, and $\underline{A}$ and $\underline{B}$ as conjugate axes.

B.4   Elastic Field with Added Conditions

We can assume friction proportional to the velocity in the elastic field so that

$$\frac{d^2\underline{x}}{dt^2} = -\mu^2\underline{x} - 2\lambda^2\frac{d\underline{x}}{dt}$$

$$\underline{0} = \frac{d^2\underline{x}}{dt^2} + 2\lambda^2\frac{d\underline{x}}{dt} + \mu^2\underline{x} \qquad\qquad\text{(B.4.1)}$$

with solutions

$$\underline{x} = e^{-\lambda^2 t}\left[\underline{C}_1 e^{\sqrt{\lambda^4 - \mu^2}t} + \underline{C}_2 e^{-\sqrt{\lambda^4 - \mu^2}t}\right], \qquad \lambda^4 - \mu^2 \neq 0. \qquad \text{(B.4.2)}$$

and

$$\underline{x} = (\underline{C}_1 + \underline{C}_2 t)e^{-\lambda^2 t}, \qquad \lambda^4 - \mu^2 = 0 \qquad \text{(B.4.3)}$$

If, in the first case, we have

$$\lambda^4 - \mu^2 < 0,$$

then, for real solution,

$$\underline{x} = e^{-\lambda^2 t}[\underline{A}\cos\omega t + \underline{B}\sin\omega t], \qquad \omega = \sqrt{\lambda^4 - \mu^2}$$

and the motion is that of the elliptical motion of the elastic field multiplied by the damping factor $\underline{e}^{-\lambda^2 t}$ which approaches zero as $t \to \infty$. Hence the motion is a kind of spiral motion. If $\underline{A}$ and $\underline{B}$ are in the same straight line, we get an ordinary damped vibration.

If, in the first case,

$$\lambda^4 - \mu^2 > 0,$$

we have

$$\underline{x} = \underline{C}_1 e^{-\lambda^2(1+\sqrt{1-\mu^2/\lambda^4})t} + \underline{C}_2 e^{-\lambda^2(1-\sqrt{1-\mu^2/\lambda^4})t},$$

and again $\underline{x} \to \underline{0}$ as $t \to \infty$.

We can also impose as a further condition a periodic force so that

$$\frac{d^2 x}{dt^2} + 2\lambda^2 \frac{dx}{dt} + \mu^2 \underline{x} = a\sin\omega t\underline{e}$$

where $\underline{e}$ is a constant unit vector. We have already the solution of the homogeneous part. A particular solution is obviously of the form

$$\underline{x} = a(m\sin\omega t + n\cos\omega t)\underline{e}.$$

We compute $\frac{dx}{dt}$ and $\frac{d^2 x}{dt^2}$ and find

$$\underline{e}\left\{a[m(\mu^2 - \omega^2) - 2\lambda^2 n]\sin\omega t + a[m(2\lambda^2\omega) + (\mu^2 - \omega^2)n]\cos\omega t\right\} = a\sin\omega t\underline{e},$$

so that we have two conditions on m and n

$$m(\mu^2 - \omega^2) - 2\lambda^2 n = 1, \qquad m(2\lambda^2\omega) + (\mu^2 - \omega^2)n = 0,$$

whence

$$n = \frac{2\lambda^2\omega}{(\omega^2 - \mu^2)^2 - 4\lambda^4\omega}, \qquad m = \frac{\omega^2 - \mu^2}{(\omega^2 - \mu^2)^2 - 4\lambda^4\omega}$$

and the general solution

$$\underline{x} = e^{-\lambda^2 t}\left[\underline{C}_1 e\sqrt{\lambda^4 - \mu^2}t + \underline{C}_2 e - \sqrt{\lambda^4 - \mu^2}t\right] + a(m\sin\omega t + n\cos\omega t)\underline{e}$$

and $\underline{x} \to \infty$ as $\lambda \to 0$ and $\omega^2 = \mu^2$ in the well-known case.

B.5　General Formulas for Central Fields

Since, for any central field of force, as we have seen,

$$\underline{x} \times \frac{d\underline{x}}{dt} = 2\underline{k}$$

where $\underline{k}$ is constant, the orbits are always in one plane, and we can introduce polar coordinates. Let us derive certain formulas in polar coordinates. Since

$$\underline{x} = |\underline{x}|\cos\theta\underline{e}_1 + |\underline{x}|\sin\theta\underline{e}_2$$

$$\frac{d\underline{x}}{dt} = \left(\frac{d|\underline{x}|}{dt}\cos\theta - |\underline{x}|(\sin\theta)\frac{d\theta}{dt}\right)\underline{e}_1 + \left(\frac{d|\underline{x}|}{dt}\sin\theta + |\underline{x}|(\cos\theta)\frac{d\theta}{dt}\right)\underline{e}_2$$

$$v^2 = \frac{d\underline{x}}{dt}\cdot\frac{d\underline{x}}{dt} = \left(\frac{d|\underline{x}|^2}{dt^2}\right)^2 + |\underline{x}|^2\left(\frac{d\theta}{dt}\right)^2 \tag{B.5.1}$$

and

$$\left|\underline{x} \times \frac{d\underline{x}}{dt}\right| = \pm|\underline{x}|^2\frac{d\theta}{dt} \tag{B.5.2}$$

Further

$$\underline{F} = m\varphi(|\underline{x}|)\frac{\underline{x}}{|\underline{x}|} \tag{B.5.3}$$

if we restrict ourselves to forces which are some function of the distance. For simplicity of notation, let $|\underline{x}| = r$. Then the energy equation becomes

$$v^2 - v_0^2 = 2\int_{\underline{x}_0}^{\underline{x}}\varphi(|\underline{x}|)\frac{\underline{x}}{|\underline{x}|}\cdot d\underline{x} = 2\int_{r_0}^{r}\varphi(r)dr \tag{B.5.4}$$

and

$$\left(\frac{dr}{dt}\right)^2 + r^2\left(\frac{d\theta}{dt}\right)^2 = 2\int_{r_0}^{r}\varphi(r)dr + v_0^2 \tag{B.5.5}$$

But, from (B.5.2),

$$\left(\frac{d\theta}{dt}\right)^2 = \frac{4k^2}{r^4}$$

and

$$\left(\frac{dr}{dt}\right)^2 = 2\int_{r_0}^r \varphi(r)dr + v_0^2 - \frac{4k^2}{r^2} \tag{B.5.6}$$

From this equation we can get

$$dt = \frac{\pm dr}{\left[2\int_{r_0}^r \varphi(r)dr + v_0^2 - \frac{4k^2}{r^2}\right]^{\frac{1}{2}}} \tag{B.5.7}$$

Or we can take

$$\left(\frac{dr}{d\theta}\right)^2 \left(\frac{d\theta}{dt}\right)^2 = 2\int_{r_0}^r \varphi(r)dr + v_0^2 - \frac{4k^2}{r^2}$$

$$\left(\frac{d\theta}{dr}\right)^2 = \frac{\frac{4k^2}{r^4}}{2\int_{r_0}^r \varphi(r)dr + v_0^2 - \frac{4k^2}{r^2}}$$

$$d\theta = \frac{\pm 2k\ dr}{r^2\left[2\int_{r_0}^r \varphi(r)dr + v_0^2 - \frac{4k^2}{r^2}\right]^{\frac{1}{2}}} \tag{B.5.8}$$

Example. Suppose a particle follows the orbit $\theta = \frac{1}{r}$. To find the law of force which produces such as orbit, that is, to find $\varphi(r)$. Using (B.5.8),

$$\frac{d\theta}{dr} = -\frac{1}{r^2} = \frac{\pm 2k}{r^2\left[2\int_{r_0}^r \varphi(r)dr + v_0^2 - \frac{4k^2}{r^2}\right]^{\frac{1}{2}}}$$

$$\frac{1}{r^4} = \frac{4k^2}{r^4\left[2\int_{r_0}^r \varphi(r)dr + v_0^2 - \frac{4k^2}{r^2}\right]}$$

$$4k^2 = 2\int_{r_0}^r \varphi(r)dr + v_0^2 - \frac{4k^2}{r^2}$$

Taking the derivative with respect to r,

$$\varphi(r) = \frac{-4k^2}{r^3}.$$

Conversely, we can start with

$$\varphi(r) = -\frac{4k^2}{r^3},$$

and using (B.5.8), we get

$$\frac{d\theta}{dr} = \frac{\pm 2k}{r^2 \left[v_0^2 - \frac{4k^2}{r_0^2}\right]^{\frac{1}{2}}}, \qquad \theta = \frac{\mp 2k}{r \left[v_0^2 - \frac{4k^2}{r_0^2}\right]^{\frac{1}{2}}} + C,$$

Obviously, we have now to add the initial conditions to get precisely $\theta = \frac{1}{r}$, for they were built into that formula. Thus the initial conditions are

$$4k^2 = v_0^2 - \frac{4k^2}{r_0^2}, \qquad v_0^2 = 4k^2 \left[1 + \frac{1}{r_0^2}\right],$$

and

$$\theta_0 = \frac{\mp 2k}{r_0 \left[v_0^2 - \frac{4k^2}{r_0^2}\right]^{\frac{1}{2}}}.$$

We thus obtain

$$\theta = \pm \frac{1}{r}.$$

Whether we have positive or negative sign will depend on which way the spiral turns, whether θ increases with r or decreases with r.

B.6   Note on Dimensions

The axioms of Newton, as we have given them, introduce three fundamental and irreducible physical dimensions: distance, time, and mass which we denote with the symbols, $L, T,$ and M. Then we can characterize the dimensions of the different entities we have introduced in this chapter, except for action which will appear later.

$$\text{velocity} = \text{LT}^{-1}$$
$$\text{acceleration} = \text{LT}^{-2}$$
$$\text{force} = \text{MLT}^{-2}$$
$$\text{momentum} = \text{MLT}^{-1}$$
$$\text{energy} = \text{ML}^2\text{T}^{-2}$$
$$\text{action} = \text{ML}^2\text{T}^{-1}$$
$$\text{angular velocity} = \text{T}^{-1}$$

Obviously any equation which is to be valid must have the same dimensions throughout. For, otherwise, if one changes units of measurement, the equation is not invariant.

Chapter 5

The Displacement and Kinematics of a Rigid Body

In the preceding chapters, we have implicitly used rigid bodies as frames of reference for the moving particles. They were always fixed in absolute space. But now we should like to consider frames of reference moving in any manner with respect to absolute space, and to consider the motions of particles with respect to these. We shall want to know how the acceleration vectors transform with respect to a rigid system of coordinates moving in any manner with respect to another system.

A   The Displacement of a Rigid Body.

If a set of points transforms so that the distance between any two remains the same without reflection, then this constitutes the displacement of a rigid set. - Analytically, this means that for any two points given by position - vectors $\underline{x}_i$ and $\underline{x}_j$ from some arbitrary fixed origin 0

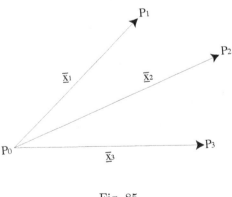

$$(\underline{x}_i - \underline{x}_j)^2 = (\underline{x}'_i - \underline{x}'_j)^2 \qquad (1.1)$$

Fig. 85

where $\underline{x}'_i$ are the transformed $\underline{x}_i$ or the $\underline{x}_i$ after displacement; and, for any

187

four non-coplanar points P_0, P_1, P_2, P_3, where

$$\overline{x}_1 = \overrightarrow{P_0P_1}, \qquad \overline{x}_2 = \overrightarrow{P_0P_2}, \qquad \overline{x}_3 = \overrightarrow{P_0P_3}$$

$\overline{x}_1 \cdot \overline{x}_2 \times \overline{x}_3$ and $\overline{x}_1' \cdot \overline{x}_2' \times \overline{x}_3'$ have the same sign.

We can think of the whole 3-space of points as being rigidly displaced. Then every orthonormal set $\{\underline{e}_i\}$ goes into an orthonormal set $\{\underline{e}_i'\}$ so that (see III.A.9.5)

$$\underline{e}_i' = t_i^j \underline{e}_j, \qquad \det(t_i^j) = 1.$$

Under such transformations, the inner product (here the dot product) is invariant. Or we can immediately deduce it from what we have

$$(\overline{x}_i - \overline{x}_j)^2 = (\overline{x}_i' - \overline{x}_j')^2$$
$$(\overline{x}_i)^2 - 2\overline{x}_i \cdot \overline{x}_j + (\overline{x}_j)^2 = (\overline{x}_i')^2 - 2\overline{x}_i' \cdot \overline{x}_j' + (\overline{x}_j')^2$$
$$\overline{x}_i \cdot \overline{x}_j = \overline{x}_i' \cdot \overline{x}_j'. \tag{1.2}$$

Again, since from

$$(\overline{x}_i \times \overline{x}_j)^2 = (\overline{x}_i' \times \overline{x}_j')^2 \tag{1.3}$$

we get, by interchange of dot and cross,

$$\overline{x}_i \cdot [\overline{x}_j \times (\overline{x}_i \times \overline{x}_j)] = \overline{x}_i' \cdot [\overline{x}_j' \times (\overline{x}_i' \times \overline{x}_j')]$$
$$\overline{x}_i \cdot [(\overline{x}_j)^2 \overline{x}_j - (\overline{x}_j \cdot \overline{x}_i)\overline{x}_j] = \overline{x}_i' \cdot [(\overline{x}_j')^2 \overline{x}_i' - (\overline{x}_j' \cdot \overline{x}_i')\overline{x}_j'],$$

therefore, by reversing steps, (1.1) is invariant.

Further

$$[\underline{a}, \underline{b}, \underline{c}]\underline{d} = [\underline{c} \cdot (\underline{a} \times \underline{b})]\underline{d} - (\underline{c} \cdot \underline{d})(\underline{a} \times \underline{b}) + (\underline{d} \cdot \underline{c})(\underline{a} \times \underline{b})$$
$$= \underline{c} \times [\underline{d} \times (\underline{a} \times \underline{b})] + (\underline{d} \cdot \underline{c})(\underline{a} \times \underline{b})$$
$$= \underline{c} \times [(\underline{d} \cdot \underline{b})\underline{a} - (\underline{d} \cdot \underline{a})\underline{b}] + (\underline{d} \cdot \underline{c})(\underline{a} \times \underline{b})$$
$$[\underline{a}, \underline{b}, \underline{c}]\underline{d} = (\underline{d} \cdot \underline{a})(\underline{b} \times \underline{c}) + (\underline{d} \cdot \underline{b})(\underline{c} \times \underline{a}) + (\underline{d} \cdot \underline{c})(\underline{a} \times \underline{b}) \tag{1.4}$$

If we dot this with $\underline{e} \times \underline{f}$, we have

$$[\underline{abc}][\underline{def}] = (\underline{d} \cdot \underline{a})(\underline{b} \times \underline{c}) \cdot (\underline{e} \times \underline{f}) + (\underline{d} \cdot \underline{b})(\underline{c} \times \underline{a}) \cdot (\underline{e} \times \underline{f})$$
$$+ (\underline{d} \cdot \underline{c})(\underline{a} \times \underline{b}) \cdot (\underline{e} \times \underline{f})$$
$$= (\underline{d} \cdot \underline{a})[\underline{b} \cdot (\underline{c} \times (\underline{e} \times \underline{f}))] + (\underline{d} \cdot \underline{b})[\underline{c} \cdot (\underline{a} \times (\underline{e} \times \underline{f}))]$$
$$+ (\underline{d} \cdot \underline{c})[\underline{a} \cdot (\underline{b} \times (\underline{e} \times \underline{f}))]$$
$$= (\underline{d} \cdot \underline{a})[(\underline{b} \cdot \underline{e})(\underline{c} \cdot \underline{f}) - (\underline{c} \cdot \underline{e})(\underline{f} \cdot \underline{b})]$$
$$+ (\underline{d} \cdot \underline{b})[(\underline{c} \cdot \underline{e})(\underline{a} \cdot \underline{f}) - (\underline{c} \cdot \underline{f})(\underline{a} \cdot \underline{e})]$$
$$+ (\underline{d} \cdot \underline{c})[(\underline{a} \cdot \underline{e})(\underline{b} \cdot \underline{f}) - (\underline{b} \cdot \underline{e})(\underline{a} \cdot \underline{f})]$$

$$[abc][def] = \begin{vmatrix} a \cdot d & b \cdot d & c \cdot d \\ a \cdot e & b \cdot e & c \cdot e \\ a \cdot f & b \cdot f & c \cdot f \end{vmatrix} \tag{1.5}$$

Using (1.5) with substitution of $\overline{x}_i, \overline{x}_j, \overline{x}_k$ for $\underline{a}, \underline{b}, \underline{c}$ and $\underline{d}, \underline{e}, \underline{f}$, and likewise with substitution of $\overline{x}_i', \overline{x}_j', \overline{x}_k'$

$$\left[\overline{x}_i \overline{x}_j \overline{x}_k\right]^2 = \left[\overline{x}_i' \overline{x}_j' \overline{x}_k'\right]^2$$

since the dot products are invariant. But we have assumed the signs of $\left[\overline{x}_i \overline{x}_j \overline{x}_k\right]$ and $\left[\overline{x}_i' \overline{x}_j' \overline{x}_k'\right]$ remain the same; therefore

$$\left[\overline{x}_i \overline{x}_j \overline{x}_k\right] = \left[\overline{x}_i' \overline{x}_j' \overline{x}_k'\right] \tag{1.6}$$

By using (1.4) we can prove that the displacement of any three non-collinear points determines the displacement of any fourth. Thus given the displacements of P_0, P_1, P_2, we can find the displacement of P_3. We have

$$\left[\overline{x}_1', \overline{x}_2', \overline{x}_1' \times \overline{x}_2'\right]\overline{x}_3' = \left[\overline{x}_1' \times \overline{x}_2'\right]^2 \overline{x}_3' = \left(\overline{x}_3' \cdot \overline{x}_1'\right)\left[\overline{x}_2' \times \left(\overline{x}_1' \times \overline{x}_2'\right)\right]$$
$$+ \left(\overline{x}_3' \cdot \overline{x}_2'\right)\left[\left(\overline{x}_1' \times \overline{x}_2'\right)\right] \times \overline{x}_1' + \left(\overline{x}_3' \cdot \overline{x}_1' \times \overline{x}_2'\right)\left(\overline{x}_1' \times \overline{x}_2'\right).$$

or

$$\left[\overline{x}_1 \times \overline{x}_2\right]^2 \overline{x}_3' = \left(\overline{x}_3 \cdot \overline{x}_1\right)\left[\overline{x}_2' \times \left(\overline{x}_1' \times \overline{x}_2'\right)\right] + \left(\overline{x}_3 \cdot \overline{x}_1\right)\left[\left(\overline{x}_1' \times \overline{x}_2'\right) \times \overline{x}_1'\right]$$
$$+ \left(\overline{x}_3 \cdot \overline{x}_1 \times \overline{x}_2\right)\left(\overline{x}_1' \times \overline{x}_2'\right) \times \overline{x}_1' \tag{1.7}$$

For we are given the relative displacement, $\overline{x}_1'$ and $\overline{x}_2'$.

The displacement of a rigid body is a <u>pure translation</u> if and only if every point has the same displacement vector $\underline{d}$. It is to be noticed that a displacement is given by the initial and terminal positions without regard to the intermediate positions, that is, to the paths by which the displacement is made.

The displacement is a pure rotation if and only if there is a point of the body (or rigidly connected with it) which remains fixed. Analytically, if the x_i are from the fixed point,

$$\left(\underline{x}_i - \underline{x}_j\right)^2 = \left(\underline{x}_i' - \underline{x}_j'\right)^2 \tag{1.8}$$

because of rigidity, and, because of pure rotation,

$$x_i^2 = \left(\underline{x}_i\right)^2 \text{ or } \left(\underline{x}_i\right)^2 - \left(\underline{x}_i'\right)^2 = 0 \tag{1.9}$$

From these last two equations, we can derive a condition for pure rotation of a rigid body which will be useful in the sequel. From (1.8) we get

$$0 = (\underline{x}_i - \underline{x}_j)^2 - (\underline{x}_i' - \underline{x}_j')^2 \equiv [(\underline{x}_i - \underline{x}_j) - (\underline{x}_i' - \underline{x}_j')] \cdot [(\underline{x}_i - \underline{x}_j) + (\underline{x}_i' - \underline{x}_j')]$$
$$\equiv [(\underline{x}_i - \underline{x}_i') - (\underline{x}_j - \underline{x}_j')] \cdot [(\underline{x}_i + \underline{x}_i') - (\underline{x}_j + \underline{x}_j')].$$

And from (1.9) we get

$$0 = (\underline{x}_i - \underline{x}_i') \cdot (\underline{x}_i + \underline{x}_i').$$

Hence from both

$$0 = (\underline{x}_i - \underline{x}_i') \cdot (\underline{x}_j + \underline{x}_j') + (\underline{x}_j - \underline{x}_j') \cdot (\underline{x}_i + \underline{x}_i') \qquad (1.10)$$

Theorem V.1.1. Any displacement of a rigid body can be reduced to a pure displacement followed by a pure rotation about any given point of the body.

Proof: Let P_0 be the given point, and P_0' after displacement. Then let P_i' be P_i after displacement. The $\underline{x}_i$ and $\underline{x}_i'$ are taken from an absolute origin 0. Then we give all P_i the translation

$$\underline{d} = \overrightarrow{P_0 P_0'}$$

so that

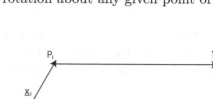

Fig. 86

$$\underline{x}_0' - \underline{x}_0 = \underline{d} = \hat{\underline{x}}_i - \underline{x}_i$$

where $\hat{\underline{x}}_i$ is position-vector of $\hat{P}_i$, the point P_i after translation $\underline{d}$. P_i' is P_i after the given total displacement of the rigid body.

Now let

$$\underline{x}_i - \underline{x}_0 = \underline{X}_i, \qquad \underline{x}_i' - \underline{x}_0' = \underline{X}_i'.$$

Then the displacement of any point P_i can be expressed as

$$\underline{x}_i' - \underline{x}_i = (\underline{x}_i' - \underline{x}_0') - (\underline{x}_i - \underline{x}_0')$$
$$= (\underline{x}_i' - \underline{x}_0') - (\underline{x}_i - \underline{x}_0) + (\underline{x}_0' - \underline{x}_0) = (\underline{X}_i' - \underline{X}_i) + \underline{d}$$

Hence the displacement of any point P_i is composed of the displacement $\underline{d}$ for all points, independent of i, and the displacement $\underline{X}_i$ to $\underline{X}_i'$.

Because of the conditions of rigidity

$$\underline{X}_i^2 = (\underline{X}_i')^2$$
$$(\underline{X}_j - \underline{X}_i)^2 = (\underline{X}_j' - \underline{X}_i')^2$$

so that this satisfies (1.8) and (1.9), the conditions of pure rotation. *Q.E.D.*

Suppose the point O to be the one fixed point of a rigid body, and any point P displaced to P'. Letting

$$\overrightarrow{OP} = \underline{x}, \quad \overrightarrow{OP'} = \underline{x}',$$

since

$$|\underline{x}| = |\underline{x}'|,$$

we construct a plane through

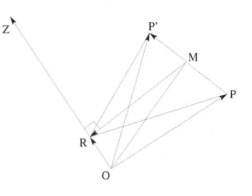

Fig. 87

O and M the midpoint of PP', normal to PP'. Take OZ anywhere in that plane through O and draw MR normal to OZ. Then

$$\overrightarrow{PP'} = \underline{x}' - \underline{x}, \qquad \overrightarrow{OM} = \frac{1}{2}(\underline{x}' + \underline{x})$$

$$\underline{x}' - \underline{x} = \lambda \left[\underline{e} \times \frac{1}{2}(\underline{x}' + \underline{x}) \right]$$

where $\underline{e}$ is a unit vector in OZ. This is so because $\overrightarrow{PP'}$ is normal to $\overrightarrow{OM}$ and $\underline{e}$, both vectors in the plane. We have now to evaluate λ. But

$$|\overrightarrow{PM}| = \frac{1}{2}|\underline{x}' - \underline{x}| = |\overrightarrow{RM}| \tan\left(\frac{\theta}{2}\right), \qquad 0 < \theta < \pi,$$

where θ is the angle PRP'. And also

$$|\underline{x}' - \underline{x}| = \lambda|\overrightarrow{OM}| \sin ROM = \lambda|\overrightarrow{OM}| \frac{|\overrightarrow{RM}|}{|\overrightarrow{OM}|} = \lambda|\overrightarrow{RM}|,$$

and therefore

$$\lambda = 2\tan\left(\frac{\theta}{2}\right),$$

so that we have Rodrigues' Formula

$$\underline{x}' - \underline{x} = 2\left(\tan\frac{\theta}{2}\right)\underline{e} \times \frac{1}{2}(\underline{x}' + \underline{x}), \qquad 0 < \theta < \pi, \tag{1.11}$$

Theorem V.1.2. (Euler's Theorem). Given any displacement of two points P_1 and P_2 of a rigid body about a fixed point O of that body, the displacement can be effected about an axis of rotation through O, the same for all points P of the rigid body. It is necessary that the points be such that

$$(\underline{x}_1' - \underline{x}_1) \neq \lambda(\underline{x}_2' - \underline{x}_2).$$

Proof: The intuitive view is here simple. We know that, for each of the two given points, there is, by Rodrigues' Formula, a plane of such axes. The intersection of these two planes gives the axis common to both. We then have to show that this is the same for all others.

Analytically, we will show there is one and only one $\underline{Z}$ such that

$$\underline{x}_1' - \underline{x}_1 = \underline{Z} \times \frac{1}{2}(\underline{x}_1' + \underline{x}_1), \qquad \underline{x}_2' - \underline{x}_2 = \underline{Z} \times \frac{1}{2}(\underline{x}_2' + \underline{x}_2),$$

$$\underline{x}' - \underline{x} = \underline{Z} \times \frac{1}{2}(\underline{x}' + \underline{x})$$

where these are displacements of a rigid body in pure rotation about a point O.

A. First, we find the $\underline{Z}$ satisfying the first two displacements. If such a $\underline{Z}$ exists, it must be normal both to $\underline{x}_1' - \underline{x}_1$ and $\underline{x}_2' - \underline{x}_2$. Hence, if it exists,

$$\underline{Z} = \lambda(\underline{x}_2' - \underline{x}_2) \times (\underline{x}_1' - \underline{x}_1).$$

Therefore

$$\underline{x}_1' - \underline{x}_1 = \lambda\left[(\underline{x}_2' - \underline{x}_2) \times (\underline{x}_1' - \underline{x}_1)\right] \times \frac{1}{2}\left(\underline{x}_1' + \underline{x}_1\right),$$

$$\underline{x}_2' - \underline{x}_2 = \lambda\left[(\underline{x}_2' - \underline{x}_2) \times (\underline{x}_1' - \underline{x}_1)\right] \times \frac{1}{2}\left(\underline{x}_2' + \underline{x}_2\right).$$

Then, expanding the triple cross products and using (1.8) and (1.9), we get

$$\underline{x}_1' - \underline{x}_1 = \frac{1}{2}\lambda\left[(\underline{x}_2' - \underline{x}_2) \cdot (\underline{x}_1' + \underline{x}_1)\right](\underline{x}_1' - \underline{x}_1),$$

$$\underline{x}_2' - \underline{x}_2 = \frac{1}{2}\lambda\left[-(\underline{x}_2' + \underline{x}_2) \cdot (\underline{x}_1' - \underline{x}_1)\right](\underline{x}_2' - \underline{x}_2),$$

so that it is necessary that

$$\lambda = \frac{2}{(\underline{x}_2' - \underline{x}_2) \cdot (\underline{x}_1' + \underline{x}_1)} = \frac{-2}{(\underline{x}_2' + \underline{x}_2) \cdot (\underline{x}_1' - \underline{x}_1)}$$

which is so, since, by (1.10), for a rigid body in pure rotation, these are equal. Since the steps are reversible, the condition is both necessary and sufficient. Hence

$$\underline{Z} = \lambda\left[(\underline{x}_2' - \underline{x}_2) \times (\underline{x}_1' - \underline{x}_1)\right],$$

$$\lambda = \frac{2}{(\underline{x}_2' - \underline{x}_2) \cdot (\underline{x}_1' + \underline{x}_1)} = \frac{-2}{(\underline{x}_2' + \underline{x}_2) \cdot (\underline{x}_1' - \underline{x}_1)} \tag{1.12}$$

B. We now show that, for any other point $\underline{x}$ rigidly connected with $\underline{x}_1$ and $\underline{x}_2$, such a $\underline{Z}$ exists and conversely.

Suppose $\underline{x}$ represents a point rigidly connected with the body in pure rotation which consists of $\underline{x}_1$ and $\underline{x}_2$. Then

$$(\underline{x}' - \underline{x}) \cdot (\underline{x}' + \underline{x}) = 0$$
$$(\underline{x}' - \underline{x}) \cdot (\underline{x}_1' + \underline{x}_1) = -(\underline{x}' + \underline{x}) \cdot (\underline{x}_1' - \underline{x}_1),$$
$$(\underline{x}' - \underline{x}) \cdot (\underline{x}_2' + \underline{x}_2) = -(\underline{x}' + \underline{x}) \cdot (\underline{x}_2' - \underline{x}_2).$$

But

$$(\underline{x}' - \underline{x}) \cdot (\underline{x}' + \underline{x}) = \lambda \underline{Z} \times (\underline{x}' + \underline{x}) \cdot \frac{1}{2}(\underline{x}' + \underline{x}) = 0 \tag{1.13}$$

and

$$(\underline{x}' - \underline{x}) \cdot (\underline{x}_1' + \underline{x}_1) = \lambda \underline{Z} \times (\underline{x}' + \underline{x}) \cdot \frac{1}{2}(\underline{x}_1' + \underline{x}_1) \tag{1.14}$$

since with one of the values of λ, when expanded,

$$\lambda[(\underline{x}_2' - \underline{x}_2) \times (\underline{x}_1' - \underline{x}_1)] \times (\underline{x}' + \underline{x}) \cdot (\underline{x}_1' + \underline{x}_1) = -(\underline{x}' + \underline{x}) \cdot (\underline{x}_1' - \underline{x}_1),$$

and, with the other value of λ,

$$(\underline{x}' - \underline{x}) \cdot (\underline{x}_2' + \underline{x}_2) = \lambda \underline{Z} \times (\underline{x}' + \underline{x}) \cdot \frac{1}{2}(\underline{x}_2' + \underline{x}_2) \tag{1.15}$$

Putting (1.13), (1.14), and (1.5) together, we have

$$(\underline{x}' + \underline{x}) \cdot [(\underline{x}' - \underline{x}) - \lambda \underline{Z} \times \frac{1}{2}(\underline{x}' + \underline{x})] = 0$$
$$(\underline{x}_1' + \underline{x}_1) \cdot [(\underline{x}' - \underline{x}) - \lambda \underline{Z} \times \frac{1}{2}(\underline{x}' + \underline{x})] = 0$$
$$(\underline{x}_2' + \underline{x}_2) \cdot [(\underline{x}' - \underline{x}) - \lambda \underline{Z} \times \frac{1}{2}(\underline{x}' + \underline{x})] = 0$$

Therefore, if $(\underline{x}' + \underline{x}), (\underline{x}_1' + \underline{x}_1)$, and $(\underline{x}_2' + \underline{x}_2)$ are neither coplanar nor zero-vectors, then

$$(\underline{x}' - \underline{x}) - \lambda \underline{Z} \times \frac{1}{2}(\underline{x}' + x) = \underline{0},$$

since in 3-space it is not possible for a non-zero vector to be normal to three non-coplanar non-zero vectors.

Conversely, if $\underline{x}$ represents point whose displacement is represented by $\underline{x}'$ so that

$$\underline{x}' - \underline{x} = \lambda[(\underline{x}_1' - \underline{x}_1) \times (\underline{x}_2' - \underline{x}_2)] \times \frac{1}{2}(\underline{x}' + \underline{x})$$

where λ has the values of (1.12), then we see, using the values of λ,

$$(\underline{x}' - \underline{x}) \cdot (\underline{x}' + \underline{x}) = 0$$

$$(\underline{x}' - \underline{x}) \cdot (\underline{x}_1' + \underline{x}_1) = \lambda[(\underline{x}_2' - \underline{x}_2) \times (\underline{x}_1' - \underline{x}_1)] \times \frac{1}{2}(\underline{x}' + \underline{x}) \cdot (\underline{x}_1' + \underline{x}_1)$$

$$= \lambda[-\frac{1}{2}(\underline{x}' + \underline{x}) \cdot (\underline{x}_1' - \underline{x}_1)][(\underline{x}_2' - \underline{x}_2) \cdot (\underline{x}_1' + \underline{x}_1)]$$

$$= -(\underline{x}' + \underline{x}) \cdot (\underline{x}_1' - \underline{x}_1),$$

and

$$(\underline{x}' - \underline{x}) \cdot (\underline{x}_2' + \underline{x}_2) = \lambda[(\underline{x}_2' - \underline{x}_2) \times (\underline{x}_1' - \underline{x}_1)] \times \frac{1}{2}(\underline{x}' + \underline{x}) \cdot (\underline{x}_2' + \underline{x}_2)$$

$$= -(\underline{x}' + \underline{x}) \cdot (\underline{x}_2' - \underline{x}_2).$$

The first expresses the fact that the point P_λ represented by $\underline{x}$, is rigidly attached to O. The last two express the fact that P is rigidly connected with P_1 and P_2, represented by $\underline{x}_1$ and $\underline{x}_2$. Hence it is rigidly connected with the rigid body.

Example. Suppose a unit cube is rotated about a fixed axis $00'$ joining two opposite vertices so that it is displaced into itself with the least angle of rotation θ about $00'$. That is, vertex $2 \to$ vertex $2', 1 \to 1'$, and, given this, we shall find the angle θ and the displacements of the other vertices.

Choosing the orthonormal set $\{\underline{e}_i\}$ about 0 as in figure, we have

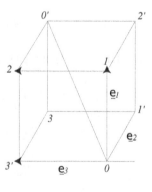

Fig. 88

$$\underline{x}_1 = \underline{e}_1, \quad \underline{x}_1' = \underline{e}_2, \quad \underline{x}_2 = \underline{e}_1 + \underline{e}_3, \quad \underline{x}_2' = \underline{e}_1 + \underline{e}_2.$$

Then, by (1.12),

$$\underline{Z} = \lambda[(\underline{x}_2' - \underline{x}_2) \times (\underline{x}_1' - \underline{x}_1)],$$

$$\lambda = \frac{2}{(\underline{x}_2' - \underline{x}_2) \cdot (\underline{x}_1' + \underline{x}_1)} = \frac{-2}{(\underline{x}_2' + \underline{x}_2) \cdot (\underline{x}_1' - \underline{x}_1)} = 2\left(\tan\frac{\theta}{2}\right)\underline{k}$$

where $\underline{k}$ is unit vector in direction $\overrightarrow{OO'}$. Hence

$$\underline{k} = \frac{1}{3}(\underline{e}_1 + \underline{e}_2 + \underline{e}_3)$$

where the sense will have to be decided from

$$\underline{Z} = \lambda[(\underline{e}_2 - \underline{e}_3) \times (\underline{e}_2 - \underline{e}_1)] = \lambda(\underline{e}_1 + \underline{e}_2 + \underline{e}_3)$$

$$\lambda = \frac{2}{(\underline{e}_2 - \underline{e}_3) \cdot (\underline{e}_1 + \underline{e}_2)} = \frac{-2}{(2\underline{e}_1 + \underline{e}_2 + \underline{e}_3) \cdot (\underline{e}_2 - \underline{e}_1)} = \frac{2}{1}.$$

Hence

$$\underline{Z} = 2(\underline{e}_1 + \underline{e}_2 + \underline{e}_3) = 2\left(\tan\frac{\theta}{2}\right)\frac{1}{\sqrt{3}}(\underline{e}_1 + \underline{e}_2 + \underline{e}_3)$$

$$\tan\frac{\theta}{2} = \sqrt{3}, \qquad \frac{\theta}{2} = \arctan\sqrt{3}, \qquad \theta = 120°.$$

Further, since

$$\underline{x}_3' - \underline{x}_3 = \underline{Z} \times \frac{1}{2}(\underline{x}_3' + \underline{x}_3),$$

$$\underline{x}_3' = \underline{x}_3 + (\underline{e}_1 + \underline{e}_2 + \underline{e}_3) \times (\underline{x}_3' + \underline{x}_3)$$

$$= (\underline{e}_2 + \underline{e}_3) + (\underline{e}_1 + \underline{e}_2 + \underline{e}_3) \times [\underline{x}_3' + (\underline{e}_2 + \underline{e}_3)]$$

which gives us three scalar equations for finding the components of $\underline{x}_3'$. Solving,

$$\underline{x}_3' = \underline{e}_1 + \underline{e}_3,$$

that is, vertex 3 → vertex 2.

It is to be noted that Euler's Theorem shows that all points of the rigid body are displaced about the axis $\underline{Z}$ by the same angle θ, where θ is the angle subtending the displacement in the plane through the original position of point and in displaced position, normal to the axis $\underline{Z}$.

In general, finite displacements do not form an Abelian group. Therefore, in general, the order of two finite displacements cannot be interchanged. But we shall see that, for small displacements of pure rotation which we define as such that $\theta_1\theta_2$ is negligible with respect to θ_1 and θ_2 where θ_1 and θ_2 are the angles of rotation, they are interchangeable. This is important to establish the commutativity of the addition of angular velocities of a rigid body.

Suppose a displacement of pure rotation of any point P of a rigid body

$$\underline{x}' - \underline{x} = \tan\left(\frac{\theta}{2}\right)\underline{k} \times (\underline{x}' + \underline{x})$$

or

$$\underline{x}' - \tan\left(\frac{\theta}{2}\right)\underline{k} \times \underline{x}' = \underline{x} + \tan\left(\frac{\theta}{2}\right)\underline{k} \times \underline{x} \tag{1.16}$$

Then, crossing with $\underline{k}$,

$$\underline{k} \times \underline{x}' - \tan\left(\frac{\theta}{2}\right)\underline{k} \times (\underline{k} \times \underline{x}') = \underline{k} \times \underline{x} + \tan\left(\frac{\theta}{2}\right)\underline{k} \times (\underline{k} \times \underline{x})$$

$$\underline{k} \times \underline{x}' - \tan\left(\frac{\theta}{2}\right)[(\underline{k} \cdot \underline{x}')\underline{k} - \underline{x}'] = \underline{k} \times \underline{x} + \tan\left(\frac{\theta}{2}\right)[(\underline{k} \cdot \underline{x})\underline{k} - \underline{x}] \tag{1.17}$$

But, from (1.16) , we get

$$\underline{k} \cdot \underline{x}' - \underline{k} \cdot \underline{x} = 0,$$

so that we have from (1.17)

$$\underline{k} \times \underline{x}' + \tan\left(\frac{\theta}{2}\right)\underline{x}' = \underline{k} \times \underline{x} - \tan\left(\frac{\theta}{2}\right)\underline{x} + 2(\underline{k} \cdot \underline{x})\tan\left(\frac{\theta}{2}\right)\underline{k} \tag{1.18}$$

Eliminating $\underline{k} \times \underline{x}'$ between (1.16) and (1.18)

$$\underline{x}' + \tan^2\left(\frac{\theta}{2}\right)\underline{x}' = 2\tan\left(\frac{\theta}{2}\right)(\underline{k} \times \underline{x}) + \left[1 - \tan^2\left(\frac{\theta}{2}\right)\right]\underline{x}$$

$$+ 2(\underline{k} \cdot \underline{x})\tan^2\left(\frac{\theta}{2}\right)\underline{k}$$

$$\underline{x}'\left(\frac{1}{\cos^2\frac{\theta}{2}}\right) = \underline{x}\frac{\cos^2\frac{\theta}{2} - \sin^2\frac{\theta}{2}}{\cos^2\frac{\theta}{2}} + 2\frac{\sin\frac{\theta}{2}}{\cos\frac{\theta}{2}}(\underline{k} \times \underline{x}) + 2(\underline{k} \cdot \underline{x})\frac{\sin^2\frac{\theta}{2}}{\cos^2\frac{\theta}{2}}$$

$$\underline{x}' = \underline{x}\cos\theta + (\underline{k} \times \underline{x})\sin\theta + 2(\underline{k} \cdot \underline{x})\sin^2\frac{\theta}{2}\underline{k}.$$

Neglecting θ^2, that is, letting

$$\cos\theta = 1, \qquad \sin\theta = \theta,$$

we have

$$\underline{x}' - \underline{x} = (\underline{k} \times \underline{x})\theta \ldots \tag{1.19}$$

If we then consider two successive small displacements θ_1 and θ_2, given by (1.19), we get

$$\underline{x}'' - \underline{x}' = (\underline{k}_2 \times \underline{x}')\theta_2, \qquad \underline{x}' - \underline{x} = (\underline{k}_1 \times \underline{x})\theta_1.$$

Eliminating $\underline{x}'$,

$$\underline{x}'' - \underline{x} = (\underline{k}_1 \times \underline{x})\theta_1 + (\underline{k}_2 \times \underline{x})\theta_2 + \underline{k}_2 \times (\underline{k}_1 \times \underline{x})\theta_2\theta_1 \qquad (1.20)$$

$$\underline{x}'' - \underline{x} \approx (\underline{k}_1\theta_1 + \underline{k}_2\theta_2) \times \underline{x} \qquad (1.21)$$

so that we see, by (1.20) and (1.21), that the displacements of pure rotation are only permutable if we neglect θ_1^2, θ_2^2, and $\theta_1\theta_2$.

B   The Kinematics of a Rigid Body.

Suppose, as above, that $\underline{x}$ represents any point of a rigid body with one point O, the origin, fixed, and $\underline{x}'$ the position vector of the point after displacement, then, by Euler's Theorem, for all points of the body there exists one and only one $\underline{Z}$ such that

$$\underline{x}' - \underline{x} = \underline{Z} \times \frac{1}{2}(\underline{x}' + \underline{x}),$$

$$\underline{Z} = \lambda[(\underline{x}_2' - \underline{x}_2) \times (\underline{x}_1' - \underline{x}_1)] = 2 \tan\frac{\theta}{2}\underline{k}$$

$$\lambda = \frac{2}{(\underline{x}_2' - \underline{x}_2) \cdot (\underline{x}_1' + \underline{x}_1)} = \frac{-2}{(\underline{x}_2' + \underline{x}_2) \cdot (x_1' - x_1)}$$

for a given displacement of the body determined by $\underline{x}_1, \underline{x}_1', \underline{x}_2, \underline{x}_2'$. Consider the displacement as time-dependent so that we have $\underline{x}$ at t and $\underline{x}'$at t',

$$\underline{x}' - \underline{x} = \Delta\underline{x}, \ t' - t = \Delta t.$$

We suppose also that the displacement follows a smooth curve from $\underline{x}$ to $\underline{x}'$, and we consider different displacements along the curves as $\Delta t \to 0$ and $\underline{x}' - \underline{x} \to \underline{0}$. For these different displacements, we have different values of θ (or $\Delta\theta$ as we shall write it) and different values of $\underline{k}$. Under these assumptions, $\underline{x}' - \underline{x}$ and $\frac{1}{2}(\underline{x}' + \underline{x})$ are continuous vector-functions of t. Hence $\underline{Z}$ is a continuous function of t. Therefore, if

$$\lim_{\Delta t \to 0} \frac{\underline{x}' - \underline{x}}{\Delta t} = \frac{d\underline{x}}{dt}$$

exists, then

$$\lim_{\Delta t \to 0} \frac{\underline{Z} \times \frac{1}{2}(\underline{x}' + \underline{x})}{\Delta t} = \left(\lim_{\Delta t \to 0}\frac{\underline{Z}}{\Delta t}\right) \times \underline{x}$$

exists, and we have also the existence of

$$\lim_{\Delta t \to 0} \frac{Z}{\Delta t} = \lim_{\Delta t \to 0} 2 \frac{\tan \frac{\Delta \theta}{2}}{\Delta t} k = \lim_{\Delta t \to 0} \left(\frac{2 \sin \frac{\Delta \theta}{2}}{\frac{\Delta \theta}{2}} \right) \left(\frac{\frac{\Delta \theta}{2}}{\Delta t} \right) \left(\frac{1}{\cos \frac{\Delta \theta}{2}} k \right)$$

$$\lim_{\Delta t \to 0} \frac{\Delta \theta}{\Delta t} k = \omega.$$

If k is a constant vector, then we have the case of rotation about a fixed axis in the direction k and through O, and

$$\lim_{\Delta t \to 0} \frac{\Delta \theta}{\Delta t} k = \frac{d\theta}{dt} k.$$

The angular velocity vector whose direction is normal to the plane of rotation, sense determined by the right-hand-screw, and whose magnitude is the angular speed of any point about that axis. If k is not a constant vector, then it approaches a limit k which is called the **instantaneous axis** of rotation, and its coefficient which we still designate as $\frac{d\theta}{dt}$, the angular velocity at the time t.[1]

We write, therefore, Poisson's Formula,

$$\frac{dx}{dt} = \omega \times x \tag{2.1}$$

We can immediately generalize this for a displacement of a rigid body with no point fixed. By Theorem V.1.1., we can express the displacement of any points $x \to x'$ by the translation $x'_0 - x_0$ of a given point x_0 plus a rotation of x about that point:

$$x' - x = x'_0 - x_0 + 2 \tan \frac{\Delta \theta}{2} k \times \frac{1}{2} \left[(x' - x'_0) + (x - x_0) \right].$$

Dividing by Δt as before and going to the limit, we have

$$\frac{dx}{dt} = \frac{dx_0}{dt} + \omega_0 \times (x - x_0) \tag{2.2}$$

This naturally raises the question whether the ω_0 would be different for different choices of x_0 as we seem to indicate by our notation. Now we have, according to this notation, for x_1,

$$\frac{dx}{dt} = \frac{dx_1}{dt} + \omega_1 \times (x - x_1) \tag{2.3}$$

---

[1] Anyone desiring a more meticulous proof of the existence of ω should see Milne, Vectorial Mechanics, pp. 163-4.

and, since by Euler's Theorem, with respect to $\underline{x}_0$, all points have the same $\underline{\omega}$, therefore, by (2.2),

$$\frac{d\underline{x}_1}{dt} = \frac{d\underline{x}_0}{dt} + \underline{\omega}_0 \times (\underline{x}_1 - \underline{x}_0) \tag{2.4}$$

so that

$$\frac{d\underline{x}}{dt} = \left[\frac{d\underline{x}_0}{dt} + \underline{\omega}_0 \times (\underline{x}_1 - \underline{x}_0)\right] + \underline{\omega}_1 \times (\underline{x} - \underline{x}_1),$$

and, therefore, using (2.2),

$$\underline{\omega}_0 \times (\underline{x}_1 - \underline{x}_0) + \underline{\omega}_1 \times (\underline{x} - \underline{x}_1) = \underline{\omega}_0 \times (\underline{x} - \underline{x}_0),$$
$$(\underline{\omega}_0 - \underline{\omega}_1) \times (\underline{x} - \underline{x}_1) = \underline{0}.$$

But $(\underline{x} - \underline{x}_1)$ is arbitrary so that

$$\underline{\omega}_0 - \underline{\omega}_1 = \underline{0}.$$

Hence $\underline{\omega}$ is a function of t, but not of $\underline{x}$ nor of $\underline{x}_0$ for a given rigid body. We can simply write $\underline{\omega}$ without any subscript in (2.2).

It follows that, in the case of a rigid body with one point fixed where

$$\frac{d\underline{x}}{dt} = \underline{\omega} \times \underline{x},$$

$$\nabla \times \frac{d\underline{x}}{dt} = \begin{vmatrix} \underline{e}_1 & \underline{e}_2 & \underline{e}_3 \\ \frac{\partial}{\partial x} & \frac{\partial}{\partial y} & \frac{\partial}{\partial z} \\ (\omega_2 z - \omega_3 y) & -(\omega_1 z - \omega_3 x) & (\omega_1 y - \omega_2 x) \end{vmatrix} = 2\underline{\omega}$$

In the case of rotation about a fixed point, $\underline{\omega}$ as we have seen, represents the instantaneous axis of rotation along which at that instant, all points have zero velocity. For, if $\underline{x}$ is parallel to $\underline{\omega}$ through 0,

$$\frac{d\underline{x}}{dt} = \underline{\omega} \times \underline{x} = \underline{0}$$

and the velocity $\frac{d\underline{x}}{dt}$ of any point $\underline{x}$ not on the axis is normal to $\underline{\omega}$. For

$$\frac{d\underline{x}}{dt} \cdot \underline{\omega} = \underline{\omega} \times \underline{x} \cdot \underline{\omega} = 0.$$

But, in the general case of (2.2), there need not ever be such a line of points with zero velocity. The analogous set of points here is those which have no component of their velocities normal to $\underline{\omega}$, that is, whose velocities are parallel to $\underline{\omega}$. To find such a set, we suppose we know some $\underline{x}_0$ and $\frac{d\underline{x}_0}{dt}$. We want to find in terms of these the set of points $\underline{x}$ where

$$\underline{\omega} \times \frac{d\underline{x}}{dt} = \underline{0}.$$

If $\underline{x}$ exists, then

$$\underline{\omega} \times \frac{d\underline{x}}{dt} = \underline{\omega} \times \frac{d\underline{x}_0}{dt} + \underline{\omega} \times [\underline{\omega} \times (\underline{x} - \underline{x}_0)] = \underline{0}$$

$$= \underline{\omega} \times \frac{d\underline{x}_0}{dt} + [\underline{\omega} \cdot (\underline{x} - \underline{x}_0)]\underline{\omega} - \underline{\omega}^2(\underline{x} - \underline{x}_0) = \underline{0}$$

$$\underline{\omega}^2(\underline{x} - \underline{x}_0) = \underline{\omega} \times \frac{d\underline{x}_0}{dt} + [\underline{\omega} \cdot (\underline{x} - \underline{x}_0)]\underline{\omega}$$

$$(\underline{x} - \underline{x}_0) = \left(\underline{\omega} \times \frac{d\underline{x}_0}{dt}\right)/\underline{\omega}^2 + \frac{[\underline{\omega} \cdot (\underline{x} - \underline{x}_0)]}{\underline{\omega}^2}\underline{\omega} \qquad (2.5)$$

In spite of the fact, that we have $\underline{x} - \underline{x}_0$ involved in the right-hand side of (2.5), we shall see that this equation actually gives us the position of all points $\underline{x}$ from $\underline{x}_0$ having the property we want. For, if we substitute (2.5) back in

$$\frac{d\underline{x}}{dt} = \frac{d\underline{x}_0}{dt} + \underline{\omega} \times (\underline{x} - \underline{x}_0)$$

to see if $\frac{d\underline{x}}{dt}$ is actually parallel to $\underline{\omega}$, we get

$$\frac{d\underline{x}}{dt} = \frac{d\underline{x}_0}{dt} + \underline{\omega} \times \frac{1}{\omega^2}\left[\underline{\omega} \times \frac{d\underline{x}_0}{dt} + \lambda\underline{\omega}\right]$$

$$= \frac{d\underline{x}_0}{dt} + \frac{\underline{\omega} \cdot \frac{d\underline{x}_0}{dt}}{\omega^2}\underline{\omega} - \frac{d\underline{x}_0}{dt} + \underline{0}$$

where λ is any arbitrary expression so that

$$\frac{d\underline{x}}{dt} = \frac{\underline{\omega} \cdot \frac{d\underline{x}_0}{dt}}{\omega^2}\underline{\omega} \qquad (2.6)$$

if and only if

$$\underline{x} - \underline{x}_0 = \frac{1}{\omega^2}\left[\underline{\omega} \times \frac{d\underline{x}_0}{dt}\right] + \lambda\underline{\omega}, \quad -\infty < \lambda < \infty \qquad (2.7)$$

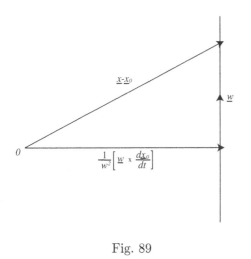

Fig. 89

This last expression represents a straight line of points in the direction $\underline{w}$ through the point given by the vector $\frac{1}{w^2}\left[\underline{w} \times \frac{d\underline{x}_0}{dt}\right]$ from O, the point given by $\underline{x}_0$. Such a straight line of points is called the **instantaneous axis** of the rigid body at that given time. We see, when $\frac{d\underline{x}_0}{dt}$ is normal to w, then $\frac{d\underline{x}}{dt} = \underline{0}$, which is the special case. Such is also the case when the displacements of all the points are parallel to a given plane. For then $\underline{Z}$ is perpendicular to two non-parallel displacements parallel to the same plane so the $\underline{Z}$ is always normal to the plane and therefore so is $\underline{w}$.

Example. Suppose a wheel rolls in a straight line without slipping at constant angular velocity or so that its center moves at a constant linear velocity $\underline{v}_0$ where the center has position-vector $\underline{x}_0$. Since it does not slip, letting a be the radius,

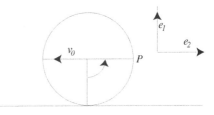

Fig. 90

$$s = a\theta, \quad \frac{ds}{dt} = a\frac{d\theta}{dt}, \quad \underline{v}_0 = -\underline{e}_1 a\frac{d\theta}{dt} = -\underline{e}_1 aw.$$

Now we wish to find $\underline{x}$ such that $\frac{d\underline{x}}{dt} = \underline{0}$, since this is the special case of plane motion and the velocity of all points on the instantaneous axis is zero, But, by (2.7)

$$\underline{x} - \underline{x}_0 = [w\underline{e}_3 \times (-\underline{e}_1 aw)]\frac{1}{w^2} = -a\underline{e}_2 + \lambda w\underline{e}_3.$$

That is, the line through the point of contact normal to the plane of the wheel.

Again to find the velocity of the point P in the figure, by (2.2), $\underline{x}$ is position-vector of P,

$$\frac{d\underline{x}}{dt} = -\underline{e}_1 aw + w\underline{e}_3 \times (a\underline{e}_1)$$

$$= aw \times (-\underline{e}_1 + \underline{e}_2).$$

C   Motion with Respect to Moving Rigid Frames.

Suppose F is a fixed rigid frame with axes $\{\underline{e}_i\}$ with $\overline{F}$ a rigid frame with origin $\underline{x}_{\overline{0}}$ and axes $\{\overline{\underline{e}}_i\}$ moving with respect to F. Let $\underline{x} = \overline{\underline{x}} + \underline{x}_{\overline{0}}$ be the position-vector of any particle moving with respect to $\overline{F}$ in any way. Then:

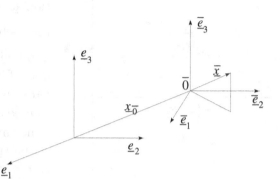

Fig. 91

$$\underline{x} = \underline{x}_{\overline{0}} + \sum_i (\overline{\underline{x}} \cdot \overline{\underline{e}}_i), \ i = 1, 2, 3.$$

and

$$\begin{aligned}
\frac{d\underline{x}}{dt} &= \frac{d\underline{x}_{\overline{0}}}{dt} + \sum_i (\overline{\underline{x}} \cdot \overline{\underline{e}}_i) \frac{d\overline{\underline{e}}_i}{dt} + \sum_i \frac{d(\overline{\underline{x}} \cdot \overline{\underline{e}}_i)}{dt} \overline{\underline{e}}_i, \\
&= \frac{d\underline{x}_{\overline{0}}}{dt} + \sum_i (\overline{\underline{x}} \cdot \overline{\underline{e}}_i) \underline{\omega} \times \overline{\underline{e}}_i + \sum_i \frac{d(\overline{\underline{x}} \cdot \overline{\underline{e}}_i)}{dt} \overline{\underline{e}}_i
\end{aligned} \tag{3.1}$$

since one can consider the position-vector $\underline{x}_{\overline{0}} + \overline{\underline{e}}_i$ of the endpoint of each axis-vector in $\overline{F}$ and apply (2.2). But

$$\sum_i (\overline{\underline{x}} \cdot \overline{\underline{e}}_i) \underline{\omega} \times \overline{\underline{e}}_i = \sum_i \underline{\omega} \times (\overline{\underline{x}} \cdot \overline{\underline{e}}_i) \overline{\underline{e}}_i = \underline{\omega} \times \overline{\underline{x}}$$

where $\underline{\omega}$ is, of course, the angular velocity of $\overline{F}$ with respect to the fixed frame F. Moreover, we can easily interpret the last term in (3.1), for $\sum_i \dfrac{d(\overline{\underline{x}} \cdot \overline{\underline{e}}_i)}{dt} \overline{\underline{e}}_i$ is the velocity of $\overline{\underline{x}}$ relative to $\overline{F}$, that is, as seen by an observer fixed in $\overline{F}$. We write it $\dfrac{d\overline{\underline{x}}}{dt}$[2] and we have

$$\frac{d\underline{x}}{dt} = \frac{d\underline{x}_{\overline{0}}}{dt} + \underline{\omega} \times \overline{\underline{x}} + \frac{\overline{d\underline{x}}}{dt}, \tag{3.2}$$

$$= \frac{d\underline{x}_{\overline{0}}}{dt} + \left[\underline{\omega} \times (\cdot) + \frac{\overline{d}(\cdot)}{dt} \right] \overline{\underline{x}} \tag{3.3}$$

---

[2]In many texts it is written $\frac{\partial \underline{x}}{\partial t}$ which we avoid for obvious reasons.

In the last form, we are emphasizing the operator $\underline{\omega} \times (\cdot) + \frac{\overline{d}(\cdot)}{dt}$ which will be used in the sequel. It can be applied to any free vector in $\overline{F}$. Hence, when we differentiate any vector without regard to origin, that is, as in a vector space, we can analyze it in relation to F and $\overline{F}$ as in (3.3) by first applying this operator since the first term $\frac{d x_{\overline{0}}}{dt}$ will obviously not appear for such a vector. Thus, for any vector $\underline{P}$,

$$\frac{d\underline{P}}{dt} = \underline{\omega} \times \underline{P} + \frac{\overline{d}\underline{P}}{dt} \tag{3.4}$$

where $\frac{d\underline{P}}{dt}$ is the rate of change of $\underline{P}$ with respect to F and $\frac{\overline{d}\underline{P}}{dt}$, the rate of change with respect to $\overline{F}$. In particular,

$$\frac{d\underline{\omega}}{dt} = \underline{\omega} \times \underline{\omega} + \frac{\overline{d}\underline{\omega}}{dt} = \frac{\overline{d}\underline{\omega}}{dt} \tag{3.5}$$

an important result.

Again suppose $\overline{F}$ rotates with respect to F at $\underline{\omega}$ as before but with its origin fixed at the origin of F, while $\overline{\underline{x}}$ is a position-vector fixed in $\overline{\overline{F}}$ which rotates at $\overline{\underline{\omega}}$ with respect to $\overline{F}$ and with its origin also at origin of F. Then, since $\underline{x}_0 = \underline{0}$,

$$\frac{d\overline{\underline{x}}}{dt} = \underline{\omega} \times \overline{\underline{x}} + \frac{\overline{d}\overline{\underline{x}}}{dt},$$

but

$$\frac{\overline{d}\overline{\underline{x}}}{dt} = \overline{\underline{\omega}} \times \overline{\underline{x}},$$

and

$$\frac{d\overline{\underline{x}}}{dt} = \underline{\omega} \times \overline{\underline{x}} + \underline{\omega} \times \overline{\underline{x}} = (\underline{\omega} + \overline{\underline{\omega}}) \times \overline{\underline{x}} \tag{3.6}$$
$$= (\overline{\underline{\omega}} + \underline{\omega}) \times \overline{\underline{x}}$$

since we saw that small displacements are commutative. We can, therefore, add angular velocities on a rigid body as ordinary vectors. Here, of course, $\underline{x}$ and $\overline{\underline{x}}$ coincide.

Example.[3] A rough sphere of radius a is pressed between two parallel planes which are rotating at constant angular velocities $\underline{\omega}_1$ and $\underline{\omega}_2$ normal to the planes, but coincident. We wish to find the velocity of C, the center of the sphere.

---

[3] Cf. Milne, **Vectorial Mechanics**, pp. 190-191.

The roughness of the sphere means that the velocities of boards at A and B are equal to the velocities of the sphere at A and B respectively. Hence, taking position-vectors from a fixed origin O, by (2.1) and (2.2),

$$\omega_1 \underline{e} \times (\underline{x}_A - \underline{x}_1) = \frac{d\underline{x}_C}{dt} + \underline{\Omega} \times (\underline{x}_A - \underline{x}_C) \tag{3.7}$$

$$\omega_2 \underline{e} \times (\underline{x}_B - \underline{x}_2) = \frac{d\underline{x}_C}{dt} + \underline{\Omega} \times (\underline{x}_B - \underline{x}_C) \tag{3.8}$$

where $\underline{x}_A, \underline{x}_B, \underline{x}_C$ are the position-vectors of the two points of contact and the center of the sphere respectively, $\underline{x}_1, \underline{x}_2$ are the position-vectors of the fixed points of the two boards respectively, and $\underline{\Omega}$ is the angular velocity of the sphere. The vector $\underline{e}$ is, of course, normal to the boards.

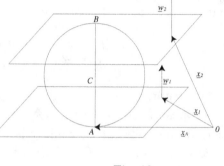

Fig. 92

Adding the two equations, we eliminate Ω, since

$$\underline{x}_A - \underline{x}_C = -\underline{e}a, \qquad \underline{x}_B - \underline{x}_C = \underline{e}a,$$

we get

$$\frac{d\underline{x}_C}{dt} = \frac{1}{2}[\omega_1 \underline{e} \times \underline{x}_A + \omega_2 \underline{e} \times \underline{x}_B - (\omega_1 \underline{e} \times \underline{x}_1 + \omega_2 \underline{e} \times \underline{x}_2)].$$

But

$$\underline{e} \times \underline{x}_A = \underline{e} \times \underline{x}_B = \underline{e} \times \underline{x}_C$$

so that

$$\frac{d\underline{x}_C}{dt} = \frac{1}{2}\left[(\omega_1 + \omega_2)\underline{e} \times \left\{\underline{x}_C - \frac{\omega_1\underline{x}_1 + \omega_2\underline{x}_2}{\omega_1 + \omega_2}\right\}\right].$$

If now we suppose $\underline{\bar{x}}_A, \underline{\bar{x}}_B, \underline{\bar{x}}_C, \underline{\bar{x}}_1$, and $\underline{\bar{x}}_2$ are the vector components of $\underline{x}_A$ etc. parallel to the plane of the boards, then

$$\frac{d\underline{\bar{x}}_C}{dt} = \frac{1}{2}\left[(\omega_1 + \omega_2)\underline{e} \times \left\{\underline{\bar{x}}_C - \frac{\omega_1\underline{\bar{x}}_1 + \omega_2\underline{\bar{x}}_2}{\omega_1 + \omega_2}\right\}\right];$$

and, since

$$\frac{d\left(\frac{\omega_1\underline{\bar{x}}_1 + \omega_2\underline{\bar{x}}_2}{\omega_1 + \omega_2}\right)}{dt} = \underline{0},$$

we can write

$$\frac{d\bar{x}_C}{dt} = \frac{d}{dt}\left[\frac{\omega_1\bar{x}_1 + \omega_2\bar{x}_2}{\omega_1 + \omega_2}\right] + \frac{1}{2}\left[(\omega_1 + \omega_2)\underline{e} \times \left\{\bar{x}_C - \frac{\omega_1\bar{x}_1 + \omega_2\bar{x}_2}{\omega_1 + \omega_2}\right\}\right]$$

which is of the general form

$$\frac{d\bar{x}_C}{dt} = \frac{d\bar{x}_O}{dt} + \underline{\omega} \times [\bar{x}_C - \bar{x}_O]$$

where

$$\bar{x}_O = \frac{\omega_1\bar{x}_1 + \omega_2\bar{x}_2}{\omega_1 + \omega_2}, \qquad \underline{\omega} = \frac{1}{2}(\omega_1 + \omega_2)\underline{e}.$$

Hence the velocity of C corresponds to an angular velocity $\frac{1}{2}(\omega_1 + \omega_2)\underline{e}$ of C about a fixed point in the plane through C parallel to the boards which has the position-vector $\frac{\omega_1\bar{x}_1 + \omega_2\bar{x}_2}{\omega_1 + \omega_2}$ from $\overline{O}$, the normal projection of O on that plane.

Subtracting, on the other hand, (3.7) from (3.8), we get

$$\omega_2\underline{e} \times (\bar{x}_C - \bar{x}_2) - \omega_1\underline{e} \times (\bar{x}_C - \bar{x}_1) = 2\underline{\Omega} \times \underline{e}a.$$

Crossing with $\underline{e}$, since

$$\underline{e} \cdot (\bar{x}_C - \bar{x}_2) = \underline{e} \cdot (\bar{x}_C - \bar{x}_1) = 0,$$

we get

$$\omega_2(\bar{x}_C - \bar{x}_2) - \omega_1(\bar{x}_C - \bar{x}_1) = 2a\underline{\Omega} - 2(\underline{\Omega} \cdot \underline{e})a\underline{e}$$

which shows that $\underline{\Omega}$ has the completely determined vector-component parallel to the plane given by the left-hand side divided by $2a$ and an indeterminate component normal to the plane, $(\underline{\Omega} \cdot \underline{e})\underline{e}$.

We can now pass from the velocity of a particle with respect to moving frames immediately to their accelerations. For, from (3.2),

$$\frac{d^2x}{dt^2} = \frac{d}{dt}\left(\frac{dx_{\overline{O}}}{dt}\right) + \frac{d}{dt}\left(\underline{\omega} \times \bar{x} + \frac{d\bar{x}}{dt}\right)$$

$$= \frac{d^2x_{\overline{O}}}{dt^2} + \left[\underline{\omega}() + \frac{\overline{d}()}{dt}\right]\left(\underline{\omega} \times \bar{x} + \frac{d\bar{x}}{dt}\right)$$

$$= \frac{d^2x_{\overline{O}}}{dt^2} + \underline{\omega} \times (\underline{\omega} \times \bar{x}) + \underline{\omega} \times \frac{\overline{d\bar{x}}}{dt} + \frac{\overline{d\underline{\omega}}}{dt} \times \bar{x} + \underline{\omega} \times \frac{\overline{d\bar{x}}}{dt} + \frac{\overline{d}^2\bar{x}}{dt^2}$$

$$= \frac{d^2x_{\overline{O}}}{dt^2} + \underline{\omega} \times (\underline{\omega} \times \bar{x}) + \frac{d\underline{\omega}}{dt} \times \bar{x} + 2\underline{\omega} \times \frac{\overline{d\bar{x}}}{dt} + \frac{\overline{d}^2\bar{x}}{dt^2} \qquad (3.9)$$

using (3.3) and (3.5). It is clear that the acceleration of a point $\underline{x}$ **fixed** in $\overline{F}$, that is, momentarily coinciding with $\underline{x}$ in $\overline{F}$, is

$$\frac{d^2 \underline{x_{\overline{O}}}}{dt^2} + \underline{\omega} \times (\underline{\omega} \times \underline{x}) + \frac{d\underline{\omega}}{dt} \times \underline{x}$$

where $\frac{d^2 \underline{x_{\overline{O}}}}{dt^2}$ is the acceleration of the origin $\overline{O}$. It is also obvious that $\frac{\overline{d}^2 \underline{x}}{dt^2}$ is the acceleration of $\underline{x}$ with respect to an observer fixed in $\overline{F}$. The surprising term is what is called the **Coriolis acceleration**

$$2\underline{\omega} \times \frac{\overline{d}\underline{x}}{dt}.$$

Example. (1) Suppose a vertical circle of radius a rotates at constant angular velocity $\underline{\omega}$ about its vertical diameter, and $\overline{e}_2$ and $\overline{e}_3$ are taken in the plane of the circle as in the figure with $\overline{e}_1$ normal to them. Then $\underline{\omega} = \omega\overline{e}_3$ is fixed in absolute space as well as in the plane of the circle. A particle of mass m is free to move under gravity without friction on the circle. Let φ be the angle the radius to m makes with the vertical from the bottom. Here $\overline{O}$ coincides with O and $\underline{x}$ with $\underline{x}$. Hence (3.9) becomes

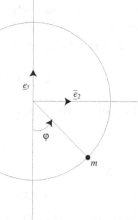

$$\frac{d^2 \underline{x}}{dt^2} = \underline{\omega} \times (\underline{\omega} \times \underline{x}) + 2\underline{\omega} \times \frac{\overline{d}\underline{x}}{dt} + \frac{\overline{d}^2 \underline{x}}{dt^2},$$

Fig. 93

where

$$\underline{x} = -a\cos\varphi\overline{e}_3 + a\sin\varphi\overline{e}_2$$

$$\frac{\overline{d}\underline{x}}{dt} = (+a\sin\varphi\overline{e}_3 + a\cos\varphi\overline{e}_2)\dot{\varphi},$$

$$\frac{\overline{d}^2 \underline{x}}{dt^2} = (+a\cos\varphi\overline{e}_3 - a\sin\varphi\overline{e}_2)(\dot{\varphi})^2$$
$$+ (a\sin\varphi\overline{e}_3 + a\cos\varphi\overline{e}_2)\ddot{\varphi},$$

$$\underline{\omega} \times (\underline{\omega} \times \underline{x}) = -a\omega^2\sin\varphi\overline{e}_2,$$

$$\underline{\omega} \times \frac{\overline{d}\underline{x}}{dt} = -a\omega\dot{\varphi}\cos\varphi\overline{e}_1.$$

By Newton's Axiom,

$$\underline{P} - mg\bar{\underline{e}}_3 = -maw^2 \sin \varphi \bar{\underline{e}}_2 - 2mwa\dot{\varphi} \cos \varphi \bar{\underline{e}}_1$$
$$+ ma \left[(\cos \varphi(\dot{\varphi})^2 + \sin \varphi \ddot{\varphi} \right] \bar{\underline{e}}_3 + ma[-\sin \varphi(\dot{\varphi})^2 + \cos \varphi \ddot{\varphi}] \bar{\underline{e}}_2$$

where $\underline{P}$ is the force of constraint normal to the curve. Hence

$$\underline{P} = P_1 \bar{\underline{e}}_1 - \lambda \bar{\underline{x}} = P_1 \bar{\underline{e}}_1 - \lambda(-a \cos \varphi \bar{\underline{e}}_3 + a \sin \varphi \bar{\underline{e}}_2)$$

We get three equations

$$P_1 = -2maw\dot{\varphi} \cos \varphi, \tag{3.10}$$
$$-\lambda a \sin \varphi = m[-aw^2 \sin \varphi - a \sin \varphi(\dot{\varphi})^2 + a \cos \varphi \ddot{\varphi}] \tag{3.11}$$
$$\lambda a \cos \varphi = [g + a \cos \varphi(\dot{\varphi})^2 + a \sin \varphi \ddot{\varphi}]m \tag{3.12}$$

We can eliminate λ and $(\dot{\varphi})^2$ together from the last two equations by multiplying the first by $\cos \varphi$ and the second by $\sin \varphi$, and adding:

$$0 = g \sin \varphi - aw^2 \sin \varphi \cos \varphi + a\ddot{\varphi} \tag{3.13}$$

We see that, corresponding to the Coriolis acceleration, we have the Coriolis force

$$P_1 \bar{\underline{e}}_1 = -2maw\dot{\varphi}(\cos \varphi)\bar{\underline{e}}_1$$

the force normal to the plane of the circle which keeps it on the circle. We can find the other component of $\underline{P}$ by evaluating λ in terms of $\cos \varphi$ and $\sin \varphi$ by substituting in (3.12) and (3.11) for $\ddot{\varphi}$ and $(\dot{\varphi})^2$ from (3.13).

In a later chapter, we shall solve this same problem in another and simpler way.

Example. (2) To find the set of smooth curves fixed in a rigid frame rotated about a fixed vertical axis at constant angular velocity $\underline{w}$ such that a particle of mass m free to move without friction under gravity on anyone of these curves, will remain at rest relative to the moving frame at any point of the curve. Here we have, taking $\overline{O} = O$ on the vertical axis,

$$m\frac{d^2\bar{\underline{x}}}{dt^2} = m\underline{w} \times (\underline{w} \times \bar{\underline{x}}) = mg + \underline{P} \tag{3.14}$$

since

$$\frac{d\bar{\underline{x}}}{dt} = \underline{0} = \frac{\bar{d}^2\bar{\underline{x}}}{dt^2},$$

where $\underline{P}$ is normal to the tangent to the curve. Suppose

$$\overline{\underline{x}} = \overline{x}(\sigma)\overline{\underline{e}}_1 + \overline{y}(\sigma)\overline{\underline{e}}_2 + \overline{z}(\sigma)\overline{\underline{e}}_3,$$

$$\frac{\overline{d\underline{x}}}{d\sigma} = \frac{d\overline{x}}{d\sigma}\overline{\underline{e}}_1 + \frac{d\overline{y}}{d\sigma}\overline{\underline{e}}_2 + \frac{d\overline{z}}{d\sigma}\overline{\underline{e}}_3,$$

then

$$\underline{P} \cdot \frac{\overline{d\underline{x}}}{d\sigma} = 0.$$

But

$$m\underline{\omega} \times (\underline{\omega} \times \overline{\underline{x}}) = m\omega\overline{\underline{e}}_3 \times (\omega\overline{\underline{e}}_3 \times \overline{\underline{x}}) = -m\omega^2(\overline{x}\,\overline{\underline{e}}_1 + \overline{y}\,\overline{\underline{e}}_2).$$

Therefore, dotting (3.15) with $\dfrac{\overline{d\underline{x}}}{d\sigma}$, we have

$$0 = g\frac{d\overline{z}}{d\sigma} - \omega^2\left(\overline{x}\frac{d\overline{x}}{d\sigma} + \overline{y}\frac{d\overline{y}}{d\sigma}\right),$$

$$C = g\overline{z} - \frac{\omega^2}{2}(\overline{x}^2 + \overline{y}^2)$$

which is a paraboloid of revolution. In other words, such a curve must lie on a paraboloid of revolution with the fixed vertical axis of rotation as its axis. This corresponds to the well-known bucket-of-water experiment where, if a bucket of water is rotated rapidly the water whirls finally with the bucket, rising in the shape of a paraboloid of revolution. Newton argued that in this experiment we have a privileged case where we can actually observe an absolute motion since in a purely relative motion there would be no reason (from point of view of Newton's system) for the water to rise in a paraboloid. The proponents of General Relativity, however, allege the possible influence of the rotation of the stars with respect to the bucket.

D   Applications of Relative Accelerations to the Rotating Earth

If $\overline{F}$ moves at constant linear velocity with respect to F, then $\underline{\omega} = \underline{0}$, $\dfrac{d^2\underline{x}_O}{dt^2} = \underline{0}$, and (3.9) becomes

$$\frac{d^2\underline{x}}{dt^2} = \frac{\overline{d}^2\overline{\underline{x}}}{dt^2},$$

that is, the absolute acceleration and the relative acceleration are equivalent. This means that the laws of force of Newton are the same for any frame of reference moving at a constant velocity with respect to the absolute as they are for the absolute frame.

If further we take $\overline{F}$ as a sphere rotating at constant velocity about its absolutely fixed center, and $\overline{x}$ to a point fixed in the surface of that sphere, then (3.9) becomes

$$\frac{d^2\overline{x}}{dt^2} = \underline{\omega} \times (\underline{\omega} \times \overline{x}) = w\underline{e} \times (w\underline{e} \times \overline{x})$$
$$= w^2\underline{e} \times (\underline{e} \times \overline{x}) \qquad (4.1)$$

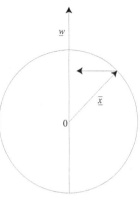

Fig. 94

is a vector pointing to the axis and normal to it.

Now suppose a rotating sphere as before but let $\overline{O}$ be on the surface of the sphere and let $\{\overline{e}_i\}$ be a set of axes on the surface of the sphere, $\overline{e}_3$ outward along the radius, $\overline{e}_1$ tangent to the great circle through the two poles pointing to the South Pole as in the figure. Then using (3.9) and (4.1), we have

$$\frac{\overline{d}^2\overline{x}}{dt^2} = \frac{d^2x}{dt^2} - \underline{\omega} \times (\underline{\omega} \times x_{\overline{O}}) - 2\underline{\omega} \times \frac{\overline{dx}}{dt} - \underline{\omega} \times (\underline{\omega} \times \overline{x}) \qquad (4.2)$$

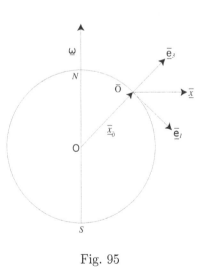

Fig. 95

If we now suppose a particle of mass m suspended at $\overline{O}$ by a string, and the string is cut, then the particle at rest is subject to the force

$$m\underline{g} = m\frac{dx^2}{dt^2}, \qquad \underline{g} = -g\overline{e}_3,$$

assuming what we will prove later, that the force exerted by a homogeneous sphere of mass M acts as if the whole mass were concentrated in the center. The same is true if the density varies as some function of the distance from the center. Then (4.2) becomes

$$\frac{\overline{d}^2\overline{x}}{dt^2} = \underline{g} - \underline{\omega} \times (\underline{\omega} \times \overline{x}_{\overline{0}}) = \overline{g}, \qquad (4.3)$$

where $\overline{g}$ is the acceleration as judged by an observer fixed on the surface of the rotating sphere. The case of the rotating Earth approximates that of the sphere whose density varies as a function of the distance from the center where we neglect the motion of the Earth about the sun, since for a short space of time it can be considered as moving at a constant velocity with respect to the system of the fixed stars, so large is its orbit relative to its rotation on its axis. Hence the plumb line falls, not along $\overline{O}O$ in the figure, but dips in the northern hemisphere towards S.

Example. Suppose a train moving north at $x_{\overline{O}} = \underline{0}$ with velocity, relative to observer fixed at $\overline{O}$, of $\dfrac{\overline{d\underline{x}}}{dt}$. Then we have from (4.2) and (4.3)

$$\frac{\overline{d^2\underline{x}}}{dt^2} = \overline{\underline{g}} - 2\underline{\omega} \times \frac{\overline{d\underline{x}}}{dt}$$

which means that the rails must exert a force equal to $m \times \left(\overline{\underline{g}} - 2\underline{\omega} \times \dfrac{\overline{d\underline{x}}}{dt} \right)$ on the train to keep it going towards the North and to keep it from sinking into the Earth. The force $2m\underline{\omega} \times \dfrac{\overline{d\underline{x}}}{dt}$ acts in the direction from out of the plane of the paper if $\dfrac{\overline{d\underline{x}}}{dt}$ is taken in sense $-\overline{e}_1$. This is the well-known effect of the right-hand rail's wearing faster than the other in the northern hemisphere; the same effect appears in river banks.

Example of Free Fall. Suppose a particle is thrown on the surface of the rotating sphere (Earth) with initial position $\overline{\underline{x}}_1$ and initial velocity $\overline{\underline{v}}_1$ with respect to the axes at rest on the surface of the sphere. We suppose the area of projection to be small enough so that the force with respect to frame fixed on surface is constantly $\overline{\underline{g}}$ and we shall always neglect $\underline{\omega} \times (\underline{\omega} \times \overline{\underline{x}})$ which is always very small for the rotation of the Earth. But we are not neglecting $\underline{\omega} \times (\underline{\omega} \times \overline{\underline{x}}_0)$ which is included in $\overline{\underline{g}}$. Then we have

$$\frac{d^2\underline{x}}{dt^2} = \underline{\omega} \times (\underline{\omega} \times x_{\overline{0}}) + 2\underline{\omega} \times \frac{\overline{d\underline{x}}}{dt} + \frac{\overline{d^2\underline{x}}}{dt^2}$$

$$\underline{g} = \underline{\omega} \times (\underline{\omega} \times x_{\overline{0}}) + 2\underline{\omega} \times \frac{\overline{d\underline{x}}}{dt} + \frac{\overline{d^2\underline{x}}}{dt^2}$$

$$\overline{\underline{g}} = 2\underline{\omega} \times \frac{\overline{d\underline{x}}}{dt} + \frac{\overline{d^2\underline{x}}}{dt^2} \tag{4.4}$$

Now $\overline{\underline{g}}$ is a constant in frame $\overline{F}$; and $\underline{\omega}$ is given as constant in frame F, so then is it in $\underline{F}$ since

$$\frac{d\underline{\omega}}{dt} = \frac{\overline{d\underline{\omega}}}{dt}.$$

Therefore we integrate in $\overline{F}$ with respect to t,

$$\underline{g}t = 2\underline{\omega} \times \overline{\underline{x}} + \frac{d\overline{\underline{x}}}{dt} + \underline{C} \tag{4.5}$$

where $\underline{C}$ is constant with respect to $\overline{F}$. Using initial conditions,

$$\underline{0} = 2\underline{\omega} \times \overline{\underline{x}}_1 + \overline{\underline{v}}_1 + \underline{C},$$
$$\underline{C} = -2\underline{\omega} \times \overline{\underline{x}}_1 - \overline{\underline{v}}_1.$$

Substituting for $\frac{d\overline{\underline{x}}}{dt}$ in (4.4) from (4.5)

$$\underline{g} = 2\underline{\omega} \times [\underline{g}t - 2\underline{\omega} \times \overline{\underline{x}} - \underline{C}] + \frac{d^2\overline{\underline{x}}}{dt^2}$$

Integrating again in $\overline{F}$, and neglecting again $\underline{\omega} \times (\underline{\omega} \times \overline{\underline{x}})$,

$$\underline{g}t = \underline{\omega} \times \underline{g}t^2 - 2\underline{\omega} \times \underline{C}t + \frac{d\overline{\underline{x}}}{dt} + \underline{C}_1,$$
$$\underline{C}_1 = \overline{\underline{v}}_1.$$

Integrating again,

$$\frac{1}{2}\underline{g}t^2 = \underline{\omega} \times \underline{g}\frac{t^3}{3} - \underline{\omega} \times \underline{C}\frac{t^2}{2} + \overline{\underline{x}} + \underline{C}_1 t + \underline{C}_2$$

$$\overline{\underline{x}} = \frac{1}{2}\underline{g}t^2 + \overline{\underline{v}}_1 t + \overline{\underline{x}}_1 - \underline{\omega} \times \left[\underline{g}\frac{t^3}{3} - \overline{\underline{v}}_1\frac{t^2}{2}\right] \tag{4.6}$$

Larmor Precession. Suppose a particle of mass m moving under a force $\underline{f}$ with respect to F, a fixed frame. Now let m be perturbed by a small force $\alpha\frac{d\underline{x}}{dt} \times \underline{e}$, where $\underline{e}$ is a constant unit vector so that

$$m\frac{d^2\underline{x}}{dt^2} = \underline{f} + \alpha\frac{d\underline{x}}{dt} \times \underline{e} \tag{4.7}$$

Suppose $\overline{F}$ moving with angular velocity $\underline{\omega} = \omega\underline{e}$; let $\underline{x} = \overline{\underline{x}}$, that is, let the origin of $\overline{F}$ be fixed at origin of F. Then (4.7) becomes

$$m\underline{\omega} \times (\underline{\omega} \times \underline{x}) + 2m\underline{\omega} \times \frac{d\overline{\underline{x}}}{dt} + m\frac{d^2\overline{\underline{x}}}{dt^2} = \underline{f} + \alpha\left(\frac{d\overline{\underline{x}}}{dt} + \underline{\omega} \times \underline{x}\right) \times \underline{e},$$

$$m\frac{d^2\overline{\underline{x}}}{dt^2} = -(2m\omega + \alpha)\underline{e} \times \frac{d\overline{\underline{x}}}{dt} - (m\omega^2 + \alpha\omega)(\underline{e} \times (\underline{e} \times \underline{x})) + \underline{f}.$$

We can eliminate the $\underline{e} \times \dfrac{\overline{d}\underline{x}}{dt}$ term by choosing

$$\omega = -\frac{\alpha}{2m}$$

so that

$$m\frac{\overline{d}^2\underline{x}}{dt^2} = +\frac{\alpha^2}{4m}\underline{e} \times (\underline{e} \times \underline{x}) + \underline{f}.$$

The first term on the right is a force directed to the axis $\underline{e}$ as seen above. If the force $\underline{f}$ rotates with the frame $\overline{F}$, we can see that the addition of the force of perturbation causes a precession about $\underline{e}$ of the whole system at an angular velocity $-\frac{\alpha}{2m}$ with the addition of a centripetal force. Larmor applied this to the perturbation of electrons by a magnetic field (see Milne, **Vectorial Mechanics**, pp.253-55.

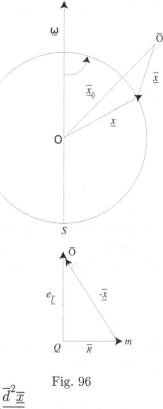

The Foucault Pendulum. We consider a long pendulum of length ℓ swinging through a small angle at the surface of the rotating Earth. The point from which the pendulum hangs is taken as $\overline{O}$, the origin of frame $\overline{F}$, the rotating Earth. Then from $\overline{O}$ to the pendulum bob of mass m is $\overline{\underline{x}}$ and from O to m is $\underline{x}$. O will not be exactly the center of the Earth for reasons we have seen. $O\overline{O}$ will be in the direction of $\overline{g}$. In detail, we have

$$-\overline{\underline{x}} = \ell\overline{\underline{n}}, \qquad \overrightarrow{Q\overline{O}} = \ell\underline{e}, \qquad \underline{\omega} = \omega\underline{e}$$

and the vector from Q to m is taken as $\overline{\underline{r}}$ where we assume the angle is so small the $\overline{\underline{r}}$ is always normal to $\ell\underline{e}$. Hence

$$\overline{\underline{r}} = \ell\underline{e} - \ell\underline{n}$$

and

$$\overline{\underline{r}} \cdot \overline{\underline{e}} = 0.$$

By Axiom 2, neglecting $\underline{\omega} \times (\underline{\omega} \times \overline{\underline{x}})$,

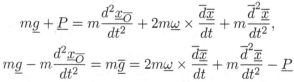

Fig. 96

$$m\underline{g} + \underline{P} = m\frac{d^2\underline{x_{\overline{O}}}}{dt^2} + 2m\underline{\omega} \times \frac{\overline{d}\overline{\underline{x}}}{dt} + m\frac{\overline{d}^2\overline{\underline{x}}}{dt^2},$$

$$m\underline{g} - m\frac{d^2\underline{x_{\overline{O}}}}{dt^2} = m\overline{g} = 2m\underline{\omega} \times \frac{\overline{d}\overline{\underline{x}}}{dt} + m\frac{\overline{d}^2\overline{\underline{x}}}{dt^2} - \underline{P}$$

where $\underline{P}$ is force of constraint in direction $\underline{\bar{n}}$. Hence

$$-m\overline{g}\underline{e} + P\underline{\bar{n}} = 2mw\underline{e} \times \frac{\overline{d}\underline{x}}{dt} + m\frac{\overline{d}^2\underline{x}}{dt^2}.$$

But

$$\underline{x} = -\ell\underline{\bar{n}} = \overline{r} - \ell\underline{\bar{e}} \tag{4.8}$$

$$\frac{\overline{d}\underline{x}}{dt} = \frac{\overline{d}\overline{r}}{dt}, \quad \frac{\overline{d}^2\underline{x}}{dt^2} = \frac{\overline{d}^2\overline{r}}{dt^2}$$

since $\underline{\bar{e}}$ is constant with respect to $\frac{\overline{d}}{dt}$. Hence

$$-\overline{g}\underline{e} + \frac{P}{m}\underline{\bar{n}} = 2w\underline{e} \times \frac{\overline{d}\overline{r}}{dt} + \frac{\overline{d}^2\overline{r}}{dt^2} \tag{4.9}$$

and to eliminate P, we cross (4.9) with $\underline{\bar{n}}$ so that

$$-\overline{g}\underline{e} \times \underline{\bar{n}} = 2w\left(\underline{e} \times \frac{\overline{d}\overline{r}}{dt}\right) \times \underline{\bar{n}} + \frac{\overline{d}^2\overline{r}}{dt^2} \times \underline{\bar{n}} \tag{4.10}$$

But, from (4.8)

$$\underline{\bar{n}} = \underline{\bar{e}} - \frac{1}{\ell}\overline{r}$$

so that (4.10) becomes

$$\frac{\overline{g}}{\ell}\underline{\bar{e}} \times \overline{r} = 2w\left[(\underline{\bar{e}} \cdot \underline{e})\frac{\overline{d}\overline{r}}{dt} - \left(\underline{e} \cdot \frac{\overline{d}\overline{r}}{dt}\right)\underline{e} + \frac{1}{\ell}\left(\overline{r} \cdot \frac{\overline{d}\overline{r}}{dt}\right)\underline{e} - \frac{1}{\ell}(\overline{r} \cdot \underline{e})\frac{\overline{d}\overline{r}}{dt}\right]$$
$$+ \frac{\overline{d}^2\overline{r}}{dt^2} \times \underline{\bar{e}} - \frac{1}{\ell}\frac{\overline{d}^2\overline{r}}{dt^2} \times \overline{r}$$

We cross this equation with $\underline{\bar{e}}$, and expand the triple products to get

$$-\frac{\overline{g}}{\ell}\overline{r} = 2w\left[(\underline{\bar{e}} \cdot \underline{e}) - \frac{1}{\ell}(\overline{r} \cdot \underline{e})\right]\underline{\bar{e}} \times \frac{\overline{d}\overline{r}}{dt} + \frac{2w}{\ell}\left(\overline{r} \cdot \frac{\overline{d}\overline{r}}{dt}\right)\underline{\bar{e}} \times \underline{e} + \frac{\overline{d}^2\overline{r}}{dt^2}.$$

We shall now drop two terms: (1) the term $\frac{2w}{\ell}(\overline{r} \cdot \underline{e})\underline{\bar{e}} \times \frac{\overline{d}\overline{r}}{dt}$ since w is small, ℓ large, $|\overline{r}|$ very small, and $\left|\frac{\overline{d}\overline{r}}{dt}\right|$ very small. For we assumed $\overline{r}$ is normal to $\underline{\bar{e}}$ so that

$$\left|\frac{\overline{d}\overline{r}}{dt}\right| = [2\overline{g}\overline{h}]^{\frac{1}{2}}$$

where $\bar{h}$ is negligible. (2) the term $\frac{2\omega}{\ell}\left(\bar{r}\cdot\frac{d\bar{r}}{dt}\right)\bar{e}\times\underline{e}$ for the same reasons. Therefore, there remains

$$-\frac{\bar{g}}{\ell}\bar{r} = 2\omega(\bar{e}\cdot\underline{e})\bar{e}\times\frac{d\bar{r}}{dt} + \frac{d^2\bar{r}}{dt^2} \tag{4.12}$$

$$= 2\bar{\omega}\times\frac{d\bar{r}}{dt} + \frac{d^2\bar{r}}{dt^2}.$$

If we add the negligible $\bar{\omega}\times(\bar{\omega}\times\bar{r})$ to this equation so that

$$-\frac{\bar{g}}{\ell}\bar{r} = \bar{\omega}\times(\bar{\omega}\times\bar{r}) + 2\bar{\omega}\times\frac{d\bar{r}}{dt} + \frac{d^2\bar{r}}{dt^2}, \tag{4.13}$$

we see immediately that we have

$$-\frac{\bar{g}}{\ell}\bar{r} = \frac{d^2\bar{r}}{dt^2} = \omega\times(\omega\times\bar{r}) + 2\bar{\omega}\times\frac{d\bar{r}}{dt} + \frac{d^2\bar{r}}{dt^2}$$

where the absolute acceleration is with respect to the fixed stars approximately and represents a simple pendulum with respect to them while the other representation is that of a rigid frame rotating at angular velocity

$$\bar{\omega} = \omega(\underline{e}\cdot\bar{e})\bar{e},$$

which is a rotation about $\bar{e}$ from west through south to east in the northern hemisphere so that the pendulum appears to an observer on Earth to rotate from east through south to west at a speed less then the Earth's about its poles according to the latitude. Thus at latitude 45° north

$$|\bar{\omega}| = \frac{|\omega|}{\sqrt{2}}.$$

E   The Euler Angles.

For a later study of the top we introduce here the Euler angles for the motion of a rigid body with one point fixed (or for one point moving arbitrarily on a fixed plane). We shall find these angles by three successive rotations of an orthonormal set. In the original position, we shall designate them as $\{\underline{e}_i\}$, after the first rotation as $\{\bar{e}_i\}$, after the second as $\{\bar{\bar{e}}_i\}$, after the third as $\{\hat{\underline{e}}_i\}$.

First, rotate $\underline{e}_1$ to $\bar{e}_1$ around $\bar{e}_3$, by angle ψ. Then,

$$\bar{e}_i = a_i^j \underline{e}_j, \quad \bar{e}_3 = \underline{e}_3$$

where

$$a_i^j = \bar{e}_i \cdot \underline{e}_j.$$

Then the matrix of this transformation is

$$[A_\psi] = \begin{pmatrix} \cos\psi & \cos\left(\frac{\pi}{2}+\psi\right) & 0 \\ \cos\left(\frac{\pi}{2}-\psi\right) & \cos\psi & 0 \\ 0 & 0 & 1 \end{pmatrix}$$

Second, rotate $\bar{e}_3$ to $\bar{\bar{e}}_3$ around $\bar{e}_1$, by angle θ. Then

$$\bar{\bar{e}}_i = \bar{a}_i^j \bar{e}_j, \qquad \bar{\bar{e}}_1 = \bar{e}_1,$$

where

$$\bar{a}_i^j = \left(\bar{\bar{e}}_i \cdot \bar{e}_j\right).$$

Then the matrix of the transformation is

$$[A_\theta] = \begin{pmatrix} 1 & 0 & 0 \\ 0 & \cos\theta & \cos\left(\frac{\pi}{2}+\theta\right) \\ 0 & \cos\left(\frac{\pi}{2}\theta\right) & \cos\theta \end{pmatrix}$$

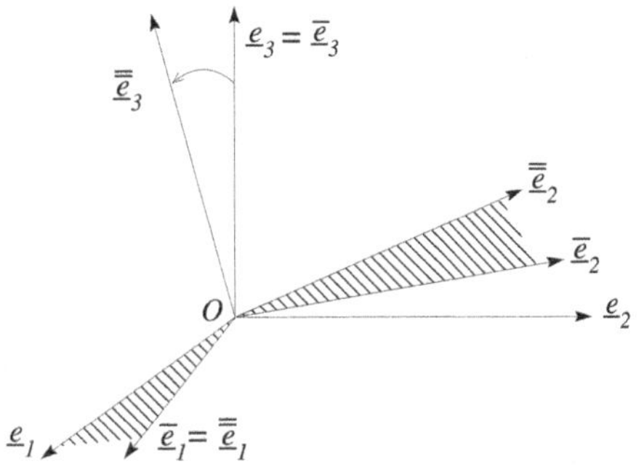

Fig. 97

Third, rotate $\bar{\bar{e}}_1$ to $\hat{e}_1$ about $\bar{\bar{e}}_3$ by angle ϕ. Then $\hat{e}_i = \bar{\bar{a}}_i^j \bar{\bar{e}}_j, \hat{e}_3 = \bar{\bar{e}}_3$ where $\bar{\bar{a}}_i^j = \left(\hat{e}_i \cdot \bar{\bar{e}}_j\right).$ Then the matrix of the transformation is

$$[A_\varphi] = \begin{pmatrix} \cos\varphi & \cos\left(\frac{\pi}{2}+\varphi\right) & 0 \\ \cos\left(\frac{\pi}{2}-\varphi\right) & \cos\varphi & 0 \\ 0 & 0 & 1 \end{pmatrix}$$

The three angles of rotation ψ, θ, φ are thus seen to be independent of each other and it is also clear that by variation of these angles the rigid body with point O fixed can be brought to any position. It is important for us to be able to express $\underline{w}$, the angular velocity of the rigid body, in terms of the orthonormal system $\hat{e}_i$ which moves as a rigid frame $\bar{F}$. But we have immediately

$$\underline{w} = \dot{\psi}\bar{e}_3 + \dot{\theta}\bar{\bar{e}}_1 + \dot{\varphi}\hat{e}_3$$

where these do not form an orthonormal set. We can transform $\bar{e}_3$ and $\bar{\bar{e}}_1$ into the $\hat{e}_i$ by taking the inverses of the transformations. We have only to take the transposes, since they are orthogonal,

$$[A_\psi]^{-1} = \begin{pmatrix} \cos\psi & \sin\psi & 0 \\ -\sin\psi & \cos\psi & 0 \\ 0 & 0 & 1 \end{pmatrix}$$

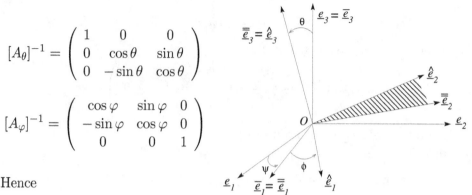

$$[A_\theta]^{-1} = \begin{pmatrix} 1 & 0 & 0 \\ 0 & \cos\theta & \sin\theta \\ 0 & -\sin\theta & \cos\theta \end{pmatrix}$$

$$[A_\varphi]^{-1} = \begin{pmatrix} \cos\varphi & \sin\varphi & 0 \\ -\sin\varphi & \cos\varphi & 0 \\ 0 & 0 & 1 \end{pmatrix}$$

Hence

$$\bar{e}_3 = \left(\bar{a}_3^j\right)^{-1} \bar{\bar{e}}_j = \sin\theta \bar{\bar{e}}_2 + \cos\theta \bar{\bar{e}}_3 \qquad\qquad \text{Fig. 98}$$

$$\bar{\bar{e}}_2 = \left(\bar{\bar{a}}_2^j\right)^{-1} \hat{e}_j = \sin\varphi \hat{\underline{e}}_1 + \cos\varphi \hat{\underline{e}}_2$$

$$\bar{\bar{e}}_1 = \cos\varphi \hat{\underline{e}}_1 - \sin\varphi \hat{\underline{e}}_2, \qquad \bar{\bar{e}}_3 = \hat{e}_3.$$

Therefore

$$\underline{\omega} = \dot{\psi}[\sin\theta \bar{\bar{e}}_2 + \cos\theta \bar{\bar{e}}_3] + \dot{\theta}[\cos\varphi \hat{\underline{e}}_1 - \sin\varphi \hat{\underline{e}}_2] + \dot{\varphi}\hat{\underline{e}}_3$$

$$= \dot{\psi}\sin\theta[\sin\varphi \hat{\underline{e}}_1 + \cos\varphi \hat{\underline{e}}_2] + \dot{\psi}\cos\theta \hat{\underline{e}}_3 + \dot{\theta}[\cos\varphi \hat{\underline{e}}_1 - \sin\varphi \hat{\underline{e}}_2] + \dot{\varphi}\hat{\underline{e}}_3$$

$$\underline{\omega} = [\dot{\theta}\cos\varphi + \dot{\psi}\sin\theta\sin\varphi]\hat{\underline{e}}_1 + [\dot{\psi}\sin\theta\cos\varphi - \dot{\theta}\sin\varphi]\hat{\underline{e}}_2$$

$$+ [\dot{\psi}\cos\varphi + \dot{\varphi}]\hat{\underline{e}}_3$$

Chapter 6

Emission and Quasi-elastic Dynamical Theories of Light

The attempt to bring the phenomena of light under the general laws of mechanics was already made by Kepler. We shall start with the theories developed in the seventeenth century, on the one hand by Descartes as realized by Newton, and on the other by Huygens, the first a particle or emission theory and the second a wave theory. It is the development and deepening of these theories in the nineteenth century which will finally lead to the recognition of the inadequacy of the assumptions concerning the measurement of time and space in the classical mechanics as we have seen it unfold in the first five chapters of this book.

The principal phenomena of light that seventeenth century theories set themselves to explain were (1) the old law of reflection, (2) the new law of refraction, and (3) the phenomenon of double refraction in Iceland spar. The phenomena of interference as seen in Newton's rings and in diffraction were imaginatively investigated but not integrated mechanically into the theories themselves. The same is true of the different refrangibility of coherent light of different colors with the analysis of white light into the spectrum which it entails.

Contrary to the assumptions of the Greeks and Medievals, the scientists of the seventeenth century assumed that light traveled with a finite speed even before this was confirmed by the following ingenious method of Römer in 1675.

Jupiter moves very slowly about the sun S compared to the Earth E. Consider the innermost moon of Jupiter and compute the time from an emergence from Jupiter's shadow to the next emergence, $42\frac{1}{2}$ hours, during which time

the Earth has moved very little from its position at E so that one can take this
as   the   moon's   periodic   time   about   Jupiter.
It is noticed that the emergence from
the shadow is independent of the observer on the Earth. Then one com-
putes the time and the number of
such emergences of the moon as the
Earth moves from the position E on
the side of its orbit near Jupiter to the
position E' on the farther side. Let m
be the number of times that the moon
is seen to emerge. Then $m(42\frac{1}{2})$ hours
is the time it would take if the Earth
had remained at E. But the time observed turns out to be longer so that,
if t is the time observed,

$$t - m(42\frac{1}{2})$$

is the time in hours it takes the light
to travel the distance $(JE' - JE)$.
On this basis it was computed that
light takes about 22 minutes to tra-

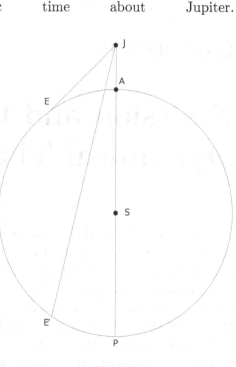

Fig. 99

verse the distance AP, which comes to a speed of about 228,000 kilometers
per second. Subsequent measurements gave 8.2 seconds or a speed of 303,200
kilometers per second. Modern measurements give 498.58 seconds and a speed
of 299,860 kilometers per second.

A   The Emission Theory of Descartes and Newton

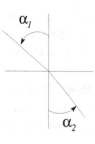

Fig. 100

The theory of Descartes is not strictly an emission theory,
but in the hands of Newton it becomes such, and here we are
not interested in the historical details but in the development
of the ideas which lead to the present state of the science.
Hence we give the theory in the form it leaves the hands of
Newton.

The first task of this theory was to predict the law of the
bending of light, or refraction of light, in passing from one
medium to another, using the principles of mechanics. It is
usually reported that Snell's Laws of Sines, that is, the law

that the ratio of the sine of the angle of incidence and of the sine of the angle of refraction is constant for two given media and a given coherent light independent of the angle of incidence, where the refracted ray lies in the same plane as the incident ray and normal (the plane of incidence), - that this law was found experimentally and then explained by Descartes. But as so often has happened in physics it was the other way around. Descartes deduced the law from his principles and then verified it experimentally by a hyperbolical glass built according to the principles.

The argument of Newton (and Descartes) is that light is made up of particles emitted by an agitated body which have a constant velocity in any given medium, but which are affected by forces acting in a small zone about the surface dividing the two media in a direction normal to the surface. The argument is that within the homogeneous medium the particles of the medium are acting on each other with the same forces in all directions; but, near the surface, there is a difference represented by the difference of the media. Obviously these forces will vary only as the depth along the normal. Hence, if the plane dividing the two media is a plane through the x-axis normal to the paper and to the y-axis, then

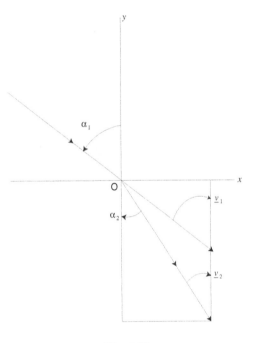

Fig. 101

$$|\underline{v}_2|^2 - |\underline{v}_1|^2 = \int_{-y_1}^{y_2} f(y)\underline{e}_2 \cdot (dx\underline{e}_1 + dy\underline{e}_2)$$

$$= \int_{-y_1}^{y_2} f(y)dy = U \qquad (1.1)$$

where $\underline{v}_1$ is velocity in first medium and $\underline{v}_2$ is velocity in second medium and $f(y)$ is some function the exact knowledge of which is not essential to the argument. The integral will monotonically increase or decrease according to some law. This is the scalar equation. Further since the forces act only in the direction of y, the x-components of $\underline{v}_1$ and $\underline{v}_2$ remain the same so that from this vector equation

$$|\underline{v}_1| \sin \alpha_1 = |\underline{v}_2| \sin \alpha_2$$

$$\frac{\sin \alpha_1}{\sin \alpha_2} = \frac{|\underline{v}_2|}{|\underline{v}_1|} = \mu \tag{1.2}$$

where $|\underline{v}_1|$ is independent of the angle α_1. But U, in (1.1), is also independent of the angles. Therefore $|\underline{v}_2|$ is also independent of the angles and so also μ. [1]

Having found this formula (1.2), Descartes verified it by constructing a hyperbolical glass as follows.

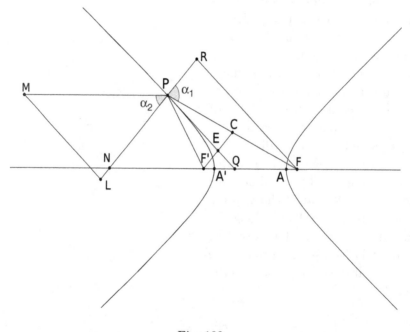

Fig. 102

Consider a solid piece of glass formed by an hyperbola rotated about its axis, and consider a plane section of it. Draw the tangent PQ at P intersecting the axis at Q, and draw the normal PL. If F' and F are the two foci, it is a well-known property of the hyperbola that

$$\angle F'PQ = \angle FPQ.$$

---

[1] This is essentially the explanation of Newton in **Opticks**, Bk.I, Part I, pg.79-80, London, 1931 and in **Principia**, Bk.1, Sect. XIV.

Take PM parallel to the axis with $PM = PF$ and construct ML normal to PL, and FR normal to PL. Then

$$\frac{ML}{FR} = \frac{\sin \alpha_1}{\sin \alpha_2}.$$

But it is obvious triangles PLM, FPN, and PRF are similar respectively to triangles FRN, FCF', and $PF'E$ where $F'C$ has been constructed parallel to PL. But, from a well-known property of the hyperbola

$$PF - PF' = 2a$$

and from the equality of the angles

$$PF' = PC.$$

Therefore

$$\frac{CF}{PC} = \frac{A'A}{PF}.$$

Since, from triangles FCF', and FPN, we have

$$\frac{CF}{FF'} = \frac{PF}{NF}.$$

Therefore, substituting in the previous ratio, we get

$$\frac{(PF)(PF')}{(NF)(PC)} = \frac{AA'}{FF'}.$$

Again, from triangles NRF and $PF'E$,

$$\frac{PF'}{NF} = \frac{F'E}{NR},$$

and again, substituting in the preceding ratio,

$$\frac{F'E}{NR} \cdot \frac{PF}{PC} = \frac{AA'}{FF'}.$$

Again, from triangles $PF'E$ and PML,

$$\frac{F'E}{PL} = \frac{PF}{PM} = \frac{PC}{PF},$$

so that

$$\frac{PL}{NR} = \frac{AA}{FF'}.$$

But, from the triangles PLM and FRN,

$$\frac{ML}{FR} = \frac{PL}{NR} = \frac{AA}{FF'}$$

or

$$\frac{\sin \alpha_1}{\sin \alpha_2} = \frac{2a}{2ae} = \frac{1}{e} = \mu.$$

If one considers light from a distance falling perpendicularly on the glass as MP, then it will fall on the focus F, if the eccentricity of the hyperbola is equal to the inverse of the refractive index μ from glass to air. And conversely if the light focuses at F, then it has obeyed Snell's Law. Thus did Descartes find and check this law.

But we can get more from (1.1). For

$$\frac{\sin \alpha_1}{\sin \alpha_2} = \mu = \frac{|v_2|}{|v_1|} = \sqrt{1 + \frac{U}{|v_1|^2}} \tag{1.3}$$

where U is the ordinary force function divided by $2m$. We can distinguish several cases.

a. When U is positive, $\alpha_1 > \alpha_2$ and the light bends down in the second medium as is the case from air to glass. It is to be noticed that, according to this theory, $|v_2|$, the speed of light in glass, is greater than the speed of light in air. This disturbed some contemporaries.

b. When $U < 0$, but $|U| < |v_1|^2 \cos^2 \alpha_1$, then

$$0 < \sin \alpha_1 < \frac{\sin \alpha_1}{\sin \alpha_2} < 1,$$

from which we conclude that

$$\frac{\pi}{2} > \alpha_2 > \alpha_1 > 0$$

and the light in the second medium turns up.

c. When $U < 0$, but $|U| = |v_1|^2 \cos^2 \alpha_1$, then

$$\sin \alpha_1 = \frac{\sin \alpha_1}{\sin \alpha_2},$$

so that

$$\alpha_2 = \frac{\pi}{2}$$

and the light just skims along the surface between the two media.

d. When $U < 0$, but $|U| > |\underline{v}_1|^2 \cos^2 \alpha_1$, then

$$\frac{\sin \alpha_1}{\sin \alpha_2} < \sin \alpha_1$$

so that

$$\sin \alpha_2 > 1$$

and refraction is impossible. This is the case of total reflection. That the angle of refraction is equal to the angle of incidence is argued from the fact that the function $f(y)$ is constant for a given depth in the zone about the plane between the two media so that the curve in and out of this zone is symmetrical with respect to the normal to the boundary between the media.

It is to be noted that this theory provides for to-tal transmission of the light into the second medium or total reflection. But the ordinary experience of looking down on one's own image in water is con-vincing evidence that even when $\alpha_1 = 0$, there is

Fig. 103

partial reflection and partial transmission. An even more telling experiment was performed by Hooke and Newton where a thin spherical shaped glass with large radius is placed on a plane piece. If a light falls on it, there appear al-ternate rings of light and dark, of reflection and transmission about the point of contact. The emission theory had no ready solution for the phenomena of partial reflection.

This law of refraction gave Newton a means of analyzing out light with different refrangibil-ity. Letting sunlight fall through a narrow aperture on a prism which produced an elongated im-age in colors running from red as the least refracted through yel-low, green, blue, to violet as the most refracted. The reason for this dispersion can be seen by considering a ray entering at P with angle α_1 and one part being refracted at angle α_2 so as to be parallel to the base of the prism

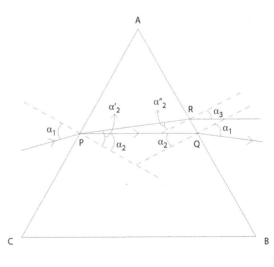

Fig. 104

BC, then continuing to Q where
it is refracted again at α_1. Sup-
pose another part is refracted less by angle α_2', continuing to R where the
incident angle is α_2'' and the refracted angle α_3. Then

$$\frac{\sin \alpha_1}{\sin \alpha_2} = \mu, \qquad \frac{\sin \alpha_1}{\sin \alpha_2'} = \mu'$$

where

$$\alpha_2' > \alpha_2, \qquad \mu' < \mu.$$

But, as exterior angle of a triangle,

$$\frac{\pi}{2} - \alpha_2'' > \frac{\pi}{2} - \alpha_2, \qquad \alpha_2'' < \alpha_2.$$

Therefore

$$\frac{\sin \alpha_2}{\sin \alpha_1} = \frac{1}{\mu}, \qquad \frac{\sin \alpha_2''}{\sin \alpha_3} = \frac{1}{\mu'}, \qquad \frac{1}{\mu'} > \frac{1}{\mu}$$

and

$$\sin \alpha_2 \sin \alpha_3 < \sin \alpha_1 \sin \alpha_2'', \qquad \alpha_3 < \alpha_1.$$

Thus the light, after passing through the prism, is spread out. In this way,
Newton is able to define monochromatic or coherent light by its index of refraction and by its inability to be broken down any further. All this, it is to be
remarked, is with respect to a given observer. What further complications will
arise from comparing indices of refraction for different observers will appear
later.

   In the late seventeenth century, under the influence of the views of Leibniz
who was searching for a new type of mathematical law to explain the actual
movements of particles, the following principle was devised: Consider the
possible paths along smooth curves between two points $\underline{x}_0$ and $\underline{x}_1$, and let
$v(S)$ be the speed at any point $\underline{x}$ at an arc-length s along the curve from $\underline{x}_0$.
Then we consider a new mathematical entity $v(s)ds$, the action at s, where
ds is the infinitesimal arc-length. We assert that the particle will follow that
path where

$$\int_{s_0}^{s_1} v(s)ds$$

is the least. Such a principle covers the whole history of the motion. Leibniz
considered such principles as transcending the ordinary mechanical laws and
as being superior to them. They became more important later.

   The application of the principle to the case of the emission theory of light
as it passes from one diaphanous medium to another is made mathematically

simple by the fact that we assume constant velocity in each medium and a
curve in the thin zone between the two which is so small as only to serve as
a change in direction from one medium to the other. Then we have the total
action over the complete path

$$A = v_1 s_1 + v_2 s_2,$$

$$A = v_1 (h_1^2 + x^2)^{\frac{1}{2}} + v_2 \left(h_2^2 + (a - x)^2\right)^{\frac{1}{2}},$$

$$\frac{dA}{dx} = \frac{v_1 x}{(h_1^2 + x^2)^{\frac{1}{2}}} + v_2 \frac{(a - x)}{\left(h_2^2 + (a - x)\right)^{\frac{1}{2}}} = 0,$$

$$\frac{\sin \alpha_1}{\sin \alpha_2} = \frac{v_2}{v_1}$$

From geometrical considerations, this is obviously a minimum.

B   The Wave Theory of Huygens.

In 1678, Huygens communicated to a learned society his wave theory of light
which in his hands not only predicted the laws of reflection and refraction
but also could be used to mathematicize a new phenomenon, that of double
refraction in Iceland Spar. Here one assumes that light is a wave disturbance
made by the agitations of a body, the primary light source, in an all pervading
perfectly elastic element called the aether. The wave, whose nature was not
defined, spreads in a spherical front from the light source as a center, in the
vacuum and diaphanous bodies.[2]

In addition Huygens assumed that as the initial wave-front spreads, it in
turn agitates every point of the aether which itself becomes a secondary source
of light and the center of a spherical wave so that the visible light consists of
the primary wave-front as the envelope of those parts of the small secondary
waves which move in the same sense. This is called **Huygens' Principle.** It
is this principle which shall be applied to deduce the law of refraction.

For this purpose we consider a plane-front disturbance, that is, a spherical wave-front whose center is so far away that small parts of the spherical
surface can be considered planes. We consider the section of such a front
OA in the plane of incidence. OX is the line in this plane between the two
media. It is assumed that the speeds of the waves in each medium are constant in that medium. As the points of OA reach the boundary between the

---

[2]We here take the common sense meaning of "diaphanous". It can be defined precisely
in the classical electromagnetic theory.

two media, they become centers of waves which advance at different speeds from those in the first medium. Thus O is the center of a spherical wave in the second medium whose radius

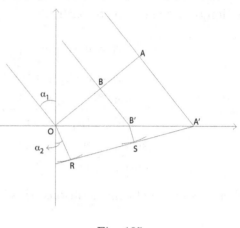

is OR traveled in the time it takes the disturbance at A to reach A' in the first medium. We draw a tangent to this circle (tangent plane to the spherical surface) from A'. Now we wish to show that $A'R$ will be the envelope or common wave front of all the waves from the points along OA'. That is, we must show that the time from B to B' plus the time from B' to S', where $B'S$ is the radius of the circle tangent also to OR is equal to the time from A to A' or from O to R.

Fig. 105

If v_1 is speed in first medium and v_2 in the second, then

$$t = \frac{OR}{v_2} = \frac{AA'}{v_1}$$

and

$$t_1 + t_2 = \frac{BB'}{v_1} + \frac{B'S}{v_2}$$

where t, t_1, t_2 are the respective times. We wish to show

$$t = t_1 + t_2.$$

But, by similar triangles,

$$\frac{B'S}{OR} = \frac{B'A'}{OA'}, \qquad \frac{BB'}{AA'} = \frac{OB'}{OA'}$$

so that

$$B'S = \frac{(B'A')(OR)}{OA'}, \qquad BB' = \frac{(OB')(AA')}{OA'}$$

$$v_1 = \frac{AA'}{t}, \qquad v_2 = \frac{OR}{t},$$

and

$$\frac{BB'}{v_1} + \frac{B'S}{v_2} = t\frac{(OB' + B'A')}{OA'} = t.$$

It has been assumed as intuitively obvious that the plane tangent to the spheres with radii $OR, B'S$ etc. touches them in a line $A'R$ which lies in plane of incidence. It follows immediately that

$$\frac{v_1}{v_2} = \frac{AA'}{OR} = \frac{AA'}{OA'} \bigg/ \frac{OR}{OA'} = \frac{\sin \alpha_1}{\sin \alpha_2} = \mu \qquad (2.1)$$

Here the ratio of the speeds is reversed with respect to the corresponding equation (1.2) in the theory of emission. And so the wave theory predicts that the speed of light in glass is less than the speed of light in air, the reverse of the prediction of the emission theory. It will only be in the middle of the nineteenth century that one will be able to decide by experiment between the two.

By the same method, that is of computing the re-formation of the wavefront, one can deduce the law of reflection.

But Huygens used this theory for an explanation of sorts of a new phenomenon, that of double refraction. Iceland crystal occurs in the form of a rhomboid whose acute angles are $78°8'$ and whose obtuse angles are $101°52'$. Let C and E be the opposite vertices, each composed of obtuse angles only. The line CE is called the axis of the crystal and the plane through CD and CE, which is normal to the face $ACBD$, is called the principal section. Then a ray of light falling on CD in the principal section (or on a line parallel in a plane parallel) is refracted at two

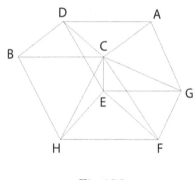

Fig.106

different angles. For if EF is made to coincide with a black line marked on paper and a line is drawn normal to it on the paper, then there appears only one line EF, but two of the line normal to it, one the result of ordinary refraction and one of extraordinary refraction.

The former is distinguished from the latter by the fact that its image remains fixed as one turns the crystal while the other does not. One can also distinguish them by observing that the extraordinary refraction appears when the incident ray is normal to the surface and that it does not obey the law of sines of the ordinary refraction. In fact, the extraordinary ray $C'T$ makes an angle

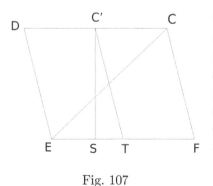

Fig. 107

$TC'S = 6°40'$ to the normal $C'S$ towards CF. Further this same refraction takes place with respect to all six faces and is therefore symmetrical with respect to the axis CE.

To account for this extraordinary refraction, Huygens supposed that it was formed by a wave-front of ellipsoids instead of the spheres of the ordinary refraction. Considered in the principal section, we would have ellipses with major axes CU in such a position that they would be tangent to FE at T at the correct angle with the normal CS. But, since this same construction would have to be true with respect to all three faces having C as vertex, therefore we are forced to suppose these ellipsoids are ellipsoids of revolution having $C'E'$ as their minor semiaxes

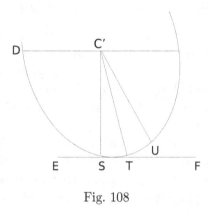

Fig. 108

of revolution parallel to the axis of the crystal CE and $C'U$ as the major semiaxis. The form of the ellipsoid is further determined by the fact that the point of contact T with EF must be such that angle $TC'S$ is $6°40'$.

The shape and position of the ellipsoids having been determined, it is now possible to predict mathematically by the same theory, what will happen to rays inclined to the normal in the extra-ordinary refraction. One has only to compute the formation of the wave-front. But to do this, we must know the relative speeds in the two media. Obviously, unlike the ordinary refraction, here theoretically because of the ellipsoids, the speed will differ in the second medium according to the inclination of the incident ray. Thus, if ZC is refracted extraordinarily into CP, then, drawing KP tangent at P and CO normal to ZC, we see that, according to Huygens' theory, the time for OK(drawn normal to OC to intersect with CD') must be the same as that for CP, CT, and CU. If we let OK represent the radius of the spherical wave in the first medium, we can fix the ratio of OK to CD' or CU by computation from an observed extraor-

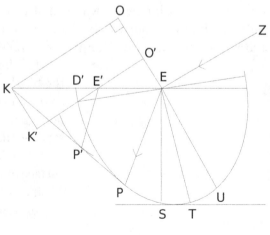

Fig. 109

dinary refraction. Once this ratio is established, the theory can
be verified by predicting other re-
fractions CP for other inclinations of ZC in the principal section and checking
further observations. In this regard, one can predict an inclination of ZC for
which there is no refraction and this too is experimentally verified. Thus, having observed a given ray ZC and its extraordinary refraction CP, we compute
OK which will, of course, remain constant for any CP, since it represents the
distance traveled by a wave-front in the first medium while the wave-fronts
travel from C to the points on the ellipsoid according to the different directions. We can now take any other incident ray ZC, draw CO normal to it
and take O so that OK is the length we have found (which is invariant) normal to CO. Having found K, we draw the tangent to the ellipsoid by known
methods. Thus we have P and CP.

It should be noted that to complete the proof of the Huygens' construction,
we have to show that for an ellipse with center C' and tangent to KP at P'
where $C'P'$ is parallel to CP and KK' is parallel to OC,

$$\frac{CP}{OK} = \frac{C'P'}{C'K'}.$$

We do not give the proof here.

Further confirmation of the theory is had in examining the extraordinary
refraction when the incident ray is not in the principal section (or in a plane
parallel to it). Then the refracted ray, according to the ellipsoidal theory,
will not lie in the plane of incidence as is obvious if one imagines ZC rotated
about C. Thus suppose ZC lies in the plane parallel to that through AB of
Fig. 106 normal to CD and to the face, then the refracted ray will lie in the
plane through CT normal to the principal section. The phenomena are found
to fit precisely in this case also the wave-fronts of the ellipsoids.

A more remarkable property is that, if two such crystals are held the one
above the other so that their principal sections coincide, then the ordinarily
refracted ray from the first is only ordinarily refracted in the second and the
extraordinarily refracted only extraordinarily refracted. But this is reversed
when the sections are normal to each other. For all positions in between, there
are four rays in the second crystal with varying degrees of brightness.

Newton, in reviewing Huygens' **Treatise on Light**, published at the end
of the century, noticed an apparent asymmetry about the direction of propagation of light, - but, bemused perhaps by his treatment of sound in the
Principia as the propagation of pulses of condensation and expansion of the
medium he felt this argued against Huygens' theory although this very prop-

erty later became one of the strongest arguments in its favor under a new interpretation.

Finally, just as the emission theory has a minimal explanation of its formulation of the refraction law, so too has the wave-theory of Huygens, one produced earlier against Descartes by Fermat where it is assumed that the path taken by the refracted ray is that of the least time between two points in two different media where the velocities are constant in each medium. Thus the t_1 and t_2 in the respective media are

$$t_1 = \frac{s_1}{v_1}, \ t_2 = \frac{s_2}{v_2},$$

$$t = t_1 + t_2 = \frac{s_1}{v_1} + \frac{s_2}{v_2} = \frac{1}{v_1}\sqrt{h_1^2 + x^2} + \frac{1}{v_2}\sqrt{h_2^2 + (a-x)^2}$$

$$\frac{dt}{dx} = \frac{x}{v_1\sqrt{h_1^2 + x^2}} - \frac{a-x}{v_2\sqrt{h_2^2 + (a-x)^2}} = 0$$

$$\frac{\sin \alpha_1}{\sin \alpha_2} = \frac{v_1}{v_2}.$$

C   Malus' Experiments on Double Refraction and Brewsters' Law.

In 1809, Malus published a memoir describing experiments which throw into remarkable relief the property of the light rays passing from one crystal to another, as given above by Huygens. These experiments and renewed investigations by Young into the phenomena of interference will prepare the way for Fresnel's construction of the quasi-elastic theory of light as an extension of Huygens' wave-theory.

Malus noticed that, if instead of using direct light, one used light reflected from water or other transparent object (not reflecting metals) at certain angles, this light behaved with respect to a doubly refracting crystal, like the ordinary rays from another. We now give a report of these results:

A. If a ray is reflected from the surface of water at an angle of 52°45′ on to a crystal of Iceland Spar (or any other doubly refracting crystal), then

(i) When the principal section of the crystal is parallel to the plane of reflection (plane of incidence) of the water, the ray is refracted by the crystal only by the ordinary law;

(ii) When the principal section is perpendicular to the plane of reflection of the water, the ray is refracted by the crystal only according to the extraordinary law;

(iii) For other positions, it will be doubly refracted as if it were an ordinary ray which had already been doubly refracted from a crystal whose principal section was parallel to the plane of reflection of the water.

Therefore, the rays reflected from the surface of the water at this angle act like the ordinary rays resulting from a double refraction in a crystal whose principal section is parallel to the water's plane of reflection.

B. In reverse, if a ray doubly refracted from a crystal is allowed to fall on a surface of water at the angle of incidence 52°45', then,

(1) When the principal section is parallel to the plane of reflection, (i) the ordinary ray is partially reflected and partially transmitted like direct light, but (ii) the extraordinary ray is totally transmitted through the water's surface;

(2) When the principal section is normal to the plane of reflection, (i) the ordinary ray is totally transmitted, while (ii) the extraordinary ray is partially reflected and partially transmitted.

Malus noticed that the angle of incidence, 52°45' for water, increased with the refractive index of the reflecting body. Brewster very soon determined the function of increase. If θ_0 is the angle of incidence, then, by Brewster's law,

$$\tan \theta_0 = \mu \tag{3.1}$$

The phenomenon of interference was destined to be connected with these experiments of Malus in an unexpected way. We shall first describe a typical interference experiment, one constructed by Fresnel, and then show the relations to Malus experiments.

We take two reflecting mirrors OQ and OQ' placed together at an angle of almost 180°, the Fresnel mirrors. Then, if we place a single light source S in front of the mirrors as in Fig. 110, then, by the law of reflection, two beams of light, one from each mirror, will be reflected as if from two virtual sources S_1 and S_2 behind the mirrors. Since the mirrors are nearly in a straight line, $S_1 S_2$ will be very small and the beams will be nearly parallel. If we take P as midpoint of $S_1 S_2$ and draw PO, it is easy to prove the following pairs of triangles congruent: SOQ and $S_1 OQ'$ and $S_2 OQ'; S_2 OP$ so that PO is normal to $S_1 S_2$. If a screen is placed normal to PO extended, then, at M_1 where it intersects OP will be bright while dark bands will appear at intervals on either side.

This appearance which is so difficult to explain in the emission theory, has a natural explanation in the wave theory. The first dark band will be caused by the fact that the difference of path-lengths is half a wave length so that the two wave trains are out of place by that much and destroy each other, whereas, the bright bands are where they reinforce each other. There is no need at this point to suppose the waves to have a form other than those of sound which act by condensation and expansion in the line of their propagation. It is to be remarked that the interference phenomenon has been produced by two beams from the same light source. This is so to insure that the fluctuations in the two beams be correlated. When they are not from the same source, the phenomenon does not, in general, occur.

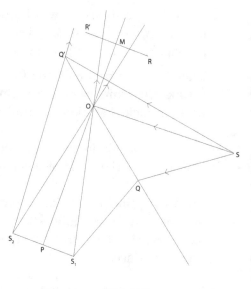

Fig. 110

Let us use the experiment of the two mirrors to measure the wave lengths of light on the assumptions which have been made, as Fresnel did. Consider the detail from the last figure, shown in Fig. 111.

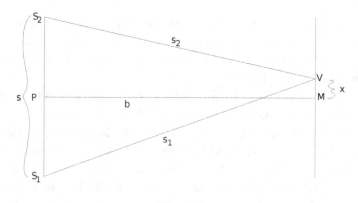

Fig. 111

Then let V be the position of the first black band from M and let

$$S_2V = s_2, \quad S_1V = s_1,$$

the path-lengths of the two beams from the source S. Now

$$s_2^2 = b^2 + \left(\frac{s}{2} - x\right)^2, \quad s_1^2 = b^2 + \left(\frac{s}{2} + x\right)^2$$

where

$$b = PM, \quad s = S_1 S_2, \quad x = VM,$$

so that

$$s_1^2 - s_2^2 = (s_1 - s_2)(s_2 + s_1) = 2xs$$

$$s_1 - s_2 = \frac{2xs}{s_1 + s_2}.$$

But, because s is very small with respect to b, we can say

$$s_1 - s_2 = \frac{\lambda}{2} \approx \frac{2xs}{2b} = \frac{xs}{b},$$

$$x \approx \frac{b\lambda}{2s}.$$

In Fresnel's experiment, $b = 7.016$m, $2s = 12.16$mm; $11x = 4.05$mm. Hence $\lambda \approx 0.000638$mm., for a red light.

But Fresnel discovered that the ordinary and extraordinary rays from the same light source will not interfere with each other. The same is true of two rays which have been passed through two piles of mica at Brewster's angle of polarization (3.1) when the planes of incidence are normal, although they do when the planes are parallel. These facts suggested to Fresnel that the waves of light consist in the vibration of the particles of the aether normal to the direction of propagation and that the polarization experienced by the ordinary and extraordinary rays consisted in the restriction of the vibrations to a single plane the one perpendicular to the other. These are the so called transverse waves as opposed to the longitudinal waves of sound. Obviously, in the case of transverse waves, polarized at right angles, the vibrations of one would not affect those of the other. The same would, of course, be true therefore for light polarized in the manner of Malus' experiment. But this will all become clearer after we have mathematicised and introduced the dynamic concepts of Fresnel. The theory, as it progresses, will also predict and discover other types of polarization.

D   Kinematics of Wave-Motion

We shall begin with the analytic representation of a plane polarized transverse wave. We suppose the wave is produced by transverse vibrations in the xy-plane in the direction of the y-axis with the direction of propagation of the wave along the x-axis. The vibration of the particles is assumed to be simple harmonic. Thus we consider

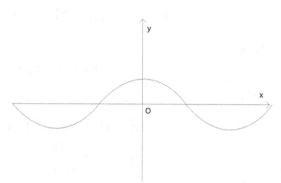

Fig. 112

$$y = a\cos\left(2\pi\omega\left(t - \frac{x}{v}\right)\right) \quad (4.1)$$

At $x = 0, t = 0, y = a$. If we keep $x = 0$, and let t vary, then at $t = \frac{1}{\omega}$, we find $y = a$ again, and this is the least period in which it does. Therefore the periodic time of vibration is (ω is called frequency)

$$PT = \frac{1}{\omega} \qquad (4.2)$$

If, on the other hand, we keep $t = 0$ and let x vary, then, at $x = \frac{v}{\omega}, y = a$ again and this is the least distance in which it does. Therefore the wave-length is

$$\lambda = \frac{v}{\omega}. \qquad (4.3)$$

It is obvious v has the dimensions of velocity. Its precise significance can be seen from

$$\frac{\lambda}{PT} = \frac{\frac{v}{\omega}}{\frac{1}{\omega}} = v, \qquad (4.4)$$

that is, v is the wave-length divided by the periodic time of vibration. This is the velocity of propagation of the disturbance. It is usual to call a the amplitude of the wave and $\omega\left(t - \frac{x}{v}\right)$ the phase.

   If we want to express a more general phase, we write

$$y = a\cos\left(2\pi\left[\omega\left(t - \frac{x}{v}\right) + \delta\right]\right) \qquad (4.5)$$

With respect to linear operations, it is usual at present to operate with a complex representation

$$y = Ae^{i2\pi\left[\omega\left(t-\frac{x}{v}\right)+\delta\right]}, A = ae^{i\alpha}.$$

If we differentiate partially with respect to time in (4.1), we have

$$\frac{\partial y}{\partial t} = -2\pi\omega a \sin\left(2\pi\omega\left(t - \frac{x}{v}\right)\right)$$

$$\left(\frac{\partial y}{\partial t}\right)^2 = 4\pi^2\omega^2 a^2 \sin^2\left(2\pi\omega\left(t - \frac{x}{v}\right)\right) \tag{4.6}$$

which is proportional to kinetic energy of the vibrating aether particle at a given point x. Disregarding the sine part which periodically varies from one to zero, we can say the kinetic energy varies as the squares of the amplitude and frequency.

Calculating further with partial derivatives with respect to t and x, we find that y of (4.1) satisfies

$$\frac{\partial^2 y}{\partial x^2} = \frac{1}{v^2}\frac{\partial^2 y}{\partial t^2} \tag{4.7}$$

which is a particular form of the wave equation.

We can immediately generalize this to a wave-front disturbance over a plane. Thus

$$\psi = a\cos\left(2\pi\omega\left(t - \frac{x \cdot n}{v}\right)\right) \tag{4.8}$$

where $\underline{n}$ is the unit vector normal to the plane

$$\underline{x} \cdot \underline{n} = \mu$$

where μ is an arbitrary constant. We may write it

$$\psi = a\cos\left(2\pi\omega\left(\frac{vt - \underline{x} \cdot \underline{n}}{v}\right)\right) \tag{4.9}$$

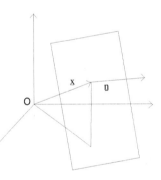

Fig. 113

Since

$$\frac{\partial \psi}{\partial x} = \frac{n_1}{v}a2\pi\omega \sin\left(2\pi\omega\left(t - \frac{x \cdot n}{v}\right)\right)$$

$$\frac{\partial^2 \psi}{\partial x^2} = -\frac{n_1^2}{v^2}4\pi^2\omega^2 \cos\left(2\pi\omega\left(t - \frac{x \cdot n}{v}\right)\right)$$

$$\frac{\partial^2 \psi}{\partial y^2} = -\frac{n_2^2}{v^2}4\pi^2\omega^2 \cos\left(2\pi\omega\left(t - \frac{x \cdot n}{v}\right)\right)$$

and correspondingly for $\frac{\partial^2 \psi}{\partial z^2}$, therefore ψ satisfies the equation

$$\left.\begin{array}{c} \dfrac{\partial^2 \psi}{\partial x^2} + \dfrac{\partial^2 \psi}{\partial y^2} + \dfrac{\partial^2 \psi}{\partial z^2} = \dfrac{1}{v^2}\dfrac{\partial^2 \psi}{\partial t^2} \\[3mm] \nabla^2 \psi = \dfrac{1}{v^2}\dfrac{\partial^2 \psi}{\partial t^2} \end{array}\right\} \qquad (4.10)$$

Suppose we consider more generally a function $\psi(\underline{x} \cdot \underline{n}, t)$ on which we put the condition that it satisfy (4.10). It obviously has a constant value over the plane

$$\underline{x} \cdot \underline{n} = \mu.$$

Let us introduce the Cartesian coordinates $0\xi, 0\eta, 0\zeta$ where 0ζ is in direction of $\underline{n}$. Then

$$\underline{x} \cdot \underline{n} = \zeta = x n_1 + y n_2 + z n_3$$

and

$$\frac{\partial \psi}{\partial x} = \frac{\partial \psi}{\partial \zeta}\frac{\partial \zeta}{\partial x} = \frac{\partial \psi}{\partial \zeta} n_1$$

$$\frac{\partial^2 \psi}{\partial x^2} = \frac{\partial^2 \psi}{\partial \zeta^2} n_1^2, \qquad \frac{\partial^2 \psi}{\partial y^2} = \frac{\partial^2 \psi}{\partial \zeta^2} n_2^2, \text{ etc.}$$

and

$$\nabla^2 \psi = \frac{\partial^2 \psi}{\partial \zeta^2}\left(n_1^2 + n_2^2 + n_3^2\right) = \frac{\partial^2 \psi}{\partial \zeta^2},$$

and with this change of variable $\psi(\underline{x} \cdot \underline{n}, t)$ satisfies

$$\frac{\partial^2 \psi}{\partial \zeta^2} = \frac{1}{v^2}\frac{\partial^2 \psi}{\partial t^2} \qquad (4.11)$$

There is a method, due to Euler, to find the general solutions of partial differential equations of the form

$$A\frac{\partial^2 \psi}{\partial x^2} + 2B\frac{\partial^2 \psi}{\partial x \partial y} + C\frac{\partial^2 \psi}{\partial y^2} = 0 \qquad (4.12)$$

of which equation (4.7) is a special case. We introduce a transformation of the variables in (4.12)

$$p = \alpha x + \beta y, \qquad q = \gamma x + \delta y \qquad (4.13)$$

where $\alpha, \beta, \gamma, \delta$ are arbitrary constants except that

$$\begin{vmatrix} \alpha & \beta \\ \gamma & \delta \end{vmatrix} \neq 0 \tag{4.14}$$

to insure invertibility. Then assuming sufficient conditions of continuity,

$$\psi(x,y) = \overline{\psi}(p,q)$$

and

$$\frac{\partial \psi}{\partial x} = \frac{\partial \overline{\psi}}{\partial p}\frac{\partial p}{\partial x} + \frac{\partial \overline{\psi}}{\partial q}\frac{\partial q}{\partial x} = \frac{\partial \overline{\psi}}{\partial p}\alpha + \frac{\partial \overline{\psi}}{\partial q}\gamma$$

$$\frac{\partial^2 \psi}{\partial x^2} = \frac{\partial^2 \overline{\psi}}{\partial p^2}\alpha^2 + 2\frac{\partial^2 \overline{\psi}}{\partial p \partial q}\alpha\gamma + \frac{\partial^2 \overline{\psi}}{\partial q^2}\gamma^2$$

$$\frac{\partial^2 \psi}{\partial y^2} = \frac{\partial^2 \overline{\psi}}{\partial p^2}\beta^2 + 2\frac{\partial^2 \overline{\psi}}{\partial p \partial q}\beta\delta + \frac{\partial^2 \overline{\psi}}{\partial q^2}\delta^2$$

$$\frac{\partial^2 \psi}{\partial x \partial y} = \frac{\partial^2 \overline{\psi}}{\partial p^2}\alpha\beta + \frac{\partial^2 \overline{\psi}}{\partial p \partial q}\gamma\beta + \frac{\partial^2 \overline{\psi}}{\partial p \partial q}\alpha\delta + \frac{\partial^2 \overline{\psi}}{\partial q^2}\gamma\delta$$

and (4.12) becomes

$$[A\alpha^2 + 2B\alpha\beta + C\beta^2]\frac{\partial^2 \overline{\psi}}{\partial p^2} + 2[A\alpha\gamma + B(\alpha\delta + \beta\gamma) + C\beta\delta]\frac{\partial^2 \overline{\psi}}{\partial p \partial q}$$

$$+ [A\gamma^2 + 2B\gamma\delta + C\delta^2]\frac{\partial^2 \overline{\psi}}{\partial q^2} = 0 \quad (4.15)$$

Our job is to choose α, β, γ, δ so that two out of three of the coefficients in (4.15) are zero while the third is not. This will involve the auxiliary equation

$$A + 2B\lambda + C\lambda^2 = 0 \tag{4.16}$$

so that three possible cases arise.

1. $B^2 - AC > 0$, real roots, hyperbolic;

2. $B^2 - AC < 0$, complex roots, elliptic;

3. $B^2 - AC = 0$, one real root, parabolic.

In the first two cases, we can let $\alpha = \gamma = 1$ and take $\beta = \lambda_1$ and $\delta = \lambda_2$ where the λ's are the distinct roots of (4.16). This makes the first and third

coefficients of (4.15) zero. We have only to show that the second coefficient is
not zero. But

$$\lambda_1 + \lambda_2 = \frac{-2B}{C}, \quad \lambda_1\lambda_2 = \frac{A}{C}$$

since

$$\lambda_1 = \frac{-B + \sqrt{B^2 - AC}}{C}, \quad \lambda_2 = \frac{-B - \sqrt{B^2 - AC}}{C}.$$

Therefore the second coefficient is

$$2[A + B(\lambda_1 + \lambda_2) + C\lambda_1\lambda_2] = 2[A - \frac{2B^2}{C} + A] = \frac{4[AC - B^2]}{C} \neq 0.$$

And we note the determinant (4.13)

$$\begin{vmatrix} 1 & \lambda_1 \\ 1 & \lambda_2 \end{vmatrix} \neq 0.$$

Hence we have

$$\frac{\partial^2 \overline{\psi}}{\partial p \partial q} = 0$$

and, integrating,

$$\frac{\partial \overline{\psi}}{\partial p} = C_1(p), \quad \overline{\psi} = \int C_1(p)dp + C_2(q)$$
$$\overline{\psi} = \overline{\psi}_1(p) + \overline{\psi}_2(q)$$
$$= \overline{\psi}_1(x + \lambda_1 y) + \overline{\psi}_2(x + \lambda_2 y) \tag{4.17}$$

where $\overline{\psi}_1$ and $\overline{\psi}_2$ are arbitrary functions determined by initial and boundary
conditions.

In the case of the wave-equation (4.7) or (4.11),

$$A = 1, \quad B = 0, \quad C = \frac{-1}{v^2}, \quad \lambda_1 = v, \quad \lambda_2 = -v$$

and

$$\overline{\psi} = \overline{\psi}_1(\zeta + vt) + \psi_2(\zeta - vt)$$
$$= \overline{\psi}_1(\underline{x} \cdot \underline{n} + vt) + \overline{\psi}_2(\underline{x} \cdot \underline{n} - vt), \tag{4.18}$$

of which

$$\psi = a \cos 2\pi\omega \frac{(-vt + \underline{x} \cdot \underline{n})}{v} + b \cos 2\pi\omega \frac{(vt + \underline{x} \cdot \underline{n})}{v}$$

is a special case. This represents two waves traveling in opposite senses.

For the third case, the reader can verify that the first and second coefficients are zero for

$$\alpha = 1, \quad \beta = \lambda_1 = \lambda_2 = \frac{-B}{C},$$

leaving γ and δ free to make the third coefficient not zero. This gives

$$\frac{\partial^2 \overline{\psi}}{\partial q^2} = 0, \qquad \overline{\psi} = \overline{\psi}_1(p) + q\overline{\psi}_2(p)$$

$$\overline{\psi} = \overline{\psi}_1(x + \lambda y) + (\gamma x + \delta y)\overline{\psi}_2(x + \lambda y).$$

That this method gives a solution of the elliptic case only if λ_1 and λ_2 are complex numbers need not worry us. We can get around this by a further linear transformation

$$\xi = \frac{p+q}{2} = x\frac{B}{C}y, \qquad \eta = \frac{p-q}{2i} = x + \frac{-B - \sqrt{B^2 - AC}}{C}y,$$

so that

$$\frac{\partial \overline{\psi}}{\partial p} = \frac{\partial \overline{\psi}}{\partial \xi}\frac{\partial \varepsilon}{\partial p} + \frac{\partial \overline{\psi}}{\partial \eta}\frac{\partial \eta}{\partial p} = \frac{1}{2}\left(\frac{\partial \overline{\psi}}{\partial \xi} + \frac{1}{i}\frac{\partial \overline{\psi}}{\partial \eta}\right)$$

$$\frac{\partial^2 \overline{\psi}}{\partial p \partial q} = \frac{1}{2}\left[\frac{\partial^2 \overline{\psi}}{\partial \xi^2}\frac{\partial \xi}{\partial q} + \frac{\partial^2 \overline{\psi}}{\partial \xi \partial \eta}\frac{\partial \eta}{\partial q} + \frac{1}{i}\frac{\partial^2 \overline{\psi}}{\partial \eta \partial \xi}\frac{\partial \xi}{\partial q} + \frac{1}{i}\frac{\partial^2 \overline{\psi}}{\partial \eta^2}\frac{\partial \eta}{\partial q}\right]$$

$$= \frac{1}{4}\left(\frac{\partial^2 \overline{\psi}}{\partial \xi^2} + \frac{\partial^2 \overline{\psi}}{\partial \eta^2}\right).$$

Thus we have the Laplace equation in two variables.

$$\nabla^2 \overline{\psi} = 0.$$

In fact, the solutions of the Laplace equation are precisely those functions which belong to an analytic function in the complex plane, that is, they satisfy the Cauchy-Riemann equations. Consider

$$\overline{\psi} = u(\xi, \eta) + iv(\xi, \eta)$$

if

$$\frac{\partial u}{\partial \xi} + \frac{\partial v}{\partial \eta} = 0, \qquad \frac{\partial u}{\partial \eta} - \frac{\partial v}{\partial \xi} = 0,$$

then

$$\frac{\partial^2 u}{\partial \xi^2} + \frac{\partial^2 v}{\partial \eta \partial \xi} = 0, \qquad \frac{\partial^2 u}{\partial \eta^2} - \frac{\partial^2 v}{\partial \xi \partial \eta} = 0$$

and

$$\frac{\partial^2 u}{\partial\xi\partial\eta} + \frac{\partial^2 v}{\partial\eta^2} = 0, \qquad \frac{\partial^2 u}{\partial\eta\partial\xi} - \frac{\partial^2 v}{\partial\xi^2} = 0$$

so that

$$\nabla^2 u = 0, \qquad \nabla^2 v = 0.$$

and, conversely, if $u(\xi, \eta)$ is given such that

$$\nabla^2 u = 0,$$

then we can construct a $v(\xi, \eta)$ such that

$$\frac{\partial v}{\partial\xi} = \frac{\partial u}{\partial\eta}, \qquad \frac{\partial v}{\partial\eta} = -\frac{\partial u}{\partial\xi}.$$

For, since $\nabla \times \left(\frac{\partial u}{\partial\eta}, -\frac{\partial u}{\partial\xi}, 0\right) = \nabla \times \left(\frac{\partial v}{\partial\xi}, \frac{\partial v}{\partial\eta}, 0\right) = 0$, therefore, in a simply connected region,

$$v(\xi, \eta) - v(\xi_0, \eta_0) = \int_{(\xi_0,\eta_0)}^{(\xi,\eta)} \left(\frac{\partial u}{\partial\eta}d\xi - \frac{\partial u}{\partial\xi}d\eta\right),$$

independently of the path. It is easy to verify that also

$$\nabla^2 v = 0.$$

It is also easy to verify that

$$u = a\log[(\xi - \xi_o)^2 + (\eta - \eta_0)^2] + b,$$

where a and b are constants, is a solution of the Laplacian.

The problem can be considerably generalized. Consider

$$A(x,y)\frac{\partial^2\psi}{\partial x^2} + 2B(x,y)\frac{\partial^2\psi}{\partial x\partial y} + C(x,y)\frac{\partial^2\psi}{\partial y^2} + f\left(x,y,\psi,\frac{\partial\psi}{\partial x},\frac{\partial\psi}{\partial y}\right) = 0.$$

Let

$$p = p(x,y), \quad q = q(x,y), \quad \frac{\partial\psi}{\partial x} = \frac{\partial\overline{\psi}}{\partial p}\frac{\partial p}{\partial x} + \frac{\partial\overline{\psi}}{\partial q}\frac{\partial q}{\partial x}, \qquad \text{etc.}$$

as before. We finally have

$$\left[A \left(\frac{\partial p}{\partial x} \right)^2 + 2B \left(\frac{\partial p}{\partial x} \frac{\partial p}{\partial y} \right) + C \left(\frac{\partial p}{\partial y} \right)^2 \right] \frac{\partial^2 \overline{\psi}}{\partial p^2}$$
$$+ \left[A \frac{\partial p}{\partial x} \frac{\partial q}{\partial x} + B \left(\frac{\partial p}{\partial x} \frac{\partial q}{\partial y} + \frac{\partial p}{\partial y} \frac{\partial q}{\partial x} \right) + C \left(\frac{\partial p}{\partial y} \frac{\partial q}{\partial y} \right) \right] \frac{\partial^2 \overline{\psi}}{\partial p \partial q}$$
$$+ \left[A \left(\frac{\partial q}{\partial x} \right)^2 + 2B \left(\frac{\partial q}{\partial x} \frac{\partial q}{\partial y} \right) + C \left(\frac{\partial q}{\partial y} \right)^2 \right] \frac{\partial^2 \overline{\psi}}{\partial q^2}$$
$$+ \overline{f} \left(x, y, \overline{\psi}, \frac{\partial \overline{\psi}}{\partial p}, \frac{\partial \overline{\psi}}{\partial q} \right) = 0,$$

where, of course,

$$\begin{vmatrix} \dfrac{\partial p}{\partial x} & \dfrac{\partial p}{\partial y} \\ \dfrac{\partial q}{\partial x} & \dfrac{\partial q}{\partial y} \end{vmatrix} \neq 0,$$

and

$$\begin{vmatrix} \alpha & \beta \\ \beta & \gamma \end{vmatrix} = \begin{vmatrix} A & B \\ B & C \end{vmatrix} \begin{vmatrix} \dfrac{\partial p}{\partial x} & \dfrac{\partial p}{\partial y} \\ \dfrac{\partial q}{\partial x} & \dfrac{\partial q}{\partial y} \end{vmatrix}^2 ,$$

α, β, and γ being the coefficients of $\frac{\partial^2 \overline{\psi}}{\partial p^2}$, $\frac{\partial^2 \overline{\psi}}{\partial p \partial q}$, and $\frac{\partial^2 \overline{\psi}}{\partial q^2}$. Then we wish as before to choose p and q so that

$$\alpha = \gamma = 0 \text{ and } \beta \neq 0,$$

that is, we must solve

$$AZ^2 + 2BZ + C = A(Z - \lambda_1)(Z - \lambda_2) = 0$$

where $\lambda_1 \neq \lambda_2$. This is possible, if and only if

$$B^2 - AC \neq 0, \qquad A \neq 0 \qquad \text{or } C \neq 0.$$

Case 1. $B^2 - AC > 0$. Then

$$A \left(\frac{\partial p}{\partial x} \middle/ \frac{\partial p}{\partial y} \right)^2 + 2B \left(\frac{\partial p}{\partial x} \middle/ \frac{\partial p}{\partial y} \right) + C = 0,$$

$$A \left(\frac{\partial q}{\partial x} \middle/ \frac{\partial q}{\partial y} \right)^2 + 2B \left(\frac{\partial q}{\partial x} \middle/ \frac{\partial q}{\partial y} \right) + C = 0;$$

$$\frac{\partial p}{\partial x} \middle/ \frac{\partial p}{\partial y} = \frac{-B + \sqrt{B^2 - AC}}{A} = \lambda_1(x, y),$$

$$\frac{\partial q}{\partial x} \middle/ \frac{\partial q}{\partial y} = \frac{-B - \sqrt{B^2 - AC}}{A} = \lambda_2(x, y).$$

This suggests that we solve

$$\frac{dy}{dx} = \frac{-\partial p}{\partial x} \middle/ \frac{\partial p}{\partial y} = -\lambda_1(x, y) \quad \text{and} \quad \frac{dy}{dx} = -\frac{\partial q}{\partial x} \middle/ \frac{\partial q}{\partial y} = -\lambda_2(x, y).$$

to have two real functions, $p(x, y)$ and $q(x, y)$, with the proper assumptions concerning A, B, and C. There will be two families of unique solutions

$$p(x, y) = \mu, \qquad q(x, y) = \nu, \qquad \mu \text{ and } \nu \text{ constants.}$$

Then we have

$$\alpha\gamma - \beta^2 = (AC - B^2) \left(\frac{\partial p}{\partial x} \frac{\partial q}{\partial y} - \frac{\partial p}{\partial y} \frac{\partial q}{\partial x} \right)^2 < 0$$

$$\beta^2 > 0,$$

So that, dividing through by β, we have

$$\frac{\partial^2 \overline{\psi}}{\partial p \partial q} = F \left(p, q, \overline{\psi}, \frac{\partial \overline{\psi}}{\partial p}, \frac{\partial \overline{\psi}}{\partial q} \right).$$

Case 2. $B^2 - AC < 0$. This is just as Case 1, if we allow complex coefficients. To avoid this, we consider a further transformation as before:

$$\xi = \frac{p + q}{2}, \qquad \eta = \frac{p - q}{2i},$$

so that we get

$$\frac{\partial^2 \overline{\psi}}{(\partial \xi)^2} + \frac{\partial^2 \overline{\psi}}{(\partial \eta)^2} = C_T \left(\xi, \eta, \frac{\partial \overline{\psi}}{\partial \xi}, \frac{\partial \overline{\psi}}{\partial \eta}, \overline{\psi} \right).$$

Case 3. $B^2 - AC = 0$. Then $\lambda_1 = \lambda_2$ and we find $p(x, y)$ so that $\alpha = 0$ and also $\beta = 0$. We can choose $q(x, y)$ at will so that $\gamma \neq 0$. Then

$$\frac{\partial^2 \overline{\psi}}{\partial q^2} = C_T \left(p, q, \overline{\psi}, \frac{\partial \overline{\psi}}{\partial p}, \frac{\partial \overline{\psi}}{\partial q} \right).$$

So far we have only dealt with a plane wave-front, suitable only for use when the light source is at a great distance. To fit the Huygens' theory of a spherical disturbance, we need to consider a function

$$\psi = \psi(r, t), \qquad r = (x^2 + y^2 + z^2)^{\frac{1}{2}}$$

where the light source is at the origin. This function is obviously constant over the sphere

$$r^2 = \lambda^2$$

at a given time. We have

$$\frac{\partial \psi}{\partial x} = \frac{\partial \psi}{\partial r} \frac{\partial r}{\partial x}, \qquad \frac{\partial^2 \psi}{\partial x^2} = \frac{\partial^2 \psi}{\partial r^2} \left(\frac{\partial r}{\partial x} \right)^2 + \frac{\partial \psi}{\partial r} \frac{\partial^2 r}{\partial x^2} \qquad \text{etc.}$$

Hence

$$\nabla^2 \psi = \frac{\partial^2 \psi}{\partial r^2} \left[\left(\frac{\partial r}{\partial x} \right)^2 + \left(\frac{\partial r}{\partial y} \right)^2 + \left(\frac{\partial r}{\partial z} \right)^2 \right] + \frac{\partial \psi}{\partial r} \left[\frac{\partial^2 r}{\partial x^2} + \frac{\partial^2 r}{\partial y^2} + \frac{\partial^2 r}{\partial z^2} \right]$$

$$= \frac{\partial^2 \psi}{\partial r^2} + \frac{\partial \psi}{\partial r} \left(\frac{2r^2}{r^3} \right) = \frac{\partial^2 \psi}{\partial r^2} + \frac{2}{r} \frac{\partial \psi}{\partial r}$$

since

$$\frac{\partial r}{\partial x} = \frac{x}{r}, \qquad \frac{\partial^2 r}{\partial x^2} = \frac{r^2 - x^2}{r^3} \qquad \text{etc.}$$

But

$$\frac{\partial^2 (r\psi)}{\partial r^2} = \frac{\partial}{\partial r} \left[\psi + r \frac{\partial \psi}{\partial r} \right] = 2 \frac{\partial \psi}{\partial r} + r \frac{\partial^2 \psi}{\partial r^2}.$$

Therefore

$$\nabla^2 \psi = \frac{1}{r} \frac{\partial^2 (r\psi)}{\partial r^2} \tag{4.19}$$

If then we are to have a solution of

$$\nabla^2 \psi = \frac{1}{v^2} \frac{\partial^2 \psi}{\partial t^2},$$

we must have a solution of

$$\frac{1}{r}\frac{\partial^2(r\psi)}{\partial r^2} = \frac{1}{v^2}\frac{\partial^2\psi}{\partial t^2}$$

$$\frac{\partial^2(r\psi)}{\partial r^2} = \frac{r}{v^2}\frac{\partial^2\psi}{\partial t^2} = \frac{1}{v^2}\frac{\partial^2(r\psi)}{\partial t^2} \tag{4.20}$$

But this is the same form as (4.7) and (4.11) with ψ replaced by $r\psi(r,t)$. Therefore the general solution is

$$r\psi = \psi_1(r + vt) + \psi_2(r - vt)$$
$$\psi = \frac{1}{r}[\psi_1(r + vt) + \psi_2(r - vt)] \tag{4.21}$$

that is, a disturbance moving to the center and from the center. It is to be noticed that the amplitude is inversely as the distance. The kinetic energy of the vibrating particle of aether is proportional to

$$\frac{1}{2}\left(\frac{\partial\psi}{\partial t}\right)^2 = \frac{1}{2}\frac{a^2}{r^2}4\pi^2\omega^2\cos^2\left(2\pi\omega\left(t - \frac{r}{v}\right)\right) \tag{4.22}$$

if one uses the special case of the solutions (4.21) representing a disturbance moving from the center. It is inversely proportional to the square of the distance from the center of disturbance. Fresnel, therefore, explains Kepler's law of the variation of illumination inversely as the square of the distance from the primary source by identifying the intensity of light with the kinetic energy of the aether particles. This led Fresnel immediately in the steps of Leibniz: Any loss of kinetic energy of light must result in heat (See Fresnel, **Theorie de Lumiere**, in Oeuvres Complètes, T.II, p.44, n.1).

In the emission theory, naturally, the intensity of light is measured by the number of particles crossing a unit surface normal to their direction in unit time.

E   The Dynamics of the Vibrating String.

The full mathematical theory of the vibrations of an elastic solid had not been completed at the time Fresnel was supposing the aether to be an elastic solid with only transverse vibrations. But the theory of the vibrations of the stretched string had been well developed by Bernoulli, d'Alembert, and Lagrange, and it is with this that Young and Fresnel will build the analogies of their theory of light. At this point, we give an elementary treatment of the problem, deferring the Lagrangian treatment until later.

Suppose we have a stretched string of homogeneous density under constant tension. We can consider the length as infinite or finite. If it is finite of length ℓ, our problem is to find a solution

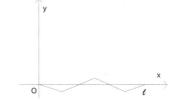

$$y = f(x,t)$$

Fig. 114

which gives the value of y for every point of the string at any distance x from 0 to $x = \ell$ at any time $t \geq 0$, when it starts from a certain position and at a certain velocity. Hence

$$f(0,t) = f(\ell,t) = 0 \tag{5.1}$$

are the boundary conditions, and the initial conditions are

$$f(x,0) = F(x), \qquad \left.\frac{\partial f(x,t)}{\partial t}\right|_{t=0} = G(x) \tag{5.2}$$

where there are boundary conditions

$$G(0) = G(\ell) = 0, \qquad \left.\frac{\partial f(x,t)}{\partial t}\right|_{x=0} = \left.\frac{\partial f(x,t)}{\partial t}\right|_{x=\ell} = 0 \tag{5.3}$$

One supposes first that he has n equal mass points m on a weightless string at equal intervals Δx. Let us consider the equation of motion of the rth mass point. We assume that the displacements of the tightly stretched string are very small, not exceeding the elastic limits, and such that S, the tension in the string, does not vary. More exactly we are neglecting the second power of θ and φ but not the first. Hence

$$\Delta \bar{x} - \Delta x = \Delta \bar{x}(1 - \cos\theta) = \Delta \bar{x}\left\{\frac{\theta^2}{2!} - \frac{\theta^4}{4!} + \ldots\right\} \approx 0 \tag{5.4}$$

The horizontal components of the forces acting on m_r are

$$-S\cos\theta + S\cos(\pi - \varphi) = -S\left[\frac{\theta^2}{2!} - \frac{(\pi-\varphi)^2}{2!} + \ldots\right] \approx 0,$$

since $(\pi - \varphi)$ is of the order of magnitude of θ. The sum of the vertical components, neglecting g which is very small with respect to S, is

$$-S\sin\theta - S\sin(\pi - \varphi) = -S[\sin\theta + \sin\varphi]$$

$$= -S\left[\frac{\theta}{1!} + \frac{(\pi-\varphi)}{1!} - \frac{\theta^3}{3!} - \frac{(\pi-\varphi)^3}{3!} + \ldots\right]$$

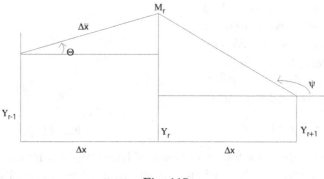

Fig. 115

which is not negligible. Hence by Newton's Axiom

$$\left. \begin{aligned} m\frac{d^2y_r}{dt^2} &= S[-\sin\theta - \sin\varphi] \\ &= S[(y_{r-1} - y_r) + (y_{r+1} - y_r)]\frac{1}{\Delta x} \end{aligned} \right\} \tag{5.5}$$

This will hold for all n mass points, with added conditions at the end points, so that

$$\left. \begin{aligned} m\frac{d^2y_1}{dt^2} &= S[(0 - y_1) + (y_2 - y_1)]\frac{1}{\Delta x} \\ m\frac{d^2y_r}{dt^2} &= S[(y_{r-1} - y_r) + (y_{r+1} - y_r)]\frac{1}{\Delta x} \\ m\frac{d^2y_n}{dt^2} &= S[(y_{n-1} - y_n) + (0 - y_n)]\frac{1}{\Delta x} \end{aligned} \right\} \tag{5.6}$$

Let us consider the limit of the rth term as $n \to \infty$ and $\Delta x \to 0$. We have

$$\lim_{\Delta x \to 0} \frac{d^2y_r}{dt^2} = \lim_{\Delta x \to 0} \frac{S}{m}[(y_{r-1} - y_r) + (y_{r+1} - y_r)]\frac{1}{\Delta x}$$

Now we have to let $m(\Delta x)$ vary as Δx varies in such a way as to keep the mass per unit length constant; this will give us finally a homogeneous density. Therefore let

$$\frac{m}{\Delta x} = \rho, \qquad m = \rho\Delta x,$$

be the constant density. Then we have

$$\lim_{\Delta x \to 0} \frac{d^2y_r}{dt^2} = \lim_{\Delta x \to 0} \frac{S}{\rho}\left[\frac{\frac{(y_{r+1} - y_r)}{\Delta x} - \frac{(y_r - y_{r-1})}{\Delta x}}{\Delta x}\right].$$

If now we remember we want y to be a continuous function of x and t, that is, at limit, $y_r = y(x, t)$ we get

$$\frac{\partial^2 y}{\partial t^2} = \frac{S}{\rho} \frac{\partial^2 y}{\partial x^2} \tag{5.7}$$

by Leibniz' well-known theorem on the limit of the second differences and the second derivative. By transfinite set theory, any continuous function determined at every rational point of an interval is determined thereby at every real point of the interval. We have only to compare (5.7) with (4.11) to see that the solution of this equation is a transverse wave with velocity of propagation

$$v^2 = \frac{S}{\rho}, \qquad v = \sqrt{\frac{S}{\rho}} \tag{5.8}$$

It is on this last equation that the Fresnel theory of light is built.

The general solution of (5.7), we have already found as

$$y = \psi_1(x + vt) + \psi_2(x - vt) \tag{5.9}$$

where v is given by (5.8). But, from the boundary conditions (5.1),

$$\psi_1(vt) + \psi_2(-vt) = 0 = \psi_1(\ell + vt) + \psi_2(\ell - vt) \tag{5.10}$$

from which, and from which alone, we can deduce the periodicity of the functions ψ_1 and ψ_2. For, with $0 \leq x \leq \ell$ and $t \geq 0$,

$$\psi_1(vt) = -\psi_2(-vt) = -\psi_2(\ell - \overline{vt + \ell}),$$
$$\psi_1(\ell + vt) = -\psi_2(\ell - vt).$$

Substituting $vt + \ell$ for vt in this last, we have

$$-\psi_2(\ell - \overline{vt + \ell}) = \psi_1(2\ell + vt) = \psi_1(vt);$$

and $vt - \ell$, we have

$$\psi_1(vt) = -\psi_2(2\ell - vt) = -\psi_2(-vt),$$

so that ψ_1 and ψ_2 are both of period 2ℓ. Translated in terms of t, this means a period of $\frac{2\ell}{v}$.

From the initial conditions (5.2) with (5.9), we get

$$\left.\begin{array}{c}\psi_1(x) + \psi_2(x) = F(x); \\ \left.\frac{\partial y}{\partial t}\right|_{t=0} = \left[\frac{\partial \psi_1}{\partial t} + \frac{\partial \psi_2}{\partial t}\right]_{t=0} = G(x), \\ \frac{\partial \psi_1(x)}{\partial x} - \frac{\partial \psi_2(x)}{\partial x} = \frac{1}{v}G(x).\end{array}\right\} \tag{5.11}$$

Since (5.10) and (5.11) define the arbitrary functions ψ_1 and ψ_2 only to within an arbitrary constant, therefore let

$$\psi_1(0) = 0, \qquad \psi_2(0) = 0.$$

Then

$$\int_0^x \left[\frac{\partial \psi_1(x)}{\partial x} - \frac{\partial \psi_2(x)}{\partial x}\right] dx = \int_0^x \frac{1}{v}G(x)dx$$
$$\psi_1(x) - \psi_2(x) = \int_0^x \frac{1}{v}G(x)dx;$$
$$\psi_1(x) + \psi_2(x) = F(x)$$

so that

$$\left.\begin{array}{c}\psi_1(x) = \frac{1}{2}\left[F(x) + \frac{1}{v}\int_0^x G(x)dx\right] \\ \psi_2(x) = \frac{1}{2}\left[F(x) - \frac{1}{v}\int_0^x G(x)dx\right]\end{array}\right\} \tag{5.12}$$

If we are dealing with an unbounded string, the boundary conditions do not hold and the periodicity of the motion cannot be deduced. The natural analogy with Fresnel's theory is with a wave traveling in one direction on a string held at one end and infinite in the direction of the wave's movement. For this we would simply have to change the boundary conditions.

There is another and deeper approach which will appear later, using the methods of Bernoulli and Lagrange, but which we will prepare for now. We can express the total kinetic energy and potential energy of the system of n mass points on the weightless elastic string. The kinetic energy T is obviously

$$T = \frac{1}{2}m \sum_r \left(\frac{dy_r}{dt}\right)^2, \tag{5.13}$$

and the potential energy,

$$
\begin{aligned}
V &= \sum_r S\left\{\left[(\Delta x)^2 + (y_r - y_{r-1})^2\right]^{\frac{1}{2}} - \Delta x\right\} \\
&\approx \frac{1}{2}\sum_r S\Delta x\left(\frac{y_r - y_{r-1}}{\Delta x}\right)^2
\end{aligned}
\tag{5.14}
$$

where S is the tension, and where one has neglected all terms beyond the second power of the binomial expansion. If one passes to the limit, then

$$
T = \frac{1}{2}\int_0^\ell \rho\left(\frac{\partial y}{\partial t}\right)^2 dx \tag{5.15}
$$

$$
V = \frac{1}{2}\int_0^\ell S\left(\frac{\partial y}{\partial x}\right)^2 dx \tag{5.16}
$$

are the kinetic and potential energies of the continuous string.

F   The Dynamic Theory of Young and Fresnel

The principal achievement of Fresnel, aided by the intuitions of Young, was to construct a dynamic theory of transverse vibrations of light, in analogy with those of the elastic string, which would predict the phenomena of diffraction, the double refraction of Iceland Spar, the experiments of Malus, the double refraction of biaxial crystals, and the formulas for partial reflection. His theory, moreover, predicted the approximately correct formula for the drag-coefficient whose existence at the time was not even

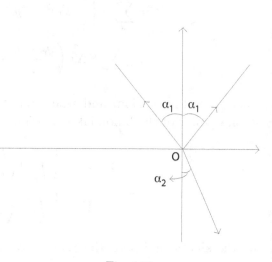

Fig. 116

suspected. When the full mathematical theory of an elastic solid is accomplished, the inadequacy of the Fresnel theory will appear. In particular, it will be seen that one cannot get rid of the longitudinal waves which have no counterpart in light. But the heuristic value and power of deduction for a wide variety of results make it an important application of the elastic field theory, however imperfect.

We shall not here treat the theory of diffraction which is derived from the Principle of Huygens. But we shall derive the others. Let us start with the theory of partial refraction and partial reflection.[3] If ψ_1 represents the disturbance of the incident wave, ψ_2 of reflected wave, and ψ_3 of refracted

---

[3]We here allow for possible differences in passing from the incident light to the reflected and refracted light as may be demanded by subsequent assumptions with respect to boundary conditions and to the conservation of kinetic energy. If we let the amplitudes take on negative values, we shall find that δ_1 and δ_2 are zero except in the important case of total reflection. Then we shall see that, by (6.11),

$$\frac{\dot{\psi}_2}{\psi_1} = \frac{-\sin(\alpha_1 - \alpha_2)}{\sin(\alpha_1 + \alpha_2)},$$

so that there is a phase jump of π from the incident to the reflected wave.

wave, we have in xy-plane,

$$\psi_1 = a \cos\left(2\pi\omega_1 \left[t - \frac{x \sin\alpha_1 - y \cos\alpha_1}{v_1} \right] \right)$$

$$\psi_2 = b \cos\left(2\pi\omega_1 \left[t - \frac{x \sin\alpha_1 + y \cos\alpha_1}{v_1} + \delta_1 \right] \right)$$

$$\psi_3 = c \cos\left(2\pi\omega_2 \left[t - \frac{x \sin\alpha_2 - y \cos\alpha_2}{v_2} + \delta_2 \right] \right)$$

We let the surface between two ordinary refracting media (excluding crystals) be through x-axis normal to the paper. At $y = 0$,

$$\psi_1 = a \cos\left(2\pi\omega_1 \left[t - \frac{x \sin\alpha_1}{v_1} \right] \right)$$

$$\psi_2 = b \cos\left(2\pi\omega_1 \left[t - \frac{x \sin\alpha_1}{v_1} + \delta_1 \right] \right)$$

$$\psi_3 = c \cos\left(2\pi\omega_2 \left[t - \frac{x \sin\alpha_2}{v_2} + \delta_2 \right] \right)$$

$$= c \cos\left(2\pi\omega_2 \left[t - \frac{x \sin\alpha_1}{v_1} + \delta_2 \right] \right)$$

since in the Huygens' wave theory

$$\frac{\sin\alpha_1}{\sin\alpha_2} = \frac{v_1}{v_2}.$$

The first assumption of Fresnel is that, in passing from one ordinary medium to a second, the elastic constant of the medium does not vary, but its density does. The variation of the elastic constant will be reserved for use in the case of doubly refracting crystals. Therefore[4]

$$\omega_1 = \omega_2. \tag{6.1}$$

We shall consider the disturbances always as analyzed into two components: (1) those normal to the plane of incidence; (2) those parallel to the plane of incidence. It is understood, of course, that all vibrations of the aether particles are normal to the direction of propagation or parallel to the plane-front of the wave.

---

[4]Since $\dfrac{d^2\psi_1}{dt^2} = -4\pi^2\omega_1^2\psi_1$, ω_1^2 is proportional to the elastic constant.

The further assumptions are that, in the case of (1), at the surface of reflection,

$$\psi_1(t) + \psi_2(t) = \psi_3(t) \tag{6.2}$$

and therefore

$$\dot{\psi}_1(t) + \dot{\psi}_2(t) = \dot{\psi}_3(t) \tag{6.3}$$

And, in the case of (2), these conditions of continuity hold only for the components parallel to the plane between the two media

$$\psi_1 \cos\alpha_1 - \psi_2 \cos\alpha_1 = \psi_3 \cos\alpha_2, \tag{6.4}$$

$$\dot{\psi}_1 \cos\alpha_1 - \dot{\psi}_2 \cos\alpha_1 = \dot{\psi}_3 \cos\alpha_2 \tag{6.5}$$

The determination of the sign $\dot{\psi}_2$ gave Fresnel some trouble. The final decision was in view of the theory of total reflection to follow. Also the fact that if the incident light is plane polarized, then, when $\alpha_1 = \pi$, the reflected light is polarized in the same plane, demands this sign convention: namely that, from the Fresnel point of view, $\dot{\psi}_1 \sin\alpha_1$ has the same sign as $\dot{\psi}_2 \sin\alpha_1$. The directions of positive $\dot{\psi}_1$ and $\dot{\psi}_2$ are given in the figure. The same thing is achieved in the electromagnetic theory of light by a different argument. One gets (6.5) again if one takes the angles from the vector of the light in the direction of

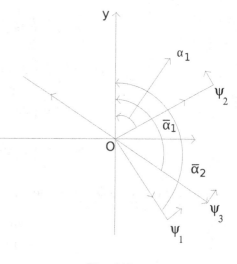

Fig. 117

its propagation as initial side to the positive y-axis as terminal side. Thus

$$\dot{\psi}_1 \cos\overline{\alpha}_1 + \dot{\psi}_2 \cos\alpha_1 = \dot{\psi}_3 \cos\overline{\alpha}_2$$

$$\overline{\alpha}_1 = \pi - \alpha_1, \quad \overline{\alpha}_2 = \pi - \alpha_2.$$

For both cases, it is assumed that kinetic energy is conserved before and after incidence so that, for (1) and (2)

$$m_1(\dot{\psi}_1)^2 = m_1(\dot{\psi}_2)^2 + m_2(\dot{\psi}_3)^2 \tag{6.6}$$

where m_1 is the mass of the aetheral wave in first medium relative to the mass of the aetheral wave in second medium. The mass of the reflected wave is the same as that of the incident wave of necessity from the theory. For both λ and ρ are same.

The assumptions (6.4) and (6.5) are strange and were immediately criticized: why should continuity only be assumed for the components parallel to the plane between the two media? And what is the idea of the mass and density of the weightless aether? It is simply a dynamic analogue of the ordinary mass.

In any case, we proceed with Fresnel's method. In accordance with the Huygens' theory

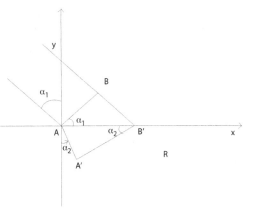

Fig. 118

$$\frac{\text{vol. of wave in } M_1}{\text{vol. of wave in } M_2} = \frac{(AB)\lambda_1}{(A'B')\lambda_2} = \frac{\lambda_1 \cos \alpha_1}{\lambda_2 \cos \alpha_2},$$

where $AB/A'B'$ is the ratio of the wave-fronts in M_1, the first medium, and M_2 the second. But

$$v_1 = \lambda_1 \omega_1, \qquad v_2 = \lambda_2 \omega_2, \qquad \omega_1 = \omega_2$$

so that

$$\frac{\lambda_1}{\lambda_2} = \frac{v_1}{v_2} = \frac{\sin \alpha_1}{\sin \alpha_2}$$

and

$$\frac{\text{vol. of wave in } M_1}{\text{vol. of wave in } M_2} = \frac{\sin \alpha_1 \cos \alpha_1}{\sin \alpha_2 \cos \alpha_2} \tag{6.7}$$

From (5.8) in the theory of the vibrating string, we have

$$\frac{v_1}{v_2} = \sqrt{\frac{S_1}{\rho_1}} \Big/ \sqrt{\frac{S_2}{\rho_2}}$$

where S is the elastic constant and ρ the density. But we have assumed

$$S_1 = S_2$$

so that

$$\frac{v_1}{v_2} = \frac{\sqrt{\rho_2}}{\sqrt{\rho_1}} = \frac{\sin \alpha_1}{\sin \alpha_2} \tag{6.8}$$

and therefore

$$\frac{m_1}{m_2} = \frac{\text{vol.}_1 \cdot \rho_1}{\text{vol.}_2 \cdot \rho_2} = \frac{\cos \alpha_1 \sin \alpha_2}{\cos \alpha_2 \sin \alpha_1} \tag{6.9}$$

It is now seen why the mass of the incident wave and the partially reflected wave are equal: the densities are the same as are the velocities of propagation since they are in the same medium.

If we now combine (6.9) and (6.6), we get for both cases (1) and (2)

$$\frac{\cos\alpha_1}{\sin\alpha_1}(\dot\psi_1^2 - \dot\psi_2^2) = \frac{\cos\alpha_2}{\sin\alpha_2}\dot\psi_3^2 \tag{6.10}$$

For (1), the vibrations normal to the plane of incidence, we divide (6.10) by (6.3) squared, to get

$$\frac{\cos\alpha_1}{\sin\alpha_1}\frac{\dot\psi_1 - \dot\psi_2}{\dot\psi_1 + \dot\psi_2} = \frac{\cos\alpha_2}{\sin\alpha_2}$$

$$\left(\frac{\dot\psi_2}{\dot\psi_1}\right)_N = \frac{\cos\alpha_1\sin\alpha_2 - \cos\alpha_2\sin\alpha_1}{\cos\alpha_1\sin\alpha_2 + \sin\alpha_1\cos\alpha_2} = \frac{\sin(\alpha_1 - \alpha_2)}{\sin(\alpha_1 + \alpha_2)}. \tag{6.11}$$

The reflection is null when $\dot\psi_2 = 0$. So there is total transmission of the light if and only if

$$\tan\alpha_2 = \tan\alpha_1, \qquad \alpha_2 = \alpha_1, \qquad \alpha_2 \neq 0, \qquad \alpha_1 \neq 0.$$

Hence for two different media, $\dot\psi_2 \neq 0$ and there is always partial reflection. For the indeterminate case where $\alpha_1 = \alpha_2 = 0$, we naturally consider

$$\left(\frac{\dot\psi_2}{\dot\psi_1}\right)_N = \lim_{\alpha_1,\alpha_2\to 0}\frac{\cos\alpha_1 - \cos\alpha_2\frac{\sin\alpha_1}{\sin\alpha_2}}{\cos\alpha_1 + \cos\alpha_2\frac{\sin\alpha_1}{\sin\alpha_2}} = \frac{1 - \mu_{12}}{1 + \mu_{12}} \tag{6.12}$$

In judging the relative intensities one must, of course, consider

$$\frac{(\dot\psi_2)^2}{(\dot\psi_1)^2} = \frac{(1 - \mu_{12})^2}{(1 + \mu_{12})^2}.$$

Thus, from air to water, where $\mu_{12} = 4/3$, we have

$$\frac{(\dot\psi_2)^2}{(\dot\psi_1)^2} = \frac{1}{49},$$

a partial reflection of about 2%. From air to crown glass where $\mu_{12} = 3/2$,

$$\frac{(\dot\psi_2)^2}{(\dot\psi_1)^2} = \frac{1}{25}.$$

For (2), the vibrations parallel to the plane of incidence, we divide (6.10) by (6.5) squared to get

$$\frac{\sin \alpha_2(\dot\psi_1 - \dot\psi_2)}{\cos \alpha_1(\dot\psi_1 + \dot\psi_2)} = \frac{\sin \alpha_1}{\cos \alpha_2}$$

$$\left(\frac{\dot\psi_2}{\dot\psi_1}\right)_P = +\frac{\tan(\alpha_1 - \alpha_2)}{\tan(\alpha_1 + \alpha_2)} = -\frac{\sin \alpha_1 \cos \alpha_1 - \sin \alpha_2 \cos \alpha_2}{\sin \alpha_1 \cos \alpha_1 + \sin \alpha_2 \cos \alpha_2} \tag{6.13}$$

Here $\dot\psi_2 = 0$, when $\tan(\alpha_1 - \alpha_2) = 0$ and $\tan(\alpha_1 + \alpha_2) \neq 0$. This is the trivial case $\alpha_1 = \alpha_2$. But $\dot\psi_2 = 0$ also when $\tan(\alpha_1 + \alpha_2) = \infty$ and $\tan(\alpha_1 - \alpha_2)$ is some finite number. Such is the case when

$$\alpha_1 + \alpha_2 = \frac{\pi}{2}, \qquad \alpha_1 = \frac{\pi}{2} - \alpha_2 \tag{6.14}$$

This is the angle of maximum polarization of Brewster's Law in (3.1), where

$$\tan \theta_0 = \frac{\sin \theta_0}{\cos \theta_0} = \frac{\sin \theta_0}{\sin \theta_1} \tag{6.15}$$

where θ_0 is the angle of incidence and θ_1 angle of refraction.
Then

$$\cos \theta_0 = \sin \theta_1, \qquad \theta_0 = \frac{\pi}{2} - \theta_1.$$

Hence, by the deduction of Fresnel, natural light which consists in vibrations of the aether in all directions normal to the direction of propagation is affected in such a way, by the passage from one ordinary medium to another, that there is always partial reflection and refraction (or transmission) for any angle of incidence of the components of vibration normal to the plane of incidence, while for the components parallel to the plane of incidence, when the angle of incidence is that of Brewster's Law, there is total transmission and no reflection. Hence, at this angle, the reflected light is plane polarized so that the vibrations are normal to the plane of incidence of the water or glass from which the light is reflected. If now we remember the results of Malus, we see that the ray of ordinary refraction is plane polarized normal to the principal section while the extraordinary ray is plane polarized parallel to the principal section. For, according to Malus' experiment, the light reflected from water at the angle of maximum polarization is refracted in Iceland crystal only ordinarily when the principal section is parallel to the plane of incidence of water, and only extraordinarily when the principal section is normal to the plane of incidence. This ties in with the observation of Fresnel that the extraordinary and ordinary rays will produce no interference patterns.

Conversely, when the rays from Iceland spar are allowed to strike the reflecting surface of the water at the appropriate angle, then, when the principal section is parallel to the plane of incidence of the water, the light of the ordinary ray is partially reflected and partially transmitted, since, according to the theory of Fresnel, the ordinary ray is polarized normal to the plane of the principal section and is hence normal to the plane of incidence of the water. Again the extraordinary ray is polarized parallel to the principal section (and normal to the direction of propagation) and hence parallel to the plane of incidence of the water and so is totally transmitted. The reader can verify the other combinations of Malus' observations.

If we now deduce in the same way the ratios of the refracted wave with the incident wave, we get

$$\left(\frac{\dot{\psi}_3}{\dot{\psi}_1}\right)_N = \frac{2\sin\alpha_2\cos\alpha_1}{\sin(\alpha_1+\alpha_2)} = \frac{2\cos\alpha_1}{\cos\alpha_1+\mu_{12}\cos\alpha_2} \tag{6.16}$$

$$\left(\frac{\dot{\psi}_3}{\dot{\psi}_1}\right)_P = \frac{2\sin\alpha_2\cos\alpha_1}{\sin(\alpha_1+\alpha_2)\cos(\alpha_1-\alpha_2)} = \frac{2\cos\alpha_1}{\mu_{12}\cos\alpha_1+\cos\alpha_2} \tag{6.17}$$

Suppose now the incident light is plane polarized so that its vibrations are in a plane making an angle β_1 with the plane of incidence. This is called the azimuth angle. Let β_2 be the azimuth angle of the reflected light and β_3 that of the refracted light. Then

$$\left(\frac{\dot{\psi}_2}{\dot{\psi}_1}\right)_N \Big/ \left(\frac{\dot{\psi}_2}{\dot{\psi}_1}\right)_P = \tan\beta_2/\tan\beta_1 = \frac{-\cos(\alpha_1-\alpha_2)}{\cos(\alpha_1+\alpha_2)} \tag{6.18}$$

and

$$\left(\frac{\dot{\psi}_3}{\dot{\psi}_1}\right)_N \Big/ \left(\frac{\dot{\psi}_3}{\dot{\psi}_1}\right)_P = \tan\beta_3/\tan\beta_1 = \cos(\alpha_1-\alpha_2);. \tag{6.19}$$

When $\alpha_1 = \frac{\pi}{2}$, we see that

$$\tan\beta_2 = \tan\beta_1.$$

So far the theory of Fresnel has only explained those appearances which were already observed and for which the theory, however far removed from the observations, was in some sense constructed. But Fresnel built, from his theory, phenomenon not yet observed which was immediately confirmed. To this end, we develop (6.11) and (6.13) for the case of total reflection. That is, when $\alpha_2 > \alpha_1$ and $\alpha_2 = \frac{\pi}{2}$ so that

$$\frac{\sin\alpha_1}{\sin\alpha_2} = \sin\alpha_1 = \mu_{12} < 1 \tag{6.20}$$

But we can write (6.11)

$$\left(\frac{\dot{\psi}_2}{\dot{\psi}_1}\right)_N = \frac{\cos\alpha_1 - \mu_{12}\cos\alpha_2}{\cos\alpha_1 + \mu_{12}\cos\alpha_2} \tag{6.21}$$

and (6.13)

$$\left(\frac{\dot{\psi}_2}{\dot{\psi}_1}\right)_P = \frac{\mu_{12}\cos\alpha_1 - \cos\alpha_2}{\mu_{12}\cos\alpha_1 + \cos\alpha_2} \tag{6.22}$$

But

$$\cos\alpha_2 = \sqrt{1 - \sin^2\alpha_2} = \sqrt{1 - \frac{\sin^2\alpha_1}{\mu_{12}^2}}, \tag{6.23}$$

which has a critical value at $\sin\alpha_1 = \mu_{12}$, is real so long as $\sin\alpha_1 < \mu_{12}$, and becomes imaginary when $\sin\alpha_1 > \mu_{12}$, so that then

$$\cos\alpha_2 = \frac{1}{\mu_{12}}\sqrt{\sin^2\alpha_1 - \mu_{12}^2} \tag{6.24}$$

Substituting in (6.21) and (6.22), we get

$$\left(\frac{\dot{\psi}_2}{\dot{\psi}_1}\right)_N = \frac{\cos\alpha_1 - i\sqrt{\sin^2\alpha_1 - \mu_{12}^2}}{\cos\alpha_1 + i\sqrt{\sin^2\alpha_1 - \mu_{12}^2}} \tag{6.25}$$

for the normal components; and

$$\left(\frac{\dot{\psi}_2}{\dot{\psi}_1}\right)_P = \frac{\mu_{12}^2\cos\alpha_1 - i\sqrt{\sin^2\alpha_1 - \mu_{12}^2}}{\mu_{12}^2\cos\alpha_1 + i\sqrt{\sin^2\alpha_1 - \mu_{12}^2}} \tag{6.26}$$

for the parallel components.

It was Fresnel who first insisted, from the principle of continuity, that these imaginary formulas be interpreted and applied physically. It is to be noted that both formulas are in the form

$$\frac{\dot{\psi}_2}{\dot{\psi}_1} = \frac{z}{\bar{z}} \tag{6.27}$$

where z is complex and $\bar{z}$ its conjugate so that representing

$$z = ae^{i\theta}, \qquad \bar{z} = ae^{-i\theta}$$

we have in both cases

$$\frac{\dot{\psi}_2}{\dot{\psi}_1} = e^{i2\theta} = \cos 2\theta + i \sin 2\theta \tag{6.28}$$

Since

$$\frac{|\dot{\psi}_2|}{|\dot{\psi}_1|} = 1,$$

the amplitudes of the incident and reflected waves are equal as we should expect. If we recall the complex representation

$$y = Ae^{i2\pi[\omega(t-\frac{x}{v})+\delta]}, \qquad \dot{y} = i2\pi\omega Ae^{i2\pi[\omega(t-\frac{x}{v})+\delta]},$$

then obviously, since ω is the same in z and $\bar{z}$ as well as t and x, the θ of (6.28) is half the difference in phase $\bar{\delta}$,

$$2\theta = (2\pi)\delta = \bar{\delta}, \qquad \theta = \frac{\bar{\delta}}{2}, \tag{6.29}$$

For the normal components,

$$\lambda \cos \theta = \cos \alpha_1, \qquad \lambda \sin \theta = \sqrt{\sin^2 \alpha_1 - \mu_{12}^2},$$

$$\tan \frac{\bar{\delta}_N}{2} = \frac{\sqrt{\sin^2 \alpha_1 - \mu_{12}^2}}{\cos \alpha_1} \tag{6.30}$$

and for the parallel components,

$$\lambda \cos \theta = \mu_{12}^2 \cos \alpha_1, \qquad \lambda \sin \theta = \sqrt{\sin^2 \alpha_1 - \mu_{12}^2},$$

$$\tan \frac{\bar{\delta}_P}{2} = \frac{\sqrt{\sin^2 \alpha_1 - \mu_{12}^2}}{\mu_{12}^2 \cos \alpha_1} \tag{6.31}$$

If now we wish the difference $\bar{\bar{\delta}}$ between the two phase-jumps which each set of components makes from incident to reflected wave, we have

$$\tan \frac{\bar{\bar{\delta}}}{2} = \frac{\tan \frac{\bar{\delta}_P}{2} - \tan \frac{\bar{\delta}_N}{2}}{1 + \tan \frac{\bar{\delta}_P}{2} \tan \frac{\bar{\delta}_N}{2}} = \frac{\cos \alpha_1 \sqrt{\sin^2 \alpha_1 - \mu_{12}^2}}{\sin^2 \alpha_1}. \tag{6.32}$$

This expression is zero at $\alpha_1 = \frac{\pi}{2}$ and at $\sin \alpha_1 = \mu_{12}$. We wish to find at what value of α_1 between these two, the tangent of the half-difference, and consequently the difference, is a maximum.

$$\frac{d}{d\alpha_1}\left(\tan \frac{\overline{\overline{\delta}}}{2}\right) = \frac{2\mu_{12}^2 - (1 + \mu_{12}^2)\sin^2 \alpha_1}{\sin^3 \alpha_1 \sqrt{\sin^2 \alpha_1 - \mu_{12}^2}} = 0.$$

$$\sin^2 \alpha_1 = \frac{2\mu_{12}^2}{1 + \mu_{12}^2} \tag{6.33}$$

$$\tan \frac{\overline{\overline{\delta}}}{2} = \frac{1 - \mu_{12}^2}{2\mu_{12}}, \text{maximum.} \tag{6.34}$$

Fresnel conceived of an experiment by which one would produce, by a sequence of total reflections of originally plane polarized homogeneous light where the normal and parallel component had equal amplitudes, a phase difference of 90° so that, according to the theory of Fresnel, the resulting light would consist of vibrations following a circular helix about the direction of propagation. Since it would by symmetrical with respect to the direction of propagation, such a light should, when passed through a crystal of Iceland spar give equal intensities of the ordinary and the extraordinary rays for different directions of the principal section contrary to what would happen with the original plane polarized light according to the experiments of Malus reported in (VI.3).

To this end, Fresnel took a block of glass where from air to glass $\mu_{21} = 1.51$, and $\mu_{12} = 0.6622$ from glass to air. For this index, the critical angle of total reflection is $\alpha_1 = 41°29'$, and the angle of incidence for maximum difference is $\alpha_1 = 51°20'$ by (6.33), and the maximum difference, by (6.34), is 45°56'. Therefore it is possible to get a difference of 45° and by two successive reflections a phase difference of 90°. Using (6.32) we solve for $\sin \alpha_1$ when $\frac{\overline{\overline{\delta}}}{2} = \frac{45°}{2}$. This gives two values

$$\alpha_1 = 48°37\frac{1}{2}', \alpha_1 = 54°37\frac{1}{3}';$$

Fresnel chose the second, for experimental reasons. Hence the glass was cut at an angle of $54°37\frac{1}{3}$, with the other faces normal to the base. Since $\alpha_1 = 54°37\frac{1}{3}' > 41°29'$, there is total reflection at the base of the glass A.

The light is plane polarized, the vibrations being in a plane through the line of propagation at an angle of 45° to the vertical plane of the paper; this makes the amplitudes of the normal and parallel components equal. If the light ray is introduced perpendicular to the face making $54°37\frac{1}{3}$, with the base, the light is reflected totally

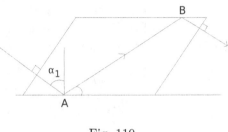

Fig. 119

at the base A at the same angle and again at the same angle at the opposite face at B. It then leaves the glass normal to the face. According to the theory of Fresnel it should be **circularly polarized** as described above. On passage through Iceland spar it behaved as Fresnel predicted. Further when the process was reversed, it produced the original plane polarized light while ordinary light would not do so. If the plane of polarization is tilted at an angle different from 45°, the amplitudes are in a constant ratio different from one and we get **elliptically polarized** light.

G   The Fresnel Theory for Production

So far the Fresnel theory has been applied to refraction and reflection between two ordinary media by using the analogy of the vibrating string, and varying the density of the aether in the two media. This development has also suggested a partial explanation of double refraction. But for a further deeper explanation of the dynamic reason for the double polarization of light in crystals, Fresnel will use a variation of the elastic constant S for different directions of vibration in the crystal. This theory will lend itself not only to the explanation of the behavior of uniaxial crystals such as Iceland Spar with one principal section but also of biaxial crystals with two principal sections.

For this purpose, Fresnel announces the following principle for which he gives a laborous and somewhat unsatisfactory proof: There exist, for every point of an elastic solid (the aether), three mutually perpendicular directions according to which any small displacement of the point (molecule of aether for Fresnel) produces a force in the direction of the displacement. For a homogeneous medium, these three axes of elasticity will be the same for all molecules in given medium (crystal). -

If, therefore, we choose these three orthogonal directions for the orthonormal basis $\{\underline{e}_1, \underline{e}_2, \underline{e}_3\}$, we have, in general, for the force exerted for unit displacement $\underline{u}$

$$\underline{F} = -(a^2 u_1 \underline{e}_1 + b^2 u_2 \underline{e}_2 + c^2 u_3 \underline{e}_3) \tag{7.1}$$

where a^2, b^2, c^2 are the elastic constants in the three directions. Then

$$\frac{\underline{u} \cdot \underline{F}}{|\underline{F}|} = \frac{-(a^2 u_1^2 + b^2 u_2^2 + c^2 u_3^2)}{|\underline{F}|} = \cos\varphi \tag{7.2}$$

$$-(a^2 u_1^2 + b^2 u_2^2 + c^2 u_3^2) = |\underline{F}| \cos\varphi = \underline{u} \cdot \underline{F} \tag{7.3}$$

is the component of $\underline{F}$ in the direction of the displacement $\underline{u}$. Now again, in analogy with the vibrating string where

$$v = \frac{\sqrt{S}}{\sqrt{\rho}},$$

S being elastic constant and ρ density, since the density in the crystal is assumed constant in agreement with the fundamental theory, absorbing the constant of proportionality into (a^2, b^2, c^2),

$$v^2 = |\underline{u} \cdot \underline{F}|, \qquad v^2 = a^2 u_1^2 + b^2 u_2^2 + c^2 u_3^2 \tag{7.4}$$

For reasons of convenience, we start with the surface.

$$\underline{x} = \frac{1}{v} \underline{u} \tag{7.5}$$

so that $|\underline{x}| = \frac{1}{v}$ always. Thus we have the velocity resulting from the anisotropic elastic medium for any direction $\underline{u}$. Given the plane-front of the light wave, we construct a plane parallel to it through the origin of this surface and consider the plane curve which is the intersection of this surface and the plane through the origin. The idea of Fresnel is to show that, omitting the component of $\underline{F}$ normal to this plane because of the assumed transverse character of the vibrations, there will be in this plane two normal directions such that the elastic force is in the direction of the displacement $\underline{u}$. These two normal directions will define the two directions of polarization of the vibrations of the light to give the ordinary and extraordinary rays with the corresponding relative velocities.

Thus the plane through the origin 0 parallel to the plane-front of light will be through the displacement $\underline{x}$. Let this plane be

$$A\xi + B\eta + C\zeta = 0 \tag{7.6}$$

Then

$$\underline{x} = \frac{1}{v}(\cos\alpha, \cos\beta, \cos\gamma) = \frac{1}{v}\underline{u} \tag{7.7}$$

and
$$A \cos \alpha + B \cos \beta + C \cos \gamma = 0 \tag{7.8}$$
$$v^2 = a^2 \cos^2 \alpha + b^2 \cos^2 \beta + c^2 \cos^2 \gamma \tag{7.9}$$

where
$$\cos^2 \alpha + \cos^2 \beta + \cos^2 \gamma = 1 \tag{7.10}$$

Then $|x|$ will attain its maximum and minimum lengths when

$$a^2 \cos \alpha \sin \alpha + b^2 \cos \beta \sin \beta \frac{d\beta}{d\alpha} + c^2 \cos \gamma \sin \gamma \frac{d\gamma}{d\alpha} = 0 \tag{7.11}$$

while, from (7.8),

$$A \sin \alpha + B \sin \beta \frac{d\beta}{d\alpha} + C \sin \gamma \frac{d\gamma}{d\alpha} = 0 \tag{7.12}$$

and, from (7.10),

$$\cos \alpha \sin \alpha + \cos \beta \sin \beta \frac{d\beta}{d\alpha} + \cos \gamma \sin \gamma \frac{d\gamma}{d\alpha} = 0. \tag{7.13}$$

Solving these last two for $\frac{d\beta}{d\alpha}$ and $\frac{d\gamma}{d\alpha}$, we have

$$\frac{d\beta}{d\alpha} = \frac{\sin \alpha(-A \cos \gamma + C \cos \alpha)}{\sin \beta(B \cos \gamma - C \cos \beta)}, \quad \frac{d\gamma}{d\alpha} = \frac{\sin \alpha(-B \cos \alpha + A \cos \beta)}{\sin \gamma(B \cos \gamma - C \cos \beta)}. \tag{7.14}$$

Substituting in (7.11), we get

$$a^2 \cos \alpha(B \cos \gamma - C \cos \beta) + b^2 \cos \beta(-A \cos \gamma + C \cos \alpha)$$
$$+ c^2 \cos \gamma(-B \cos \alpha + A \cos \beta) = 0 \tag{7.15}$$

on condition
$$B \cos \gamma - C \cos \beta \neq 0.$$

This last equation defines the principal axes as we shall now verify. For we show that the plane through either of these directions (defined by this quadratic) and the resultant force due to a displacement in either of these directions is such that the force can be resolved into a component parallel to the displacement and a component normal to the wave-front, which is to be omitted according to the fundamental assumption of transversality. Take the equation of such a plane

$$A'x + B'y + C'z = 0 \tag{7.16}$$

Through the displacement vector and force vector. Therefore it must be satisfied by

$$\frac{F}{|F|} = \frac{1}{|F|}(a^2 \cos \alpha, b^2 \cos \beta, c^2 \cos \gamma)$$

and

$$u = (\cos \alpha, \cos \beta, \cos \gamma)$$

which gives the two equations

$$\left. \begin{array}{c} A'a^2 \cos \alpha + B'b^2 \cos \beta + C'c^2 \cos \gamma = 0 \\ A' \cos \alpha + B' \cos \beta + C' \cos \gamma = 0 \end{array} \right\} \tag{7.17}$$

whence

$$\frac{B'}{A'} = \frac{(-a^2 + c^2) \cos \alpha}{(b^2 - c^2) \cos \beta}, \qquad \frac{C'}{A'} = \frac{(-b^2 + a^2) \cos \alpha}{(b^2 - c^2) \cos \gamma}.$$

We shall show what we want if we show the plane through u and F is normal to the plane parallel to the wave-front, that is, if

$$AA' + BB' + CC' = 0, \qquad A + B\frac{B'}{A'} + C\frac{C'}{A'} = 0,$$

that is

$$A + B\left(\frac{-a^2 + c^2}{b^2 - c^2}\right) \frac{\cos \alpha}{\cos \beta} + C\left(\frac{-b^2 + a^2}{b^2 - c^2}\right) \frac{\cos \alpha}{\cos \gamma} = 0$$

which is precisely condition (7.15).
But the surface

$$x = \frac{1}{v}u$$

satisfies

$$x_1^2 = \frac{u_1^2}{v^2}, \qquad x_2^2 = \frac{u_2^2}{v^2}, \qquad x_3^2 = \frac{u_3^2}{v^2}$$

so that

$$a^2 x_1^2 + b^2 x_2^2 + c^2 x_3^2 = 1$$

which is the equation of an ellipsoid. It is well known that any plane through the center of an ellipsoid intersects it in an ellipse. Hence the two directions of maximum and minimum of (7.15) are the directions of the principal axes of these ellipses and are normal to each other.

The problem now is to find a relation between the normals (A, B, C) to the planes of the incident light waves and the principal axes of the ellipses belonging to these planes. We can suppose (A, B, C) normalized so that

$$A^2 + B^2 + C^2 = 1$$

and consider a radius vector r from the origin having for each direction (A, B, C) the values of the principal axes of the corresponding ellipses so that

$$r^2 = v^2.$$

If we eliminate $(\cos\alpha, \cos\beta, \cos\gamma)$ from (7.8), (7.9), and (7.10), a tedious procedure, we get the equation of what Fresnel called the surface of elasticity

$$\frac{A^2}{r^2 - a^2} + \frac{B^2}{r^2 - b^2} + \frac{C^2}{r^2 - c^2} = 0 \qquad (7.18)$$

The theory of Fresnel is that the light is plane polarized in the crystal according to the directions of the principal axes of the ellipse belonging to the direction of propagation (A, B, C) with velocities of propagation proportional to the magnitudes of these axes. Equation (7.18) is a quadratic in r^2 or v^2 so that there are no more than two different velocities of propagation for any given direction, since $\pm v$ do not indicate different velocities but only different senses.

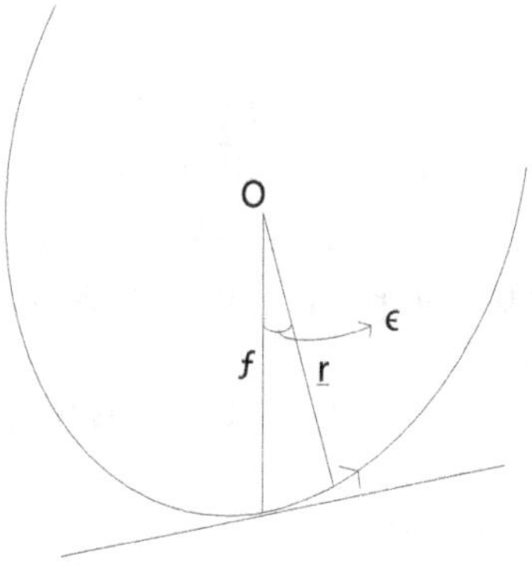

According to the Principle of Huygens, taken over by the Fresnel theory, the wave-front in the crystal will be the envelope of, or surface tangent

Fig. 120

to all these planes. Hence if r is, as before, the position-vector normal to the plane and belonging to the surface of elasticity and f is the position-vector to any point on the plane, then the equation of this plane will be, where $f = \rho[\cos\lambda, \cos\mu, \cos\nu]$,

$$A\cos\lambda + B\cos\mu + C\cos\nu = \frac{r}{\rho} = \cos\varepsilon$$

where $\rho = |f|$ and $r = \pm|r|$. And the equation of the surface tangent to all these planes at the normal distance $|r|$ will be the equation of the envelope. That is, we want the equation of the envelope to

$$f(\cos\lambda, \cos\mu; A, B) = A\cos\lambda + B\cos\mu + C\cos\nu - \frac{r}{\rho} = 0 \qquad (7.19)$$

since $\cos\nu$ and ρ are functions of $\cos\lambda$ and $\cos\mu$, and C and r are functions of A and B. For, we have in addition

$$h(A, B) \equiv A^2 + B^2 + C^2 = 1 \qquad (7.20)$$

$$g(A, B) \equiv \frac{A^2}{r^2 - a^2} + \frac{B^2}{r^2 - b^2} + \frac{C^2}{r^2 - c^2} = 0 \qquad (7.21)$$

Therefore A and B are the independent parameters. Hence the equation of the envelope must satisfy (7.19), (7.20), (7.21) and

$$\frac{\partial f}{\partial A} = 0, \quad \frac{\partial f}{\partial B} = 0, \quad \frac{\partial g}{\partial A} = 0, \quad \frac{\partial g}{\partial B} = 0, \quad \frac{\partial h}{\partial A} = 0, \quad \frac{\partial h}{\partial B} = 0 \qquad (7.22)$$

A lengthy elimination gives

$$0 = \frac{a^2 \cos^2 \lambda}{\rho^2 - a^2} + \frac{b^2 \cos^2 \mu}{\rho^2 - b^2} + \frac{c^2 \cos^2 \nu}{\rho^2 - c^2} \qquad (7.23)$$

as the equation of the wave-front in the crystal.

Startling confirmations of this theory arose from two properties of the ellipsoid.

First, if the ellipsoid has three distinct axes so that

$$a^2 x^2 + b^2 y^2 + c^2 z^2 = 1, \quad a^2 > b^2 > c^2,$$

then there exist two planes (and only two) through the y-axis which intersect the ellipsoid in circles. This is intuitively obvious since, as the plane through the y-axis turns through an angle of $90°$ from containing the x-axis to containing the z-axis, the y-axis is first the minor axis then the major axis of the ellipses of intersection. Analytically, taking any plane through y-axis

$$Lx + Nz = 0,$$

let

$$L = \pm \left(\frac{a^2 - b^2}{a^2 - c^2} \right)^{\frac{1}{2}}, \quad N = \left(\frac{b^2 - c^2}{a^2 - c^2} \right)^{\frac{1}{2}},$$

so that

$$z = \frac{L}{N} x,$$

and, with substitution in the equation of the ellipsoid,

$$y^2 = \frac{1}{b^2} \left[1 - a^2 x^2 - c^2 \frac{a^2 - b^2}{b^2 - c^2} x^2 \right]$$

Therefore the position-vector $\underline{x}$ to the intersection of the ellipsoid and plane is

$$|\underline{x}|^2 = x^2 + \frac{1}{b^2} \left[1 - a^2 x^2 - c^2 \left(\frac{a^2 - b^2}{b^2 - c^2} \right) x^2 \right] + \frac{a^2 - b^2}{b^2 - c^2} x^2 = \frac{1}{b^2}.$$

Similarly, it is easy to prove that any plane parallel to this cuts the ellipsoid in a circle. The directions normal to these two circles are called the optical axes. There can be two such directions and no more where the velocity of propagation is the same for all planes of polarization and, for all other directions, there will be two planes of polarization and hence two different velocities of propagation, each changing with the direction of the incident ray, so that neither refraction of the ray would be ordinary, but both extraordinary. Such crystals, called biaxial, were being studied by Brewster and Biot when Fresnel constructed his theory, and confirmed it. Hamilton later deduced from Fresnel's equations that, in the direction of the optical axes, the light ray gives, in the crystal, a cone of light. This phenomenon was also confirmed.

Second, if the ellipsoid is an ellipsoid of revolution so that

$$a = b, \qquad x^2 + y^2 + \bar{c}^2 z^2 = \frac{1}{a^2}, \qquad \bar{c}^2 = \frac{c^2}{a^2},$$

then any plane through the center intersects it in an ellipse one of whose axes is equal to the axis of the ellipsoid of revolution. This is shown by considering the position-vector $\underline{r}_1$ to the intersection which is normal to the z-axis, that is, to $(0,0,1)$ as well as (L, M, N) where

$$Lx + My + Nz = 0$$

is the plane through the origin. We shall show that $\underline{r}_1$ is semi-axis of the ellipse. If

$$\underline{r}_1 = (x_1, y_1, z_1),$$

$$Lx_1 + My_1 + Nz_1 = 0, \qquad\qquad\qquad z_1 = 0$$

$$Lx_1 + My_1 = 0, \qquad\qquad\qquad x_1 = \frac{-My_1}{L}$$

and

$$x_1^2 + y_1^2 = \frac{1}{a^2}, \quad y_1 = \pm \frac{L}{a\sqrt{M^2 + L^2}}, \quad x_1 = \mp \frac{M}{a\sqrt{M^2 + L^2}}.$$

Take $\underline{r}_2$ normal to $\underline{r}_1$ so that

$$\underline{r}_2 = (x_2, y_2, z_2)$$

$$Lx_2 + My_2 + Nz_2 = 0,$$

$$\frac{-M}{a\sqrt{M^2 + L^2}} x_2 + \frac{L}{a\sqrt{M^2 + L^2}} y_2 = 0.$$

$$x_2^2 + y_2^2 + \bar{c}^2 z_2^2 = \frac{1}{a^2}.$$

That is,

$$z_2 = \frac{-(Lx_2 + My_2)}{N}, \quad y_2 = \frac{M}{L}x_2, \quad z_2 = \frac{-(L^2x_2 + M^2x_2)}{LN}$$

$$x_2^2 = \frac{L^2N^2}{[N^2L^2 + N^2M^2 + \bar{c}^2(L^2 + M^2)^2]a^2}.$$

It is easy to show that $\underline{r}_1$ and $\underline{r}_2$ are the conjugate semi-axes of the ellipse. For $\underline{r}_1$ is normal to $\underline{r}_2$ and

$$\underline{x} = \underline{r}_1 \cos\varphi + \underline{r}_2 \sin\varphi$$

is the equation of an ellipse with semi-major axes $\underline{r}_1$ or $\underline{r}_2$ where

$$|\underline{r}_1| = \frac{1}{a},$$

the radius of revolution of the ellipsoid. And it can be easily shown that $\underline{x}$ satisfies the equation of the ellipsoid.

Therefore, for such an ellipsoid of revolution, the crystal is uniaxial, having one optical axis along which there is only one law of refraction, and, in any other direction, there will be two double refraction, one of which, corresponding to the direction of the axis of the ellipse always equal to $\frac{1}{a}$, is the ordinary refraction and the other extraordinary. This is the case of Iceland Spar and Huygens' account of it. Further, considering (7.23), the equation of the wave-surface, we have, for $a = b$,

$$0 = (\rho^2 - a^2)[a^2(\rho^2 - e^2)\sin^2\nu + c^2(\rho^2 - a^2)\cos^2\nu]$$

which gives two solutions

$$\rho^2 = a^2,$$

a sphere; and

$$\rho^2 = \frac{c^2}{1 + \left(\frac{c^2}{a^2} - 1\right)\cos^2\nu}$$

or, transferring to Cartesian Coordinates,

$$x^2 + y^2 + \frac{c^2}{a^2}z^2 = c^2,$$

an ellipsoid of revolution.

H   Aberration, and the Fresnel Theory.

The problem now arises, in the wave theory, whether the aether is absolutely at rest and everywhere, the perfect embodiment of Newton's absolute space, or whether it is dragged by the motion of the bodies through it, either totally or in part, so that one has still to refer to an absolute space above and beyond it. Or will the particle theory of emission fare better in this regard than the wave theory?

It was observed by Bradley in 1727 that the direction of the fixed stars changed with the yearly movement of the Earth. From the point of view of the emission or particle theory and Newtonian mechanics, the analysis of this is easily had. Intuitively, the light entering the telescope at A, instead of going directly to C, because of the motion of the telescope to the right, falls at C'. We neglect the motion of the Earth or its axis with respect to the much greater motion about the sun.

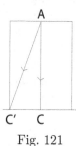

Fig. 121

Analytically, this is simply a case of the Newtonian transformation of velocities. For a stationary telescope we have

$$x = c(t_0 - t)\cos\varphi$$
$$y = c(t_0 - t)\sin\varphi.$$

For the observer moving with a telescope at velocity u in the direction of the x-axis,

$$x' = x + u(t_0 - t)$$
$$y' = y$$

where (x', y') are fixed to moving telescope. Let φ' be angle of light appearing to the observer fixed to moving telescope. Then

$$\tan\varphi' = \frac{y'}{x'} = \frac{c(t_0 - t)\sin\varphi}{c(t_0 - t)\cos\varphi + u(t_0 - t)}$$

$$\frac{\sin\varphi'}{\cos\varphi'} = \frac{\sin\varphi}{\cos\varphi + \frac{u}{c}} \tag{8.1}$$

$$\sin\varphi'\cos\varphi - \sin\varphi\cos\varphi' + \frac{u}{c}\sin\varphi' = 0$$

$$\sin(\varphi - \varphi') = \frac{u}{c}\sin\varphi' \tag{8.2}$$

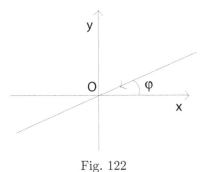

Fig. 122

where $\varphi - \varphi'$ is the angle of aberration. One can observe this angle over a half-year period as Bradley did. In this way, we shall actually observe $2(\varphi - \varphi')$ by observing φ' in two positions, φ'_1 and φ'_2, where the Earth is moving to the right in the first case and to the left in second so that, by (8.1) $\varphi'_1 < \varphi$ and $\varphi'_2 > \varphi$. We assume φ the same in both cases because of the infinite distance of the fixed star with respect to the diameter of the Earth's orbit. From another point of view, observing $2(\varphi - \varphi')$ we can use (8.2) to compute c. In fact,

$$\varphi'_2 - \varphi'_1 = 2(\varphi - \varphi'_1) = 40.89''$$

From the point of view of the wave-theory, the argument is much more complicated. Let us first consider the behavior of light in an isotropic aether for an observer moving at constant velocity $\underline{u}$ (in the direction of the positive x-axis coinciding with x'-axis) where $\underline{x}$ is with respect to fixed aether and $\underline{x}'$ with respect to frame of moving observer. The disturbance of a plane wave-front is given by

$$\psi = a \cos \left(2\pi\omega \left(t - \frac{\underline{x} \cdot \underline{n}}{v} \right) \right)$$

where v is the velocity of propagation or phase-velocity in direction $\underline{n}$, and $\underline{x} \cdot \underline{n}$ is the perpendicular distance from 0 to wave-front through P. We suppose P

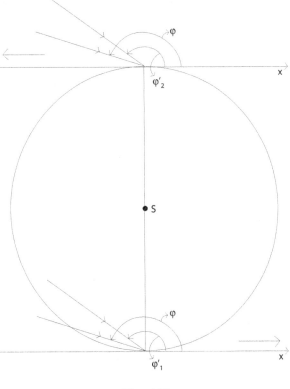

Fig. 123

fixed in aether with position-vector $\underline{x}_1$ and P' fixed in moving frame with position-vector $\underrightarrow{x}' = \overrightarrow{0'P'}$, where P' is so chosen that it will coincide with P at some time t.

If we examine the physical meaning of the phase $\omega\left(t - \frac{x \cdot n}{v}\right)$, we shall see that, within a wide range of space-time measurement conventions, this expression should be invariant under transformation. For suppose a distinguished wave passes 0 at $t_0 = 0$ and P at t, and that the observer fixed at P begins counting the wave-crests which pass by him with the arrival of the distinguished wave. Then

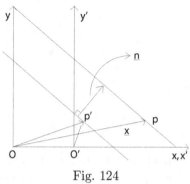

Fig. 124

$$\omega = \text{\# of waves arriving per second at } P;$$

$$\frac{x \cdot n}{v} = \text{\# of seconds from distinguished wave to pass from 0 to } P;$$

$$t - \frac{x \cdot n}{v} = \text{\# of seconds from distinguished wave's passing at } P \text{ to } t;$$

$$\omega\left(t - \frac{x \cdot n}{v}\right) = \text{\# of waves passing } P \text{ from passage of distinguished wave to } t.$$

This is the meaning of the phase expression, measured in the frame F with coordinates (x, y, z, t).

Now in F' with coordinates (x', y', z', t'), let $t' = 0$ when $0'$ is at 0, and $t'_1 = t_1$ when P' coincides with P. The expression $\omega'\left(t' - \frac{x' \cdot n'}{v'}\right)$ is the number of waves passing P' from passage of distinguished wave to t'. But, if we consider these expressions for t'_1 and t_1, we see that, since P' begins counting the waves passing him until P' coincides with P, P' counts the same waves as P. Hence, physically,

$$\omega\left(t_1 - \frac{x \cdot n}{v}\right) = \omega'\left(t'_1 - \frac{x' \cdot n'}{v'}\right)$$

for all values of the variables without much being assumed about the way space and time are measured. Therefore, we take the wave-phase as invariant.

Under Newtonian transformation, we have

$$t = t', \quad x = x' + ut', \quad y = y', \quad z = z',$$

and, for our choice of coordinates,

$$\underline{x} \cdot \underline{n} = x \cos\alpha + y \sin\alpha, \qquad \underline{x}' \cdot \underline{n}' = x' \cos\alpha' + y' \sin\alpha'.$$

Hence

$$\omega \left[t' - \frac{(x' + ut')\cos\alpha + y'\sin\alpha}{v} \right] = \omega' \left[t' - \frac{x'\cos\alpha' + y\sin\alpha'}{v'} \right]$$

and, since (x', y', z', t') are independent, their coefficients, on either side, must be equal, so that

$$\omega \left(\frac{v - u\cos\alpha}{v} \right) = \omega' \tag{8.3}$$

$$\frac{-\omega\cos\alpha}{v} = \frac{-\omega'\cos\alpha'}{v'} \tag{8.4}$$

$$\frac{-\omega\sin\alpha}{v} = \frac{-\omega'\sin\alpha'}{v'}. \tag{8.5}$$

From (8.4) and (8.5) we get

$$\tan\alpha = \tan\alpha', \quad \alpha = \alpha', \quad \underline{n} = \underline{n}', \tag{8.6}$$

We can generalize (8.3) to

$$\omega' = \omega \left(1 - \frac{\underline{u} \cdot \underline{n}}{v} \right). \tag{8.7}$$

From (8.4) or (8.5) and (8.6) we get,

$$\frac{\omega}{v} = \frac{\omega'}{v'}, \quad v' = \frac{\omega'}{\omega} v \tag{8.8}$$

Which, together with (8.7) yields

$$v' = v - \underline{u} \cdot \underline{n} \tag{8.9}$$

Furthermore

$$\lambda = \frac{v}{\omega} = \frac{v'}{\omega'} = \lambda' \tag{8.10}$$

The formula (8.7), the transformation of frequency, yields the classical Doppler effect, and (8.9) yields the transformation of phase-velocity. It should be remarked that it is natural to assume, in the wave-theory, that the velocity of propagation depends only on the medium and its velocity relative to the observer, and not on the velocity of the source.

We now proceed to develop these formulas in such a way that they can be used for observations; for we only directly observe the relative velocity of light source and observer no matter what Newtonian and Fresnel absolutes we suppose behind the scenes. For this purpose let F_0 be the frame of the light source moving at $\underline{u}_0$ with respect to the fixed aether of frame F which coincides with Newton's absolute space, and let ω be the frequency for the observer at rest in this frame of absolute aether, and ω' the frequency for the observer in F' moving at $\underline{u}$ with respect to this aether. Then, by (8.7),

$$\omega' = \omega \left[1 - \frac{\underline{u} \cdot \underline{n}}{v} \right] \tag{8.11}$$

where v is velocity of light with respect to this aether and $\underline{n}$ normal to wave-front. But, again by (8.7),

$$\omega^0 = \omega \left[1 - \frac{\underline{u}_0 \cdot \underline{n}}{v} \right] \tag{8.12}$$

where ω^0 is frequency for observer at rest in F_0. Eliminating ω, we have

$$\omega' = \omega^0 \frac{1 - \frac{\underline{u} \cdot \underline{n}}{v}}{1 - \frac{\underline{u}_0 \cdot \underline{n}}{v}}$$

$$\omega' \approx \omega^0 \left(1 - \frac{(\underline{u} - \underline{u}_0) \cdot \underline{n}}{v} - \frac{(\underline{u}_0 \cdot \underline{n})[\underline{n} \cdot (\underline{u} - \underline{u}_0)]}{v^2} \right) \tag{8.13}$$

In this formula, $\omega', \omega^0, (\underline{u} - \underline{u}_0)$ are observable. By Newtonian assumptions, $\underline{n} = \underline{n}'$ and hence is observable on this account. But $\underline{u}_0$ is not, and v is not. The Michelson-Morley experiment will later enable us to consider $v = v'$ for light in vacuum. On these conditions, we can use the formula through the first order term, and, in 1906 Stark confirmed it by the measurement of the frequency of light emitted by rapidly moving hydrogen molecules but found the second order terms too small for his techniques. The spectra of fixed stars also confirm it through the first order terms. But we shall discuss this formula more deeply later in relation to another anal-ogous one derived from another transforma-tion, the Lorentz transformation.

But we have considered the light wave and the moving observer without regard to the Principle of Huygens. If we bring this in, we shall find that we shall have to distin-guish between the direction of the primary light wave and the direction of the actually observed light ray for the observer moving with respect to the aether.

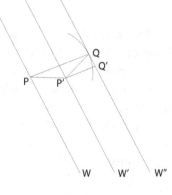

Fig. 125

In the manner of Fig. 125, let P be fixed
in the medium coinciding with moving observer P' and the wave-front W at t; and let
P' coincide with wave front W' at $t+\Delta t$ while W has moved to W'' at $t+\Delta t$. P
at t is the center of a secondary disturbance in the form of a spherical wave
having the same velocity $\underline{v}$ as the primary wave. Then PQ is normal to QW''
and

$$\overrightarrow{PQ} = v\underline{n}\Delta t$$

in fixed frame F, and

$$\overrightarrow{PP'} = \underline{u}\Delta t.$$

The phase-velocity vector for P' is

$$\overrightarrow{P'Q'} = v'\underline{n}'\Delta t,$$

where $P'Q'$ is normal to QW'' and parallel to PQ. For in the Newtonian
transformation

$$\underline{n} = \underline{n}'.$$

Hence

$$\overrightarrow{P'Q'} = v'\underline{n}\Delta t.$$

The direction and velocity of the light ray for the moving observer P' is

$$\overrightarrow{P'Q} = \overrightarrow{P'P} + \overrightarrow{PQ}$$

$$\overline{\underline{v}} = \underline{v} - \underline{u} \tag{8.14}$$

so that the ray velocity in the wave-theory obeys the law of particle velocities.
For here $\underline{v}$ is both the phase velocity and the ray velocity in the absolutely
fixed aether.

Returning to the problem of aberration, if the wave-theory is to explain
the phenomena in the manner of the particle-theory, then it would seem that
we would have to suppose the aether absolutely at rest with respect to the
fixed stars (or, since these latter were soon to appear to move with respect to
each other, as the absolute space itself). If it were dragged completely with
the Earth, there would be no aberration at all.

But Arago found in 1817 that a prism refracted the light of a fixed star in
the same way whether the Earth was moving toward the fixed star or away
from it. In either theory, the index of refraction is a function of the relative
velocities in the two media. In the emission theory, the constant velocity of
the frame would make no difference in the action of the forces. But, since

the initial speed would be different, by (1.3), the ratio of the speeds would
change. For the wave-theory of Fresnel the problem is complicated by the fact
that the velocities in different isotropic media are a function of the density
of the aether in the media. If the prism carries the aether completely with
itself keeping the relative densities of the aether inside and outside fixed, there
would be no problem, but aberration would then become impossible. If the
aether were fixed and the prism (and the Earth) moved through it without
dragging, then it is easy to show that there would be a change in the index of
refraction, assuming the ratios of aether densities remained constant.

Whatever the merits of the problem, it led Fresnel to make the natural
deduction from the axioms of his theory and so to derive a formula which was
to have a future much greater than could have been suspected at the time. It
would be natural to assume that the prism dragged just enough of the aether
to keep the higher density of the aether in it constant. Such an assumption
would not only fit the postulates of Fresnel's system, but would also effect a
compromise. Bypassing the details for the refraction experiment of Arago, we
shall develop Fresnel's theory for the two more important experiments which
took place much later.

Suppose the prism, for the sake of simplicity, moves parallel to the light
and to the Earth's motion about the sun and that end surfaces of the prism
are normal to the direction of the light. If u is velocity of Earth, ρ_1 density
of aether outside prism, ρ_2 density inside, ΔA area of prism's face, then, with
Δt a small period of time,

$$\rho_1 u(\Delta A \cdot \Delta t) = \text{amount of aether entering}$$
$$\rho_2(u - \theta u)(\Delta A \cdot \Delta t) = \text{amount of aether leaving}$$

where θ is coefficient of dragging. But, from Fresnel's theory

$$\mu_{12} = \frac{v_1}{v_2} = \frac{\rho_2^{1/2}}{\rho_1^{1/2}}$$

so that, to have a steady state of constant aether density in the prism,

$$\rho_1 u(\Delta A \cdot \Delta t) = \rho_2(u - \theta u)(\Delta A \cdot \Delta t)$$
$$\rho_1 u = \rho_1 \mu_{12}^2(u - \theta u)$$
$$\theta = 1 - \frac{1}{\mu_{12}^2} \qquad\qquad (8.15)$$

which is called the Fresnel drag-coefficient.

I   The Experiments of Fizeau and Hook.

A generation after Fresnel's death, the experiment of Fizeau was designed to test the Fresnel theory of drag for water moving with respect to the Earth, ignoring the problem of the Earth's motion with respect to an absolute aether. In this experiment, the light from a source is reflected and transmitted so as to pass around the moving water in two opposite ways as in Figure 126. The interference fringes are on the screen. They are examined first with the water going one way at 700 cms./sec., and then

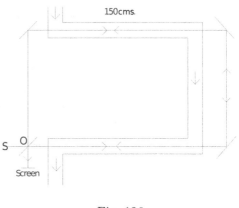

Fig. 126

the other way. This gives the effect of a velocity of 1400 cms./sec. Since the time to go around should differ with the two directions, there should be a shift in the fringes. The drag coefficient was calculated as 0.437 which should have given a shift of 0.404 of a fringe width using the formula

$$c/\mu_{12} \pm \left(1 - \frac{1}{\mu_{12}^2}\right) u$$

where c is speed of light in a vacuum. Fizeau observed 0.46. It is important to note here that the observer is fixed on the Earth and that the aether is being dragged by the water moving with respect to him.[5]

In 1868, Hoek tried to find by an analogous method the drag-coefficient of the Earth and its atmosphere with the aether absolutely at rest with respect to the fixed stars. In this experiment, the light passes around two different ways; the interference fringes are observed. Then the apparatus is rotated an angle of 180° about an axis normal to the direction of the

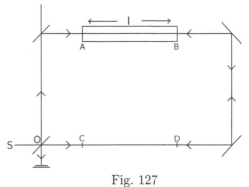

Fig. 127

---

[5]In all these experiments, the weakly silvered glass at O permits both strong reflection and transmission of light.

Earth's motion about the sun. In
the first position, the fact that the aether in the tube of water is dragged in
one direction makes the time of the passage of light clockwise shorter than
counter-clockwise. In the second position, the light that was moving clockwise
now moves counter-clockwise, and that which was moving counter-clockwise
now moves clockwise. This reversal of roles should give a shift of fringes which
is measurable. None was observed. It is to be noted here that the observer
is fixed in the frame of the moving medium which is dragging the aether,
contrary to the case of Fizeau's experiment.

We now calculate the difference in times first on the supposition that the
prism of water does not drag the aether, and secondly that it partially drags
it according to the theory of Fresnel. Since we shall consider the difference of
the times only, we need only compute the times through the prism and the
length corresponding to it on the lower branch.

A. If there is no dragging,

$$t_1(\text{time counter} - \text{clockwise}) = \frac{\ell}{c-u} + \frac{\ell}{c/\mu+u}$$

$$t_2(\text{time clockwise}) = \frac{\ell}{c+u} + \frac{\ell}{c/\mu-u}$$

$$t_2 - t_1 = \frac{\ell}{c+u} - \frac{\ell}{c-u} + \frac{\ell}{c/\mu-u} - \frac{\ell}{c/\mu+u}$$

$$= 2\ell u \left[\frac{1}{(c/\mu)^2 - u^2} - \frac{1}{c^2 - u^2} \right]$$

$$= 2\ell u \left[\frac{u^2/c^2}{1 - \frac{\mu^2 u^2}{c^2}} - \frac{1/c^2}{1 - u^2/c^2} \right]$$

$$\approx \frac{2\ell u}{c^2} \left[\mu^2 \left(1 + \frac{\mu^2 u^2}{c^2} \right) - \left(1 + \frac{u^2}{c^2} \right) \right]$$

$$\approx \frac{2\ell u}{c^2} (\mu^2 - 1),$$

omitting all terms in $\frac{u^2}{c^2}$ from the bracket. Hence the phase-difference between
two positions of the frame is

$$2\omega(t_2 - t_1) = \frac{4\ell u \omega}{c^2}(\mu^2 - 1) \tag{9.1}$$

which was judged observable.

B. If there is partial dragging,

$$t_1 = \frac{\ell}{c-u} + \frac{\ell}{c/\mu + u - \left(1 - \frac{1}{\mu^2}\right)u} = \frac{\ell}{c-u} + \frac{\ell}{c/\mu + \frac{u}{\mu^2}}$$

$$t_2 = \frac{\ell}{c+u} + \frac{\ell}{c/\mu - \frac{u}{\mu^2}}$$

$$t_2 - t_1 = 2\ell u \left[\frac{1}{c^2 - u^2/\mu^2} - \frac{1}{c^2 - u^2}\right] = \frac{2\ell u}{c^2}\left[\frac{1}{1 - u^2/c^2\mu^2} - \frac{1}{1 - u^2/c^2}\right]$$

$$\approx \frac{2\ell u}{c^2}\left[1 + \frac{u^2}{c^2\mu^2} - 1 - \frac{\mu^2}{c^2}\right] = \frac{2\ell u}{c^2}\left[\frac{u^2}{c^2\mu^2} - \frac{u^2}{c^2}\right] \tag{9.2}$$

To the order of approximation of (A), this is zero. Therefore, as long as there is no way of measuring this difference, the Fresnel theory is successful. But a new point of view, the Special Theory of Relativity, will later give these two experiments a more satisfying explanation, deducing Fresnel's coefficient from higher principles.

Boscovich had suggested, at the end of the eighteenth century, that a telescope full of water be used for the observation of aberration of stellar light. If one assumes the Earth moves through the aether without dragging, the slowing of the light would seem to call for a larger effect of aberration. But the Fresnel drag-coefficient again compensates enough to predict no change to the order of approximation of the previous experiments. And, in fact, Airy, in 1871, found no difference to this degree of approximation. Finally, in 1851, Fizeau was able to show experimentally that the velocity of light in water was less than in air as predicted by the wave theory and contrary to the emission theory.

Even though the Fresnel theory cannot fit the full mathematical treatment of the elastic solid, and sins, as we have noted, in its use of continuity conditions, yet all the formulas we have deduced with its aid from the condition of transversality, the analogy of the vibrating string and from Huygen's Principle are still valid and can be deduced from Maxell's electro-magnetic theory of light created later.

J The Michelson-Moreley Experiment.

Here the instrument is in the horizontal plane of the Earth; but the $x-$ and $y-$axes are considered as fixed in an unmovable aether with respect to the fixed stars while the instrument moves in the direction of the x-axis at the rate of the Earth a about the sun, that is, at velocity $\underline{u}$. The mirror at A is weakly-silvered so as to transmit and reflect, one ray to C which moves to C' during passage of ray, the other to B which moves to B' during the passage. Both rays meet again at A become A'' and cast interference fringes on a screen at W.

The instrument is rotated slowly in the horizontal plane so that the y-axis replaces the x-axis. The shift of the fringe pattern, if any, is measured. Let

$$AB = AC = D$$

and t_1 be the time from A to C' and t_2 the time from C' back to A''. Then

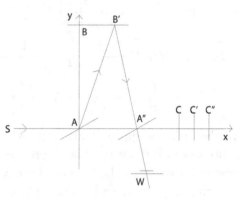

Fig. 128

$$ct_1 = D + ut_1$$
$$ct_2 = D - ut_2$$

where c is absolute velocity of light in a vacuum. Hence

$$t_1 = \frac{D}{c-u}, \quad t_2 = \frac{D}{c+u}, \quad t_1 + t_2 = \frac{2Dc}{c^2 - u^2},$$

and the distance transversed is

$$(t_1 + t_2)c = \frac{2Dc^2}{c^2 - u^2}.$$

The second path is

$$AB' = \sqrt{D^2 + x^2}, \quad x = BB',$$

where

$$\frac{x}{\sqrt{x^2 + D^2}} = \frac{u}{c}, x^2 c^2 = x^2 u^2 + D^2 u^2, x^2 = \frac{u^2 D^2}{c^2 - u^2}.$$

Hence

$$(AB')^2 = D^2 \left(\frac{1}{1 - \frac{u^2}{c^2}} \right)$$

$$AB' = D \left(1 - \frac{u^2}{c^2} \right)^{-\frac{1}{2}} \approx D \left(1 + \frac{u^2}{2c^2} + \dots \right)$$

$$2AB' \approx 2D \left(1 + \frac{u^2}{2c^2} \right).$$

Therefore the difference in distance between the two paths is approximately, that is, neglecting $\frac{u^4}{c^4}$ and higher powers,

$$\frac{2D}{1 - u^2/c^2} - 2D\left[1 + \frac{u^2}{2c^2}\right] = 2D\left[1 + \frac{u^2}{c^2}\right] - 2D\left[1 + \frac{u^2}{2c^2}\right]$$

$$= \frac{Du^2}{c^2}$$

By the shift of the instrument we get twice this difference. The phase-difference, therefore, will be

$$\omega(T_2 - T_1) = \frac{2Du^2\omega}{c^3} \tag{10.1}$$

approximately which can be compared with (9.2) and (9.1).

In the actual instrument there were more mirrors to give greater length to the paths which were 21 meters. There was observed no shift in the fringe patterns although a width of $\frac{1}{30}$ of a fringe width could be observed, and a width difference of 0.4 of a fringe was predicted. The instrument of Joos used in 1933 could observe $\frac{1}{1000}$ of a fringe width. But again no change was observed. This would seem to require that the Earth drag the aether almost totally.

Before leaving the experiment, we note that we assumed that the ray reflected at A from the moving mirror at $45°$ to the x-axis, will exactly hit the moving mirror B at B' just as if the whole instrument did not move. We here give the classical explanation for this. Suppose the mirror AR inclined at angle θ to OA, moving in the direction AP at velocity u. While light goes from O to R, R goes to R'. Let

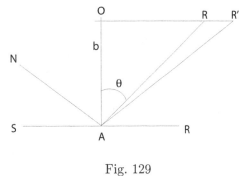

Fig. 129

$$OR = x, \quad RR' = \Delta x.$$

Therefore, for all Δx,

$$\frac{x + \Delta x}{\Delta x} = \frac{c}{u}, \quad \frac{c - u}{u} = \frac{x}{\Delta x}.$$

But

$$x = b \tan \theta$$

$$dx = b \frac{d\theta}{\cos^2 \theta} = \frac{x}{\tan \theta} \frac{d\theta}{\cos^2 \theta}$$

$$\frac{dx}{x} = \frac{d\theta}{\cos \theta \sin \theta}$$

and

$$\frac{u}{c - u} = \frac{d\theta}{\cos \theta \sin \theta}.$$

When $\theta = \frac{\pi}{4}$,

$$d\theta = \frac{1}{2} \frac{u/c}{1 - u/c} = \frac{1}{2} \frac{u}{c} \left[1 + \frac{u}{c} + \frac{u^2}{c^2} + \dots \right] = \frac{1}{2} \left[\frac{u}{c} + \frac{u^2}{c^2} + \frac{u^3}{c^3} \dots \right]$$

$$\approx \frac{1}{2} \frac{u}{c}.$$

Hence, the mirror has the effect of turning by an angle of approximately $\frac{1}{2} \frac{u}{c}$ from its fixed position. Hence the ray SA hits the mirror at $\frac{\pi}{4} + \frac{1}{2} \frac{u}{c}$ from normal AN and is reflected at $\frac{\pi}{4} + \frac{1}{2} \frac{u}{c}$ from AN. Hence it is reflected at $\frac{\pi}{2} + \frac{u}{c}$ from the original direction SA. That is, the ray is turned at an angle $\frac{u}{c}$ approximately from the direction AO as required, since

$$\frac{u}{c} = \sin(2d\theta) = 2d\theta.$$

This explanation will become obsolete in the Special Theory of Relativity.

The preceding is an explanation in terms of the wave-theory. We now add an account of this in terms of the particle theory. We shall assume the collision is elastic and that the mass of the mirror is infinitely large with respect to the mass of the light-particle, that is, $\frac{m}{M} = 0$, and that there is no friction at collision in the plane of the mirror. Then

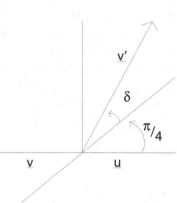

Fig. 130

1. $m\underline{v} + M\underline{u} = m\underline{v}' + M\underline{u}'$.

2. $m\underline{v}^2 + M\underline{u}^2 = m(\underline{v}')^2 + M(\underline{u}')^2$;

or

1. $mv_1 + Mu_1 = mv'_1 + Mu'_1,$ $mv_2 + Mu_2 = mv'_2 + Mu'_2$

2. $m(v_1)^2 + m(v_2)^2 + M(u_1)^2 + M(u_2)^2 = m(v'_1)^2 + m(v'_1)^2 + M(u'_2)^2.$

where u_1 and v_1 are parallel to the plane of the mirror and u_2 and v_2 are normal to u_1 and v_1. Since there is no friction from the mirror, therefore,

$$v_1 = v'_1, \qquad u_1 = u'_1$$

so that (2) becomes

$$m[(v_2)^2 - (v'_2)^2] = M[(u'_2)^2 - (u_2)^2].$$

Dividing by (1),

$$v_2 + v'_2 = u'_2 + u_2.$$

If we take M so large relative to m, that $\frac{m}{M} = 0$, we have from (1),

$$\frac{m}{M}[v_2 - v'_2] = u'_2 - u_2 = 0, \qquad u'_2 = u_2.$$

Hence $v'_2 = -v_2 + 2u_2$ and

$$\frac{v'_2}{v'_1} = \frac{-v_2 + 2u_2}{v_1} = \frac{-v_2}{v_1} + \frac{2u_2}{v_1}.$$

Since here the components of $\underline{v}$ and $\underline{u}$ are, because of the 45° angle, such that

$$v_1 = v_2, \quad u_1 = u_2, \quad \frac{v_1}{v_2} = \frac{u_1}{u_2},$$

hence

$$\tan \delta = \frac{v'_2}{v'_1} = \frac{-v_2 + 2u_2}{v_1} = -1 + \frac{2u_2}{v_1}.$$

But

$$\frac{|\underline{u}|}{|\underline{v}|} = \frac{|\underline{u}'|}{|\underline{v}'|} = \frac{[u_1^2 + u_2^2]^{\frac{1}{2}}}{[v_1^2 + v_2^2]^{\frac{1}{2}}} = \frac{u_2[1 + u_1^2/u_2^2]^{\frac{1}{2}}}{v_1[1 + v_2^2/v_1^2]^{\frac{1}{2}}} = \frac{u_2}{v_1}.$$

Therefore

$$\tan\left(\frac{\pi}{4} + \delta\right) = \frac{1 + \tan \delta}{1 - \tan \delta} = \frac{|\underline{u}'|}{|\underline{v}'|} \Big/ \left(1 - \frac{|\underline{u}'|}{|\underline{v}'|}\right).$$

Let

$$\frac{|\underline{u}|}{|\underline{v}'|} = \frac{|\underline{u}'|}{|\underline{v}'|} = \alpha.$$

Then

$$\tan\left(\frac{\pi}{4}+\delta\right) = \frac{\alpha}{1-\alpha}$$

$$\sin\left(\frac{\pi}{4}+\delta\right) = \frac{a}{[1-2(\alpha-\alpha^2)]^{\frac{1}{2}}} = \alpha[1+(\alpha-\alpha^2)+\ldots]$$

$$\approx \alpha = \frac{|\underline{u}'|}{|\underline{v}'|}.$$

Since $\underline{u}'$ is small with respect to $\underline{v}'$.

Chapter 7

The Special Theory of Relativity

A Principles of Special Relativity

[1] Physics is structured such that events are measured and described from a particular frame of reference. Because many observers who do not necessarily share a reference frame can each make measurements, the mathematics involved in reference frames moving relative to each other must be examined. Classically, one expects Galilean relativity to govern such motion. In this formalism, two reference frames, represented by three dimensional coordinate systems F and F′, share an origin at time t=0. Frame F then remains stationary while F′ moves at a constant relative velocity v.[2] A suitable choice of axes gives the following transformations that relate F′ to F:

$$x' = x - vt \tag{7.1}$$

$$y' = y \tag{7.2}$$

$$z' = z \tag{7.3}$$

$$t' = t. \tag{7.4}$$

Here the coordinate systems have the same orientation and all motion is along the x-axis; x' is the only transformed coordinate. Galileo reasonably assumed

---

[1]Chapter reconstructed by Laura Kinnaman, at the time a graduate student in the Physics Department of the University of Notre Dame, from original lecture notes taken by Thomas Banchoff and Robert Burckel, and building on an earlier transcription of these lecture notes by Paul Gibson, a 2004 Notre Dame graduate in Mathematics.

[2]Accelerating frames are by definition noninertial and shall not be considered under Galilean or Special Relativity.

that the passage of time is constant for all reference frames, so that $t' = t$ for all times. It is a simple matter to show that under Galilean transformations, Newton's laws of mechanics hold in both the F and F' systems, which implies that Newton's laws hold in the classical regime for all inertial reference frames.

The fundamental assumption of the Special Theory of Relativity

The spacetime coordinates for an object in a given reference frame can be described as a vector in four dimensional spacetime, defined by the orthogonal unit vectors $\underline{i}, \underline{j}, \underline{k}$, and $\underline{e}$. We assume that the total sum of spacetime vectors for two different observers, with reference frames moving at a constant relative velocity, will be equal between the same two events. This is Einstein's beginning hypothesis and can be formulated as

$$\Delta x\underline{i} + \Delta y\underline{j} + \Delta z\underline{k} + \Delta\tau\underline{e} = \Delta x'\underline{i'} + \Delta y'\underline{j'} + \Delta z'\underline{k'} + \Delta\tau'\underline{e'}, \qquad (7.5)$$

where $\tau = \kappa t$ and $\tau' = \kappa t'$ are called "proper time." The conversion factor κ is a constant with units of length per time; t and t' are time. Evident from this equation is that $\Delta\tau$ and $\Delta\tau'$ are not necessarily equal, because of the addition of spatial dimensions to the vector equation. In contrast, in the classical view the difference in time between two events must be the same for any two observers, regardless of relative motion between those observers.

While classical physics is governed by Galilean transformations, Special Relativity requires the development of the Lorentz transformations. As before, we shall set up two frames of reference with axes $O : x - y - z - t$ and $O' : x' - y' - z' - t'$ that coincide at $t = t' = 0$. However, for these Lorentz transformations, as shown in Fig. 7.1, the time axes are no longer necessarily parallel. Because we have set $y = y'$ and $z = z'$, (7.5) becomes simply

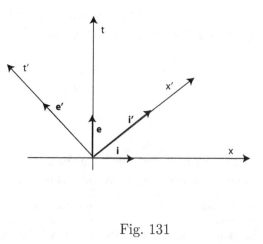

Fig. 131

$$\Delta x\underline{i} + \Delta\tau\underline{e} = \Delta x'\underline{i'} + \Delta\tau'\underline{e'}. \qquad (7.6)$$

To first solve (7.6) for x, both sides are scalar multiplied by $\underline{i}$; from Fig. 7.1

we see that $\underline{i} \cdot \underline{i}' = \cos \theta$, which simplifies the result to

$$x = x' \cos \theta - \kappa t' \sin \theta. \tag{7.7}$$

To solve (7.6) for κt, both sides are scalar multiplied by $\underline{e}$, with the simplifications that $\underline{i}' \cdot \underline{e} = \sin \theta$ and $\underline{e} \cdot \underline{e}' = \cos \theta$:

$$\kappa t = x' \sin \theta + \kappa t' \cos \theta. \tag{7.8}$$

Similarly, scalar multiplying by $\underline{i}'$ or $\underline{e}'$ gives expressions for x' and $\kappa t'$:

$$x' = x \cos \theta + \kappa t \sin \theta \tag{7.9}$$

$$\kappa t' = -x \sin \theta + \kappa t \cos \theta. \tag{7.10}$$

The assumption of $\cos \theta > 0$ is valid on physical grounds: it preserves the sense of time; that is, time moves forward in one direction. With this assumption the cosine and sine can be rewritten in terms of a positive number α:

$$\alpha = \cos \theta \tag{7.11}$$

$$\sin \theta = \sqrt{1 - \alpha^2}, \tag{7.12}$$

which in turn simplifies Eqs. 7.7 - 7.10 to

$$x = \alpha x' - \kappa \sqrt{1 - \alpha^2} t' \tag{7.13}$$

$$\kappa t = \sqrt{1 - \alpha^2} x' + \kappa \alpha t' \tag{7.14}$$

$$x' = \alpha x + \kappa \sqrt{1 - \alpha^2} t \tag{7.15}$$

$$\kappa t' = -\sqrt{1 - \alpha^2} x + \kappa \alpha t. \tag{7.16}$$

As in classical mechanics, the velocity is the time derivative of the position. However, for the Lorentz transformation, time t' is no longer a constant, and its derivative with respect to time t must also be found to completely determine the velocities. Taking the time derivative of (7.13) and (7.14) gives

$$\frac{dx}{dt} = \alpha \frac{dx'}{dt'} \frac{dt'}{dt} - \kappa \sqrt{1 - \alpha^2} \frac{dt'}{dt} \tag{7.17}$$

$$\kappa \frac{dt'}{dt} = -\sqrt{1 - \alpha^2} \frac{dx}{dt} + \kappa \alpha. \tag{7.18}$$

Combining these equations gives:

$$\frac{dx}{dt} = \frac{\alpha \frac{dx'}{dt'} - \kappa \sqrt{1 - \alpha^2}}{\alpha + \frac{\sqrt{1-\alpha^2}}{\kappa} \frac{dx'}{dt'}}. \tag{7.19}$$

Now we are in position to find physical definitions of κ and α. First, we let v_0 be the speed of F' relative to F. Hence, a particle at rest in F' has $\frac{dx'}{dt'} = 0$ and $\frac{dx}{dt} = v_0$, simplifying (7.19) to

$$v_0 = \frac{-\kappa\sqrt{1-\alpha^2}}{\alpha}. \tag{7.20}$$

At this point, to ensure that v_0 is a real number, α^2 must be less than 1. Algebraic manipulation of (7.20) gives

$$\alpha^2 = \frac{1}{1 + \frac{v_0^2}{\kappa^2}}. \tag{7.21}$$

We now show that there is a speed which is the same in both systems, which shall be called $v_f = \frac{dx'}{dt'} = \frac{dx'}{dt'}$. Applying (7.17) and then simplifying gives:

$$v_f = \frac{\alpha v_f - \kappa\sqrt{1-\alpha^2}}{\alpha + \frac{v_f\sqrt{1-\alpha^2}}{\kappa}} \tag{7.22}$$

$$v_f^2 = -\kappa^2. \tag{7.23}$$

The Michelson-Morely experiments suggest that such a speed v_f exists and is in fact c, the speed of light. Hence κ can be taken as $\kappa = ic$, which in turn gives (7.21) as

$$\alpha^2 = \frac{1}{1 - \frac{v_0^2}{c^2}}. \tag{7.24}$$

For a given velocity v_0, which evidently must always be less than c because α is real, α is a constant which shall be defined as γ_0, the Lorentz constant:

$$\alpha = \frac{1}{\sqrt{1 - \frac{v_0^2}{c^2}}} = \gamma_0, \tag{7.25}$$

where

$$\sqrt{1-\alpha^2}\,\kappa = -v_0\gamma_0 \tag{7.26}$$

and

$$\frac{\sqrt{1-\alpha^2}}{\kappa} = \frac{v_0\gamma_0}{c^2}. \tag{7.27}$$

It is now possible to rewrite Eqs. 7.13 - 7.17 in terms of c and the Lorentz constant.

$$x = \frac{x' + v_0 t'}{\sqrt{1 - \frac{v_0^2}{c^2}}} = (x' + v_0 t')\gamma_0 \tag{7.28}$$

$$t = \frac{\frac{v_0}{c^2}x' + t'}{\sqrt{1 - \frac{v_0^2}{c^2}}} = (\frac{v_0}{c^2}x' + t')\gamma_0 \tag{7.29}$$

$$x' = \frac{x - v_0 t}{\sqrt{1 - \frac{v_0^2}{c^2}}} = (x - v_0 t)\gamma_0 \tag{7.30}$$

$$t' = \frac{\frac{-v_0}{c^2}x + t}{\sqrt{1 - \frac{v_0^2}{c^2}}} = (\frac{-v_0}{c^2}x + t)\gamma_0 \tag{7.31}$$

$$\frac{dx}{dt} = \frac{\frac{dx'}{dt'} + v_0}{1 + \frac{v_0}{c^2}\frac{dx'}{dt'}} \tag{7.32}$$

Now quantities that remain invariant through these Lorentz equations can be taken to be the laws of the universe, as these quantities are the same for all observers. These equations also show that F' moving at v_0 with respect to F is exactly equivalent to F moving at $-v_0$ with respect to F', as is physically expected. It is possible to derive a relativistically invariant quantity designated s, using this equivalence in Einstein's hypothesis, (7.5):

$$x\underline{i} + y\underline{j} + z\underline{k} + ict\underline{e} = x'\underline{i}' + y'\underline{j}' + z'\underline{k}' + ict'\underline{e}'. \tag{7.33}$$

Taking the dot product of each side with itself gives

$$x^2 + y^2 + z^2 - c^2 t^2 = x'^2 + y'^2 + z'^2 - c^2 t'^2. \tag{7.34}$$

We then let $\dot{s} = |\dot{\underline{r}}|$ where $\underline{r} = x\underline{i} + y\underline{j} + z\underline{k} + ict\underline{e}$, which gives

$$\dot{s} = \frac{ds}{dt} = \sqrt{\left(\frac{dx}{dt}\right)^2 + \left(\frac{dy}{dt}\right)^2 + \left(\frac{dz}{dt}\right)^2 - c^2}. \tag{7.35}$$

Given $\tau = ict$, this velocity can be rewritten in terms of the invariant s and the proper time τ:

$$\frac{ds}{d\tau} = \sqrt{1 - \frac{(\frac{dx}{dt})^2 + (\frac{dy}{dt})^2 + (\frac{dz}{dt})^2}{c^2}}. \tag{7.36}$$

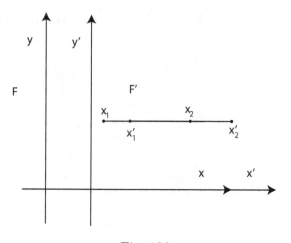

Fig. 132

The physics described by the Lorentz transformations has some interesting implications, namely length contraction and time dilation. Considering the former, suppose an observer in F examines a yardstick in the F′ frame, as shown in Fig. 2. He observes two points in F′, x_1' and x_2', which have corresponding points in F, x_1 and x_2, at time t. The Lorentz transformations give the following relationships between the coordinates in each frame:

$$x_1' = \gamma_0(x_1 - v_0 t) \tag{7.37}$$
$$x_2' = \gamma_0(x_2 - v_0 t). \tag{7.38}$$

The difference in length between them in F′ is thus

$$x_2' - x_1' = \gamma_0(x_2 - x_1). \tag{7.39}$$

However, when $|v_0| < c$, $\gamma_0 > 1$ which gives the length in F to be

$$x_2 - x_1 = \frac{1}{\gamma_0}(x_2' - x_1') < x_2' - x_1'. \tag{7.40}$$

Therefore the length of the yardstick in F is proportional to the length in F′ by $\frac{1}{\gamma_0}$, a quantity less than one. We note that the very same shrinkage would be observed by a primed observer measuring a yardstick at rest in the unprimed system. For very small velocities, $\frac{1}{\gamma_0}$ approaches 1 and length contraction is an infinitesimal effect, giving the classical result. In the other limit when v_0 approaches c, $\frac{1}{\gamma_0}$ goes to zero, which means if v_0 were greater than c, unphysical negative lengths would appear.

Considering time dilation, suppose two different events that occur at the same point x in the primed system are measured by an observer in the unprimed system. These events occur at time t_1 and t_2 in F, and at time t_1' and t_2' in F'. Again, the Lorentz transformations give the following relationships between the times, and the difference in times in each frame:

$$t_1 = \gamma_0 \left(\frac{v_0}{c^2} x' + t_1' \right) \tag{7.41}$$

$$t_2 = \gamma_0 \left(\frac{v_0}{c^2} x' + t_2' \right) \tag{7.42}$$

$$t_2 - t_1 = \gamma_0 (t_2' - t_1') > t_2' - t_1'. \tag{7.43}$$

Because the length of time between events is greater in F, (7.43) implies that moving clocks run slow when observed from the unprimed system. In other words, as v_0 increases, the passage of time slows relative to the stationary observer. However, just as a yardstick always measures a yard in the observer's own system regardless of that system's velocity relative to another, time passes normally for the observer in his own system. We once again note that at nonrelativistic velocities the time dilation will be negligible, a clear example of which is Fizeau's experiment with light in moving water. His work predicts the speed of light in water moving at v_0 to be

$$v = \frac{c}{\mu} + \left(1 - \frac{1}{\mu^2} \right) v_0, \tag{7.44}$$

where μ is the ratio of the speed of light in the medium to the speed of light in the vacuum. Using the Lorentz transformations we calculate:

$$\frac{dx}{dt} = \frac{\frac{dx'}{dt'} + v_0}{1 + \frac{dx}{dt} \frac{v_0}{c^2}} = \frac{\frac{c}{\mu} + v_0}{1 + \frac{c}{\mu} \frac{v_0}{c^2}} = \left(\frac{c}{\mu} + v_0 \right) \left(1 + \frac{v_0}{c\mu} \right)^{\frac{1}{2}}$$

$$= \left(\frac{c}{\mu} + v_0 \right) \left(1 - \frac{c}{\mu} \frac{v_0}{c^2} + \cdots \right)$$

$$= \left(\frac{c}{\mu} + v_0 - \frac{c^2}{\mu^2} \frac{v_0}{c^2} - \frac{cv_0^2}{\mu c^2} + \cdots \right), \tag{7.45}$$

which is $\approx \frac{c}{\mu} + v_0 (1 - \frac{1}{\mu^2})$ up to order $\frac{1}{c^2}$. Thus we have agreement with Fizeau's results for water moving at low velocity in comparison to the speed of light.

We now turn to another consequence of Relativity theory, the addition of vectors via a "boost". In classical physics, when a particle moves at v' in a reference frame which is also moving at v_0 relative to the observer, the velocity which the observer measures, v, is the simple vector addition of v_0

and v'. However, in relativity more caution must be taken, because of the impossibility of traveling faster than the speed of light in vacuum: an object moving at c cannot be seen to be moving at $2c$ by an observer also moving at c. Suppose a particle moves at $c - \epsilon$ in F' and F' moves at $v_0 = c - \epsilon$ with respect to the stationary frame F. Then by (7.32) we have

$$\frac{dx}{dt} = \frac{c - \epsilon + c - \epsilon}{1 + \frac{(c-\epsilon)^2}{c^2}} = \frac{2(c - \epsilon)}{\frac{2c^2 - 2c\epsilon + \epsilon^2}{c^2}} = \frac{2c^2 - 2c^2\epsilon}{2c^2 - 2c\epsilon + \epsilon^2} = \frac{c - \epsilon}{1 - \frac{\epsilon}{c} + \frac{\epsilon^2}{2c^2}}. \tag{7.46}$$

Additional algebraic manipulation with the knowledge that $c - \epsilon < c - \epsilon + \frac{\epsilon^2}{2c}$ yields

$$\frac{c - \epsilon}{1 - \frac{\epsilon}{c} + \frac{\epsilon^2}{2c^2}} < c. \tag{7.47}$$

Hence the speed of light can never be attained by adding a finite number of velocities less than c. A new dynamical system based on the speed of light must now be developed.

B Dynamics of Special Relativity

We summarize here the Lorentz transformations for position, time and velocity for a primed system F' moving with velocity v_0 with respect to an unprimed system F.

$$x = (x' + v_0 t')\gamma_0 \tag{7.48}$$

$$y = y' \tag{7.49}$$

$$z = z' \tag{7.50}$$

$$t = (\frac{v_0}{c^2} x' + t')\gamma_0 \tag{7.51}$$

$$x' = (x - v_0 t)\gamma_0 \tag{7.52}$$

$$t' = (\frac{-v_0}{c^2} x + t)\gamma_0 \tag{7.53}$$

$$\frac{dx}{dt} = \frac{\frac{dx'}{dt'} + v_0}{1 + \frac{v_0}{c^2}\frac{dx'}{dt'}} \tag{7.54}$$

$$\frac{ds}{d\tau} = \sqrt{1 - \frac{(\frac{dx}{dt})^2 + (\frac{dy}{dt})^2 + (\frac{dz}{dt})^2}{c^2}} = \frac{1}{\gamma_0} \tag{7.55}$$

We also need equations to calculate the velocities in the y and z directions; $y = y'$ but $t \neq t'$, so the chain rule of derivatives is necessary to relate $\frac{dy}{dt}$ to $\frac{dy'}{dt'}$:

$$\frac{dy}{dt} = \frac{dy'}{dt'}\frac{dt'}{dt} = \frac{dy'}{dt'}\gamma_0(-\frac{v_0}{c^2}\frac{dx}{dt} + 1) \tag{7.56}$$

$$= \frac{dy'}{dt'}\gamma_0\left[-\frac{v_0}{c^2}\left(\frac{\frac{dx'}{dt'} + v_0}{1 + \frac{dx'}{dt'}\frac{v_0}{c^2}}\right) + 1\right]. \tag{7.57}$$

Additional simplification yields the following for $\frac{dy}{dt}$ and a similar result for $\frac{dz}{dt}$

$$\frac{dy}{dt} = \frac{dy'}{dt'}\frac{1}{\gamma_0(1 + \frac{dx'}{dt'}\frac{v_0}{c^2})} \tag{7.58}$$

$$\frac{dz}{dt} = \frac{dz'}{dt'}\frac{1}{\gamma_0(1 + \frac{dx'}{dt'}\frac{v_0}{c^2})}. \tag{7.59}$$

Dynamics are concerned with momentum and energy; we shall consider the former first. Suppose a particle is moving at $\underline{u}$ in F and therefore $\underline{u}'$ in F', where

$$\underline{u} = \frac{dx}{dt}\underline{i} + \frac{dy}{dt}\underline{j} + \frac{dz}{dt}\underline{k}. \tag{7.60}$$

We are interested in the relationship $1 - \frac{u^2}{c^2}$:

$$1 - \frac{u^2}{c^2} = 1 - \frac{(\frac{dx}{dt})^2 + (\frac{dy}{dt})^2 + (\frac{dz}{dt})^2}{c^2}. \tag{7.61}$$

After algebraic manipulation we can rewrite (7.61) as

$$1 - \frac{u^2}{c^2} = \frac{\frac{1}{\gamma^2}\left[1 - \frac{(u')^2}{c^2}\right]}{\left[1 + \frac{dx'}{dt'}\frac{v_0}{c^2}\right]^2}. \tag{7.62}$$

By factoring the reciprocal of $1 - \frac{u^2}{c^2}$ into $\gamma = \frac{1}{\sqrt{1 - \frac{u^2}{c^2}}}$ and $\gamma' = \frac{1}{\sqrt{1 - \frac{(u')^2}{c^2}}}$, we obtain from (7.62)

$$\gamma^2 = \gamma'^2\gamma_0^2(1 + \frac{dx'}{dt'}\frac{v_0}{c^2})^2. \tag{7.63}$$

Since $\frac{dx'}{dt'}v_0 < c^2$ we can consider only the positive root of (7.63):

$$\gamma = \gamma'\gamma_0(1 + \frac{dx'}{dt'}\frac{v_0}{c^2}). \tag{7.64}$$

A similar calculation for $\frac{dx}{dt}\gamma$ is

$$\frac{dx}{dt}\gamma = \gamma'\gamma_0(1 + \tfrac{dx'}{dt'}\tfrac{v_0}{c^2})\left[\frac{\frac{dx'}{dt'} + v_0}{(1 + \tfrac{dx'}{dt'}\tfrac{v_0}{c^2})}\right] \tag{7.65}$$

and will yield

$$\frac{dx}{dt}\gamma = \gamma'\gamma_0(\tfrac{dx'}{dt'} + v_0). \tag{7.66}$$

Similarly we obtain

$$\frac{dx'}{dt'}\gamma' = \gamma\gamma_0(\tfrac{dx}{dt} - v_0). \tag{7.67}$$

We now state our earlier conclusion about the Lorentz transformations more formally:

Postulate: Any expression which passes through the Lorentz transformation invariant in functional form is a law of the world.

According to classical physics, mass is constant in all reference frames; every observer moving at any given velocity will measure the same mass for a given object. However, according to relativity the frame of the observer will affect the measurement. In order to investigate the relativistic effects on mass, we let

$$\sum_i M_i'\frac{dx_i'}{dt'} = \kappa', \tag{7.68}$$

where M_i' is the Newtonian inertial mass and κ' is the total momentum in F', assuming for the moment a one dimensional system. For ease of notation we set $\sum M_i' = \lambda'$. Here we assume conservation of momentum and of inertial mass in the primed system itself. The question becomes, when are these quantities conserved in F? In other words, under what conditions will these two laws remain invariant under the Lorentz transformation? We add $\sum M_i'v_0$ and multiply by γ_0 to each side of (7.68) to obtain

$$\sum_i M_i'[\tfrac{dx_i'}{dt'} + v_0]\gamma_0 = \kappa'\gamma_0 + \lambda'\gamma_0 v_0. \tag{7.69}$$

Using (7.66) we see that

$$\sum M_i'[\tfrac{dx_i'}{dt'} + v_0]\gamma_0 = \sum M_i'\frac{\gamma_i}{\gamma_i'}\frac{dx_i}{dt} = \kappa'\gamma_0 + \lambda'\gamma_0 v_0. \tag{7.70}$$

If we take

$$\frac{M_i'\gamma_i}{\gamma_i'} = M_i, \tag{7.71}$$

where M_i is the Newtonian inertial mass in F, then (7.70) becomes

$$\sum M_i \frac{dx_i}{dt} = \kappa'\gamma_0 + \lambda'\gamma_0 v_0 = \kappa. \tag{7.72}$$

Here κ is analogous to κ' and is the total momentum in F. Therefore it is sufficient for conservation of momentum in F that

$$\frac{M_i'}{\gamma_i'} = \frac{M_i}{\gamma_i} = m_i, \tag{7.73}$$

where m_i is called the "proper rest mass" of the ith particle. Only as velocities approach c will there be a measurable difference in rest and inertial masses. The proper rest mass is seen to be the same for all observers at a given time (although this rest mass might change for the particle at some point during its life). We shall prove that this is also a necessary condition for the conservation of momentum in F. Similar to the derivation of (7.72), we will now subtract $\sum M_i' v_0$ and multiply by γ_0 to each side of $\sum M_i \frac{dx_i}{dt_i} = \kappa$ to obtain

$$\sum M_i' \left[\frac{dx_i}{dt} - v_0 \right] \gamma_0 = \kappa\gamma_0 - \sum M_i\gamma_0 v_0. \tag{7.74}$$

Using (7.67) we have

$$\sum M_i' \frac{dx_i'}{dt'} \frac{\gamma_i'}{\gamma_i} = \kappa\gamma_0 - \sum M_i\gamma_0 v_0. \tag{7.75}$$

Applying (7.73) and (7.68) we finally obtain a relationship between the total momentum in each frame:

$$\kappa' = \kappa\gamma_0 - \sum M_i\gamma_0 v_0. \tag{7.76}$$

By analogy with (7.72) we see that $\sum M_i = \lambda$ is a constant in F, as we would expect. It should now be apparent that the conservation of momentum relies on (7.73) holding. Looking more closely at (7.73), we note

$$M_i = m_i\gamma_i = m_i \left[1 - \frac{u_i^2}{c^2} \right]^{-1/2}, \tag{7.77}$$

which can be rewritten with the Binomial expansion as

$$M_i c^2 = m_i c^2 + \frac{1}{2} m_i u_i^2 + \cdots \qquad (7.78)$$

Solving for the kinetic energy we have, to first order,

$$\frac{1}{2} m_i u_i^2 \approx (M_i - m_i) c^2. \qquad (7.79)$$

We see then that the kinetic energy is the difference in the energy from the inertial mass, $M_i c^2$, and the rest mass energy, $m_i c^2$.

We also note that if (7.73) holds–and now investigating the other spatial dimensions–we can write

$$M_i' \frac{dy_i'}{dt'} = M_i \frac{dy_i}{dt}, \qquad (7.80)$$

or in the case of a single particle,

$$M' \frac{dy'}{dt'} = M \frac{dy}{dt}. \qquad (7.81)$$

Given $y = y'$ and the chain rule for derivatives, we have

$$M \frac{dy}{dt} = M \frac{dy}{dt'} \frac{dt'}{dt} = M \frac{dy'}{dt'} \frac{dt'}{dt}. \qquad (7.82)$$

With (7.73) and the time derivative of (7.53) this becomes

$$M \frac{dy}{dt} = M' \frac{\gamma}{\gamma'} \frac{dy'}{dt'} \frac{dt'}{dt} = M' \frac{dy'}{dt'} \frac{\gamma \gamma_0}{\gamma'} \left[-\frac{v_0}{c^2} \frac{dx}{dt} + 1 \right]. \qquad (7.83)$$

Derivations similar to the ones that provided (7.64) give

$$\frac{\gamma'}{\gamma \gamma_0} = 1 - \frac{dx'}{dt'} \frac{v_0}{c^2}, \qquad (7.84)$$

which allows us to simplify (7.83) to

$$M \frac{dy}{dt} = M' \frac{dy'}{dt'}. \qquad (7.85)$$

Similarly, we have

$$M \frac{dz}{dt} = M' \frac{dz'}{dt'}. \qquad (7.86)$$

For this calculation and what follows, we assume that the rest mass does not vary only with time (i.e. the particle is a primitive one).

With these results we are now ready to tackle the conservation of momentum in its entirety, and from momentum deal with forces. We define the total momentum of the object as

$$M\frac{dx}{dt}\underline{i} + M\frac{dy}{dt}\underline{j} + M\frac{dz}{dt}\underline{k} + icM\underline{e} = M\frac{d\underline{r}}{dt} = \underline{p}, \tag{7.87}$$

where $\underline{r} = x\underline{i} + y\underline{j} + z\underline{k} + ict\underline{e}$ is the position vector, $M = m\gamma$ is the rest mass, and $\gamma = \frac{dt}{d\sigma}$. Hence we can rewrite the total momentum as

$$\underline{p} = m\frac{dx}{d\sigma}\underline{i} + m\frac{dy}{d\sigma}\underline{j} + m\frac{dz}{d\sigma}\underline{k} + icM\underline{e}. \tag{7.88}$$

Using Newton's Second Law we define the force on the object of proper rest mass m to be

$$\underline{F} = \frac{d\underline{p}}{dt}, \tag{7.89}$$

which can be expanded to

$$\underline{F} = \frac{d}{dt}\left(m\frac{dx}{d\sigma}\underline{i}\right) + \cdots + \frac{d}{dt}\left(mic\frac{dt}{d\sigma}\right)\underline{e} \tag{7.90}$$

$$= m\frac{d^2x}{d\sigma^2}\frac{d\sigma}{dt}\underline{i} + \cdots + mic\frac{d^2t}{d\sigma^2}\frac{d\sigma}{dt}\underline{e}. \tag{7.91}$$

Now, we know that $\left(\frac{d\sigma}{dt}\right)^2 = 1 - \dfrac{\left(\frac{dx}{dt}\right)^2 + \left(\frac{dy}{dt}\right)^2 + \left(\frac{dz}{dt}\right)^2}{c^2}$, so dividing by $\left(\frac{d\sigma}{dt}\right)^2$ yields

$$c^2 = c^2\left(\frac{dt}{d\sigma}\right)^2 - \left(\frac{dx}{d\sigma}\right)^2 - \left(\frac{dy}{d\sigma}\right)^2 + \left(\frac{dz}{d\sigma}\right)^2. \tag{7.92}$$

Furthermore, differentiating with respect to σ removes the additive constant:

$$0 = c^2\frac{dt}{d\sigma}\frac{d^2t}{d\sigma^2} - \frac{dx}{d\sigma}\frac{d^2x}{d\sigma^2} - \frac{dy}{d\sigma}\frac{d^2y}{d\sigma^2} - \frac{dz}{d\sigma}\frac{d^2z}{d\sigma^2}. \tag{7.93}$$

If we let $\underline{x}$ be the spatial part of $\underline{r}$ then (7.93) further simplifies to

$$0 = c^2\frac{dt}{d\sigma}\frac{d^2t}{d\sigma^2} - \frac{d^2\underline{x}}{d\sigma^2}\cdot\frac{d\underline{x}}{d\sigma}. \tag{7.94}$$

For a particle of Newtonian inertial rest mass m, the total energy can be written as

$$E = mc^2 + T = mc^2 + \frac{1}{2}mu^2 + \cdots, \tag{7.95}$$

where T is the kinetic energy of relativity theory. The Newtonian expression for kinetic energy can be derived by integrating $\underline{F}_s \cdot \underline{u}$ with respect to time, where $\underline{u} = \frac{dx}{dt}\underline{i} + \frac{dy}{dt}\underline{j} + \frac{dz}{dt}\underline{k}$ and $\underline{F}_s$ are the spatial components of the force $\underline{F}$.

$$\underline{F}_s \cdot \underline{u} = m\frac{d^2x}{d\sigma^2}\frac{d\sigma}{dt}\frac{dx}{dt} + \cdots + m\frac{d^2z}{d\sigma^2}\frac{d\sigma}{dt}\frac{dz}{dt}. \tag{7.96}$$

Using the chain rule and $\underline{x}$ as defined above, this expression can be simplified to

$$= m\frac{d^2\underline{x}}{d\sigma^2} \cdot \frac{d\underline{x}}{d\sigma}\left(\frac{d\sigma}{dt}\right)^2.$$

The $\underline{x}$ second derivative can be replaced by a second derivative of time using (7.94)

$$= mc^2\frac{d^2t}{d\sigma^2}\left(\frac{dt}{d\sigma}\right)\left(\frac{d\sigma}{dt}\right)^2,$$

which further simplifies to

$$= mc^2\frac{d^2t}{d\sigma^2}\frac{d\sigma}{dt}$$
$$= mc^2\frac{d}{dt}\left(\frac{dt}{d\sigma}\right); \tag{7.97}$$

the final result is

$$\underline{F}_s \cdot \underline{u} = mc^2\frac{d\gamma}{dt}. \tag{7.98}$$

Integrating (7.98) gives us the energy,

$$\int \underline{F}_s \cdot \underline{u}\,dt = mc^2\gamma + K \tag{7.99}$$

for some constant of integration K. Substituting in for γ we get

$$\int \underline{F}_s \cdot \underline{u}\,dt = mc^2\frac{1}{\left(1 - \frac{u^2}{c^2}\right)^{\frac{1}{2}}} + K. \tag{7.100}$$

Because the kinetic energy is zero when $\underline{u}$ is zero, we have

$$0 = mc^2 \frac{1}{\left(1 - \frac{0^2}{c^2}\right)^{\frac{1}{2}}} + K. \tag{7.101}$$

Solving for the constant K we have

$$K = -mc^2, \tag{7.102}$$

were mc^2 is the rest energy. We conclude that

$$\int \underline{F_s} \cdot \underline{u} dt = T = \tfrac{1}{2}mu^2 + \cdots + \cdots \tag{7.103}$$

where these last terms are the difference arising from using the relativistic mass M in expressions for F_s instead of the rest mass m.

C   Lorentz Transformations of Wave-Phase

We now wish to transform various properties of the wave equation from one inertial frame to another. To determine the Lorentz transformations that achieve this, we begin by examining $w(-t + \frac{x \cdot n}{w})$, a term in the wave equation. As before, $\underline{x}$ is the spatial part of $\underline{r}$. We let

$$\frac{\underline{x} \cdot \underline{n}w}{w} - wt = \underline{r} \cdot \underline{\kappa} \tag{7.104}$$

for some $\underline{\kappa}$. We assume that such a $\underline{\kappa}$ exists, and with the form

$$\underline{\kappa} = \frac{w}{w} \cos \alpha_1 \underline{i} + \frac{w}{w} \cos \alpha_2 \underline{j} + \frac{w}{w} \cos \alpha_3 \underline{k} - \frac{w}{ic} \underline{e}. \tag{7.105}$$

Is this $\underline{\kappa}$ invariant under Lorentz transformations? We know $\underline{r} = \underline{r}'$ and $\underline{r} \cdot \underline{\kappa} = \underline{r}' \cdot \underline{\kappa}'$, the former being another statement of Einstein's hypothesis and the later because phase is just a number, independent of the system in which the measurement is made. But if these two statements are true, and the four components of $\underline{\kappa} = (\kappa_1, \ldots, \kappa_4)$ are numbers in F and similarly the components of κ' are numbers in F', then we see that $\underline{\kappa} = \underline{\kappa}'$. This can be proven more rigorously:

For ease of computation, we work in one spatial dimension. Let

$$x_r = \sum_{s=1}^{2} \alpha_{rs} x_s' = \alpha_{rs} x_s', \tag{7.106}$$

where Einstein's summation convention is invoked for the right hand side (that is to say, any doubly repeated indices are summed over, and the explicit summation symbol can be dropped). We also define $x_1 = x$ and $x_2 = ict = \kappa t$. Now we can write

$$x_r \kappa_r = \alpha_{rs} x'_s \kappa_r = x'_s \kappa'_s \qquad (7.107)$$

where the second equality is a restatement of $\underline{r} \cdot \underline{\kappa} = \underline{r}' \cdot \underline{\kappa}'$. Rewriting (7.107) we have

$$\alpha_{rs} x'_s \kappa_r - x'_s \kappa'_s = 0, \qquad (7.108)$$

which implies

$$x'_s(\alpha_{rs}\kappa_r - \kappa'_s) = 0. \qquad (7.109)$$

Therefore, for any arbitrary x'_s we have $\alpha_{rs}\kappa_r = \kappa'_s$. A closer look at the matrix form of α_{rs} demonstrates that the inverse is also true:

$$\begin{bmatrix} a_{11} & a_{12} \\ a_{21} & a_{22} \end{bmatrix}^{-1} = \begin{bmatrix} a_{11} & -a_{12} \\ -a_{21} & a_{22} \end{bmatrix} = \begin{bmatrix} a_{11} & a_{21} \\ a_{12} & a_{22} \end{bmatrix} \qquad \alpha_{11} = \alpha_{22}, \; \alpha_{21} = -\alpha_{12}$$

Because $\alpha_{rs}^{-1} = \alpha_{rs}$, we see that $\alpha_{rs}\kappa'_s = \kappa_r$. This proves that κ is Lorentz invariant. With this understanding, we can move on to examining the transformations of other wave properties.

Now, the Newtonian transformation of the frequency ω is

$$\omega' = \omega^0 \frac{1 - \frac{\underline{u}\cdot\underline{n}}{w}}{1 - \frac{\underline{v_0}\cdot\underline{n}}{w}} = \omega^0 \left[1 - \frac{(\underline{u} - \underline{v_0}) \cdot \underline{n}}{w} - \frac{(\underline{n} - \underline{v_0})(\underline{u} - \underline{v_0}) \cdot \underline{n}}{w^2} \cdots \right] \qquad (7.110)$$

where ω^0 is the frequency observed in the rest system F of the wave source and ω' is the frequency observed by an observer having velocity $\underline{u} - \underline{v_0}$ with respect to the wave source. The velocity of the observer with respect to the (hypothetical) ether is denoted $\underline{u}$ and the velocity of the wave source is $\underline{v_0}$. We note here that $\underline{n}$ is taken to be the direction of wave propagation in both frames (proved classically to be the same for each frame). In contrast, the Lorentz transformation of ω is

$$\omega \left(t - \frac{x \cos \alpha + y \sin \alpha}{w} \right) = \omega' \left(t' - \frac{x' \cos \alpha' + y' \sin \alpha'}{w'} \right) \qquad (7.111)$$

where α is the angle between the direction of propagation and the velocity of the wave source. From (7.110), we can solve for ω^0:

$$\omega^0 = \omega' \frac{1 - \frac{\underline{v_0}\cdot\underline{n}}{w}}{1 - \frac{\underline{u}\cdot\underline{n}}{w}} = \omega' \left[1 - \frac{\underline{v_0}\cdot\underline{n}}{w} \right] \left[1 + \frac{\underline{u}\cdot\underline{n}}{w} + \left(\frac{\underline{u}\cdot\underline{n}}{w^2} \right)^2 + \cdots \right]^2$$

$$\omega^0 = \omega' \left[1 - \frac{(v_0 - u) \cdot n}{w} - \frac{(n \cdot u)(v_0 - u) \cdot n}{w^2} + \cdots \right]. \tag{7.112}$$

In (7.111) we have that $x = \overline{\gamma_0}[x' + \overline{v_0}t']$ and $t = \overline{\gamma_0}[\frac{\overline{v_0}}{c^2} + t']$, where $\overline{v_0}$ is the speed of F' with respect to F, either $\underline{u} - \overline{v_0} = \underline{v_0}$ or $\overline{v_0} - \underline{u} = \underline{v_0}$, depending on the interpretation of F and F'. With these definitions, (7.111) becomes

$$\omega' \left(t' - \frac{x' \cos \alpha' + y' \sin \alpha'}{w'} \right) =$$
$$\omega \left[\overline{\gamma_0}(\tfrac{\overline{v_0}}{c^3} x' + t') - \frac{\overline{\gamma_0}(x' + \overline{v_0}t') \cos \alpha + y' \sin \alpha}{w} \right] \tag{7.113}$$

Equating coefficients in (7.113) gives first

$$\omega' t' = \omega \left[\overline{\gamma_0} t' - \frac{\overline{\gamma_0} v_0 t' \cos \alpha}{w} \right] \tag{7.114}$$

which simplifies to

$$\omega' = \omega \overline{\gamma_0} \left[1 - \frac{\overline{v_0} \cos \alpha}{w} \right]. \tag{7.115}$$

The cosine term is the result of a dot product when the angle is known; therefore (7.115) can be generalized to

$$\omega' = \omega \overline{\gamma_0} \left[1 - \frac{\overline{v_0} \cdot n}{w} \right], \tag{7.116}$$

the relativistic Doppler effect. Equating the other terms in (7.113) now gives

$$-\frac{\omega' x' \cos \alpha'}{w'} = \omega \left[-\frac{\overline{\gamma_0} x' \cos \alpha}{w} + \frac{\overline{\gamma_0} v_0 x'}{c^2} \right]. \tag{7.117}$$

Eventually this equation simplifies to an equation relating trigonometric functions of α' and α:

$$\tan \alpha' = \frac{\frac{\sin \alpha}{w}}{\overline{\gamma_0} \left[\frac{\cos \alpha}{w} - \frac{\overline{v_0}}{c^2} \right]} = \frac{\sin \alpha (1 - \frac{\overline{v_0}^2}{c^2})^{\frac{1}{2}}}{\cos \alpha - \frac{\overline{v_0} w}{c^2}}. \tag{7.118}$$

We can compare this result with James Bradley's experiment on stellar abberation due to the changing motion of the Earth.

Now we let F' be the system in which the wave source is at rest, and F be the system in which the observer is at rest. We take the case that $\overline{v_0} = \underline{v_0} - \underline{u}$.

Here then w' corresponds to the above w^0 and w corresponds to w' from the above (classical) discussion. These redefinitions make (7.116) assume the form

$$w^0 = w' \left(1 - \frac{\overline{v_0} \cdot n}{w}\right) \left(1 - \frac{\overline{v_0}^2}{c^2}\right)^{-\frac{1}{2}}. \tag{7.119}$$

Doing a Taylor expansion on $\left(1 - \frac{\overline{v_0}^2}{c^2}\right)^{-\frac{1}{2}}$ to first order in $\frac{\overline{v_0}^2}{c^2}$, multiplying by $\left(1 - \frac{\overline{v_0} \cdot n}{w}\right)$, and substituting in the definition of $\overline{v_0}$ yields

$$w^0 = w' \left[1 - \frac{(v_0 - u) \cdot n}{w} + \frac{1}{2}\frac{(v_0 - u)^2}{c^2} + \cdots\right]. \tag{7.120}$$

Now note that the classical Doppler effect depends in the first order term on the absolute velocity of the observer (with respect to the ether) since the term $n \cdot u$ appears. Hence change in direction only of the observer with respect to the light source should give change in the Doppler effect for first order terms.

In the transverse Doppler effect where $v_0 - u = \overline{v_0} \perp n$, there is no shift classically, that is $w^0 = w'$. But relativistically we get

$$w^0 = \frac{w'\left(1 - \frac{\overline{v_0} \cdot n}{w}\right)}{\left(1 - \frac{\overline{v_0}^2}{c^2}\right)^{\frac{1}{2}}} = \frac{w'}{\left(1 - \frac{\overline{v_0}^2}{c^2}\right)^{\frac{1}{2}}} = w'\left[1 + \frac{1}{2}\frac{\overline{v_0}^2}{c^2} + \cdots\right] \tag{7.121}$$

Obviously then $w^0 > w'$ in the relativistic case. But this is just $w^0 = w'\gamma_0$. We see this is consistent with $t_2 - t_1 = (t_2' - t_1')\gamma_0$, when, because of the reversal of primes,

$$t_2 - t_1 \leftrightarrow w'$$
$$t_2' - t_1' \leftrightarrow w^0.$$

We know this is correct because w' and w^0 are in units of $seconds^{-1}$ while $t_2 - t_1$ and $t_2' - t_1'$ are in $seconds$; the transformations are consistent.

Cosmic radiation provides the following confirmation: the mean life τ' of a meson (measured in a system in which it is at rest) is 1.5×10^{-6} sec. Mesons penetrate the atmosphere to the extent of 20 kilometers from their formation to their death, a distance that without relativistic consideration would imply a speed greater than c. However, the mean life observed by us (in a system in which the meson has velocity $v_0 \approx c$) is $\tau \approx \frac{20}{c} = 7 \times 10^{-5}$ sec. Then $\frac{\tau}{\tau'} \approx 50$. Therefore from relativity theory we must have

$$\frac{1}{\left(1 - \frac{v_0^2}{c^2}\right)^{\frac{1}{2}}} = 50, \tag{7.122}$$

so $v_0 = c \left(1 - \frac{1}{2500}\right)^{\frac{1}{2}}$. This value for v_0 is confirmed experimentally.

D DeBroglie Theory of the Relation Between the Momentum Vector and the Wave Phase Vector for all Particles

The momentum vector can be written in a relativistic framework as

$$\frac{m_0}{\left(1 - \frac{u^2}{c^2}\right)^{\frac{1}{2}}} u_x \underline{i} + \cdots + \frac{i m_0 c}{\left(1 - \frac{u^2}{c^2}\right)^{\frac{1}{2}}} \underline{e} \qquad (7.123)$$

where $u^2 = u_x^2 + u_y^2 + u_z^2$. The wave phase vector is given as

$$\frac{\omega}{u} \cos \alpha_1 \underline{i} + \frac{\omega}{u} \cos \alpha_2 \underline{j} + \frac{\omega}{u} \cos \alpha_3 \underline{k} + \frac{i\omega}{c} \underline{e}; \qquad (7.124)$$

the first three terms are parallel to $\underline{n}$, as is the case when the momentum and wave phase vectors apply to the same "thing". In the case of light, for the rest mass we need $m_0 = 0$ so that $\frac{m_0}{\sqrt{1 - \frac{u^2}{c^2}}} = \frac{m_0}{0} \neq \infty$. We now make the following leap:

$$\underline{p} = h\underline{k} \qquad (7.125)$$

for some constant h. Since such a relation holds between the first three coordinates, by requiring it to hold here, we can see what this condition imposes on the fourth coordinate. This relation is Lorentz invariant (and so should be a law of the world). For $u = c$, p is indeterminate because $m = 0$ also. Now for the photon $w = c$ and we define the momentum vector in this case by $h\underline{k}$:

$$\underline{p}_{photon} = h \left[\frac{\omega}{c} \cos \alpha_1 \underline{i} + \frac{\omega}{c} \cos \alpha_2 \underline{j} + \frac{\omega}{c} \cos \alpha_3 \underline{k} + \frac{i\omega}{c} \underline{e} \right]. \qquad (7.126)$$

Equating the fourth coordinates from Eqs. 7.123 and 7.126 gives

$$iMc = \frac{i m_0 c}{\gamma} = \frac{i m_0 c}{\left(1 - \frac{u^2}{c^2}\right)^{\frac{1}{2}}} = \frac{h i \omega}{c}. \qquad (7.127)$$

Therefore $Mc^2 = h\omega$. But we know $E = Mc^2$, so for a photon, $E = h\omega$, Planck's Law. Now since de Broglie's theory is Lorentz invariant we suppose it a law of the world–that is, every particle with speed $u < c$ has a wave-vector associated with its momentum vector and a wave speed of w.

Dot multiplying the spatial parts of momentum and wave phase vectors gives

$$
\frac{m^2}{1 - \frac{u^2}{c^2}}(u_x^2 + u_y^2 + u_z^2) = \frac{m^2 u^2}{1 - \frac{u^2}{c^2}}
$$

$$
= \frac{h^2}{w^2}\left[\omega^2 \cos^2 \alpha_1 + \omega^2 \cos^2 \alpha_2 + \omega^2 \cos^2 \alpha_3\right] \quad (7.128)
$$

which impies

$$
\frac{m^2 u^2}{1 - \frac{u^2}{c^2}} = \frac{h^2 \omega^2}{w^2}. \quad (7.129)
$$

But from (7.128) we have

$$
\frac{m^2 c^2}{1 - \frac{u^2}{c^2}} = h^2 \omega^2. \quad (7.130)
$$

By dividing (7.129) by (7.130) we obtain

$$
u^2 = \frac{c^4}{w^2}, \quad (7.131)
$$

where u is the particle speed and w is the wave speed of the particle. (7.131) implies

$$
|u| = \frac{c^2}{|w|}. \quad (7.132)
$$

Hence, for all particles having $u < c$, we see that $w > c$, and this–the existence of wave velocities greater than c–has been verified by diffraction experiments from 1927. Recall $\mu = \frac{u'}{u}$ in Newtonian wave mechanics, but it is $\frac{w}{w'}$ in Huygens. So these quantities no longer conflict since the velocities being considered were not the same.

We now establish the Lorentz transformations for momentum and energy by assuming p_x transforms like x and $\frac{E}{c^2}$ transforms like t.

$$
p_x = (p_x' + v_0 \frac{E'}{c^2})\gamma_0 \quad (7.133)
$$

$$
\frac{E}{c^2} = \left(\frac{v_0}{c^2}p_x' + \frac{E'}{c^2}\right)\gamma_0 \quad (7.134)
$$

We also set $p_y = p_{y'}$ and $p_y = p_{y'}$; here $p = |p|$ is the length of the spatial part of p. Then, because p_x transforms like x and x obeys the following sort of law, we have

$$p^2 - c^2 \left(\frac{E}{c^2}\right)^2 = (p')^2 - c^2 \left(\frac{E'}{c^2}\right)^2. \tag{7.135}$$

Substituting $\dfrac{mu}{\sqrt{1-\frac{u^2}{c^2}}}$ for p' and $\dfrac{mc^2}{\sqrt{1-\frac{u^2}{c^2}}}$ for E' gives

$$p^2 - c^2 \left(\frac{E}{c^2}\right)^2 = \frac{m^2[u^2 - c^2]}{1 - \frac{u^2}{c^2}} = -mc^2. \tag{7.136}$$

These results are proven in more detail, beginning with the form of the Lorentz transformation for p_x. By (7.123), we know

$$p_x = \frac{mu_x}{\left(1 - \frac{u^2}{c^2}\right)^{\frac{1}{2}}}; \tag{7.137}$$

also

$$u_x = \frac{dx}{dt} = \frac{\frac{dx'}{dt'} + v_0}{1 + \frac{v_0}{c^2}\frac{dx'}{dt'}} = \frac{u'_x + v_0}{1 + \frac{v_0}{c^2}u_{x'}} \tag{7.138}$$

$$\gamma = \gamma'\gamma_0[1 + \frac{u'_x v_0}{c^2}]. \tag{7.139}$$

Plugging Eqs. 7.138 and 7.139 into (7.137) gives

$$p_x = \frac{m\,[u_{x'} + v_0]}{\left(1 - \frac{u^2}{c^2}\right)^{\frac{1}{2}}\left[1 + \frac{u_{x'}v_0}{c^2}\right]} = \frac{m\,[u_{x'} + v_0]}{\frac{1}{\gamma}\left[1 + \frac{u_{x'}v_0}{c^2}\right]}$$

$$= \frac{m[u_{x'} + v_0]}{\frac{1}{\gamma_0}\frac{1}{\gamma'}} = \frac{\gamma_0 m[u_{x'} + v_0]}{\left[1 - \frac{u'^2}{c^2}\right]^{\frac{1}{2}}}$$

$$= \gamma_0 \left[\frac{mu_{x'}}{\left[1 - \frac{u'^2}{c^2}\right]^{\frac{1}{2}}} + \frac{mv_0}{\left[1 - \frac{u'^2}{c^2}\right]^{\frac{1}{2}}}\right] = \gamma_0 \left[p_{x'} + \frac{mv_0}{\left(1 - \frac{u'^2}{c^2}\right)}\right]^{\frac{1}{2}}$$

$$p_x = \gamma_0 \left[p_{x'} + \frac{E'v_0}{c^2}\right], \tag{7.140}$$

which is the form we assumed earlier. The proof for the energy Lorentz transformation begins with the relativistic definition of E from (7.100):

$$E = \frac{mc^2}{\left(1 - \frac{u^2}{c^2}\right)^{\frac{1}{2}}}. \tag{7.141}$$

Using $\gamma = \gamma_0 \gamma' [1 + \frac{v_0}{c^2} \frac{dx'}{dt'}]$ we get

$$\frac{E}{c^2} = \frac{m \left(1 + \frac{v_0 u_{x'}}{c^2}\right)}{\frac{1}{\gamma_0} \left(1 - \frac{u'^2}{c^2}\right)^{\frac{1}{2}}}, \tag{7.142}$$

which simplifies to our previous assumed form, (7.134).

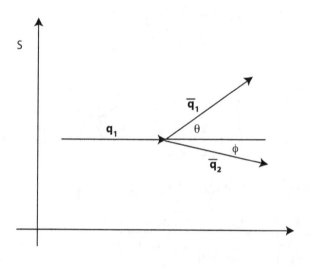

Fig. 133

To further our discussion of momentum and energy, let us consider a collision in two spatial dimensions, as shown in Fig. 3. Here the momentum of particle 1 before collision, in the S frame, is $q_1 = q_1 \underline{i}$; the momentum of particle 2 before collision in S, q_2, is 0. We define the momentum after collision for particles 1 and 2 to be $\overline{q_1}$ and $\overline{q_2}$ respectively. In the center of mass frame S', the center of mass momentum is $q' = q'_1 + q'_2 = 0$. Then we can say that S' moves at some constant speed v_0 along the x-axis. From earlier definitions we know that

$$q'_1 = \frac{mu'_1}{\left(1 - \frac{u_1^2}{c^2}\right)^{\frac{1}{2}}} \tag{7.143}$$

$$q'_2 = \frac{mu'_2}{\left(1 - \frac{u_2^2}{c^2}\right)^{\frac{1}{2}}}. \tag{7.144}$$

The conservation of both momentum and kinetic energy–there being no potential in this example–tells us that the x-component of the total p or q is

$$p'_x = 0 = q'_x$$
$$= \left[p_x - v_0 \frac{E}{c^2}\right] \gamma_0, \tag{7.145}$$

where (7.145) comes from applying the Lorentz transformation for momentum. Solving for p_x gives

$$p_x = \frac{v_0 E}{c^2}. \tag{7.146}$$

Solving this further for v_0 gives

$$v_0 = \frac{p_x c^2}{E} = \frac{c^2}{E} \left[\frac{mu_1}{\left(1 - \frac{u^2}{c^2}\right)^{1/2}}\right]. \tag{7.147}$$

Knowing $E = E_1 + E_2 = \frac{mc^2}{\left(1 - \frac{u^2}{c^2}\right)^{1/2}} + mc^2$ and setting $u = u_1$ simplifies the

v_0 expression to

$$v_0 = \frac{u_1}{1 + \left(1 - \frac{u_1^2}{c^2}\right)^{1/2}}. \tag{7.148}$$

In S', u'_2 is the speed of the particle which was at rest in S, which makes u'_2 also the speed of S with respect to S' and thus $u'_s = -v_0$. Now since primes without bars are "before collision in S'" we have, from $\underline{q}' = \underline{q}'_1 + \underline{q}'_2 = \underline{0}$,

$$\frac{u'_1}{\left(1 - \frac{u'^2_1}{c^2}\right)^{1/2}} = \frac{-u'_2}{\left(1 - \frac{u'^2_2}{c^2}\right)^{1/2}}. \tag{7.149}$$

Simplifying (7.149) gives $u_1'^2 = u_1'^2$, so $u_1' = \pm u_2'$. However, the original form of (7.149) precludes the solution $u_1' = +u_2'$, hence we have $u_1' = -u_2'$.

We investigate the effect of the conservation of kinetic energy in S': $E' = E_1' + E_2'$, which can be written

$$E_1' + E_2' = \frac{mc^2}{\left(1 - \frac{u_1'^2}{c^2}\right)^{1/2}} + \frac{mc^2}{\left(1 - \frac{u_2'^2}{c^2}\right)^{1/2}}. \tag{7.150}$$

Applying $u_1'^2 = u_1'^2 = v_0^2$ gives

$$E' = \frac{mc^2}{\left(1 - \frac{v_0^2}{c^2}\right)^{1/2}} + \frac{mc^2}{\left(1 - \frac{v_0^2}{c^2}\right)^{1/2}}, \tag{7.151}$$

which implies $E' = 2E_1' = 2E_2'$, because we see that $E_1' = E_2'$.

Returning to momentum, Fig. 3 shows us that

$$\tan\theta = \frac{\bar{p}_{1y}}{\bar{p}_{1x}} = \frac{\bar{p}_{1y}'}{\bar{p}_{1x}} \tag{7.152}$$

$$\tan\phi = \frac{\bar{p}_{2y}}{\bar{p}_{2x}} = \frac{\bar{p}_{2y}'}{\bar{p}_{2x}}. \tag{7.153}$$

But by the conservation of momentum in S', $\underline{q}' = \underline{q}_1' + \underline{q}_2' = \underline{0}$, we know $\bar{p}_{1y} = -\bar{p}_{2y}$. This in turn simplifies Eqs. 7.152 and 7.153 to the combination

$$\tan\theta\tan\phi = \frac{(\bar{p}_{1y})^2}{\bar{p}_{1x}\bar{p}_{2x}}. \tag{7.154}$$

Now $p_{1y} = p_{1y}'$ and $p_{1x} = \gamma_0(p_{1x}' + v_0 \frac{E'}{c^2})$ (similarly for p_{2y} and p_{2x}), so (7.154) becomes

$$\tan\theta\tan\phi = \frac{(\bar{p}_{1y}')^2}{\left[\bar{p}_{1x}' + v_0\frac{E_1'}{c^2}\right]\left[\bar{p}_{2x}' + v_0\frac{E_2}{c^2}c^2\right]}\left(1 - \frac{v_0^2}{c^2}\right) \tag{7.155}$$

since $\frac{1}{\gamma_0^2} = 1 - \frac{v_0^2}{c^2}$. Further manipulations yield

$$\tan\theta\tan\phi = \frac{(\bar{p}_{1y}')^2(1 - \frac{v_0^2}{c^2})}{-(\bar{p}_{1x}')^2 + \frac{v_0^2}{c^2}(\bar{E}_1')^2}$$

$$\tan\theta\tan\phi = 1 - \frac{v_0^2}{c^2}, \tag{7.156}$$

the later because $(\vec{q}_1')^2 = (\vec{p}_{1x}')^2 + (\vec{p}_{1y}')^2$. As an aside: recall $\vec{p}_{1z}' = 0$ and earlier we showed $(\vec{q}_1')^2 - \frac{(\overline{E}_1')^2}{c^2} = -mc^2$. Therefore

$$(\vec{q}_1')^2 = \frac{(\overline{E}_1')^2}{c^2} - mc^2 \tag{7.157}$$

$$= m^2 c^2 \left(\frac{1}{\left(1 - \frac{v_0^2}{c^2}\right) c^2} - 1 \right) \tag{7.158}$$

$$= \frac{(\overline{E}_1')^2 v_0^2}{c^4} \tag{7.159}$$

where $\overline{E}_1' = \frac{m^2 c^2}{1 - \frac{v_0^2}{c^2}}$. This result allows us to write

$$(\vec{q}_1')^2 = (\vec{p}_{1x}')^2 + (\vec{p}_{1y}')^2 = \frac{(\overline{E}_1')^2 v_0^2}{c^4} \tag{7.160}$$

$$(\vec{p}_{1y}')^2 = -(\vec{p}_{1x}')^2 + \frac{(\overline{E}_1')^2 v_0^2}{c^4}. \tag{7.161}$$

Returning to our $\tan\theta \tan\phi$ equation, we use Eqs. 7.156 and 7.148 to obtain

$$\tan\theta \tan\phi = 1 - \frac{u_1^2}{c^2 \left(1 + \left(1 - \frac{u_1^2}{c^2}\right)^{1/2}\right)^2}. \tag{7.162}$$

Judicious use of algebra simplifies this expression to

$$\tan\theta \tan\phi = \frac{2}{1 + \left(1 - \frac{u_1^2}{c^2}\right)^{-1/2}}. \tag{7.163}$$

The algebraic identity

$$\tan(\theta + \phi) = \frac{\tan\theta + \tan\phi}{1 - \tan\theta \tan\phi} \tag{7.164}$$

can be used to investigate different physical limits. When $u_1 \ll c$, we know $\tan\theta \tan\phi \approx 1$. In that case, $\tan(\theta + \phi) \approx \infty$ and $\theta + \phi \approx \frac{\pi}{2}$, which is the Newtonian case for collisions. When $u_1 \approx c$, we have $\tan(\theta + \phi) \ll \frac{\pi}{2}$, the extreme relativistic case.

1. Suppose a planet P moves on a circle at constant angular speed about the circle's center O. That is, $\frac{d\beta}{dt} = K$ where K is constant, independent of t. Suppose $ES \perp$ to AB intersecting circle at S & S' and observer at E. Show that the apparent angular speed $\frac{d\alpha}{dt}$ is equal to K only at S and S'. Where is the apparent slowest and fastest? Prove by methods of differential calculus.

2. Let O be center of circle as before with sun at S, and suppose P moves on circle at constant angular speed about T, not O. Show that the ratio of $\frac{d\beta}{dt}$ at A to $\frac{d\beta}{dt}$ at B is as TA to TB. If $OT = OS$, then the ratio is $\frac{SB}{SA}$. This was first discovered geometrically by Kepler and is important for his Theory. Since $(R)\beta = s$ where R is radius, β in radians and s is arc length, Therefore $\frac{d\beta}{dt}$ is proportional always to the true linear speed of the planet on its path. Do these inverse ratios from S hold for the rest of the path?

3. The cycloid is the curve traced by a point P on a circle as the circle rolls without slipping on a st. line. Find x and y as functions of a and φ where a is radius of circle and φ is angle from $O'B$ to $O'P$. These will be the parametric equations of the cycloid. This curve will be involved in the famous problem of the brachystochrone of classical mechanics.

4. Suppose a particle moves on a circle at constant speed (counterclockwise) while the center of the circle (epicycle) moves counterclockwise on a different circle, starting in the position of 1. First, without regard to speeds, find

—4 (cont.) The parametric equations of the curve traced by the particle P in terms of angles θ and φ and the radius b of epicycle and radius a of the deferent circle.

Second, suppose both movements are at constant angular speed; that is, φ = ηt and θ = μt where t is time. Find $\frac{dx}{dt}$ and $\frac{dy}{dt}$.

Third, suppose P moves on the epicycle at the same angular speed as O moves on the deferent; but P moves clockwise while O moves counterclockwise. Prove that P traces out a circle with same radius as deferent but with center on y axis at a distance b below E.

—5. Suppose a particle moves on a circle at constant speed about its center. Find $\frac{dx}{dt}$ and $\frac{dy}{dt}$, then find $\frac{d^2x}{dt^2}$. What does this represent? Can you give a simple law for its variation?

—6. A planet is seen to move about the earth irregularly, but in such a way, it always seems to move fastest in one part of the ecliptic and slower in the part opposite. The time from mean motion thru fastest to mean motion is elevated as 100 days to 75 days from mean thru slowest to mean. Find a simple theory of its path according to the Neo-Aristotelean theory of inertia.

—7. Suppose the mean sun is assumed to be always very near the center of Venus' epicycle. In Ptolemaic theory, it is always in very nearly the same st. line as seen from the earth, suppose if coincides with the center of the epicycle at the apogee and perigee of Venus' eccentric. Find the time it takes Venus to move from a fixed star with respect to an observer on the mean sun back to that that same fixed star. This is called Venus' periodic time with respect to the mean sun. Use the Ptolemaic tables.

Index